Handbook of
Abiotic and Biotic Stress Tolerance in Plants and Crop Productivity
Methods and Protocols

Handbook of
Abiotic and Biotic Stress Tolerance in Plants and Crop Productivity
Methods and Protocols

Lanee J. Allen

Editor

KOROS PRESS LIMITED
London, UK

Handbook of Abiotic and Biotic Stress Tolerance in Plants and Crop Productivity: Methods and Protocols

© 2012

Printed in 2017 for Sale in the Indian Subcontinent

Published by
Koros Press Limited
3 The Pines, Rubery B45 9FF, Rednal,
Birmingham, United Kingdom

Tel.: +44-7826-930152
Email: info@korospress.com
www.korospress.com

ISBN: 978-1-78163-009-9

Editor: Lanee J. Allen

Printed in UK

British Library Cataloguing in Publication Data
A CIP record for this book is available from the British Library

10 9 8 7 6 5 4 3 2 1

Exclusively distributed by CBS Publishers & Distributors Pvt. Ltd.

Sales & Distribution Rights only for India, Pakistan, Bangladesh, Sri Lanka, Nepal and Bhutan.This book is not to be sold outside these territories.

Contents

Preface

Changes in crop production and the impact of new food and environmental legislation are having an influence on the significance of pests and diseases which attack plants and reduce yields. Climate change will also have an impact on pests and pathogens, and may increase exposure to abiotic stresses such as drought and heat. With the expected increase in world population outstripping the land available for cropping, maximising utilisable yields by breeding for resistance to biotic and abiotic stresses is becoming an imperative. It is therefore important that plant breeders can identify the most important constraints on production in a particular crop and region. Plant production has to meet considerably mounting demands in the future. Expanding global markets and the competition of food and non-food uses require further significant progress in productivity levels. In Europe as well as globally, increased production will have to be achieved on the same or decreasing area of arable land. If global welfare is to be maintained or improved an increased efficiency per unit area is required. At the same time, climatic changes may aggravate the conditions of growth in less favourable locations. Thus, the scenario which agriculture is facing is further intensified crop rotations with a limited number of high-yielding crops for the food or raw materials market, under aggravated climatic conditions. Altogether, these developments will result in a significant increase in problems caused by biotic and abiotic stresses, which will inevitably limit yield levels. One way out will be improvement of cultivars.

Breeding programmes are currently set up to meet the new challenges. Recent biotechnological progress has opened new avenues for further and faster advances in crop breeding. Cultivars with better resistance to biotic and abiotic stress are becoming a real option. However, a number of emerging questions had to be answered. What will be the major threats in crop production systems over the next few decades? Which traits are needed and which can be expected to become available in new cultivars within the next few years? How can the new biotechnologies be helpful in producing cultivars harbouring the desired new traits? At the same time as the demand for food is intensifying, the climate is changing, with inevitable consequences for agriculture and the world's food supply. The potential

consequences have been discussed by Rosenzweig & Hillel (1995). They state that "vulnerability to climate change is systematically greater in developing countries, which in most cases are located in lower, warmer latitudes. In those regions, cereal grain yields are projected to decline under climate change scenarios, across the full range of expected warming. Agricultural exporters in middle and high latitudes ...stand to gain, as their national production is predicted to expand, and particularly if grain supplies are restricted and prices rise. Thus, countries with the lowest income may be the hardest hit."

In Europe, predicted changes in climatic conditions depend on location, with the greatest levels of warming predicted for Mediterranean and north-eastern areas, increased precipitation in northern areas (particularly in winter) and decreased precipitation in southern areas. The challenge in many areas of the world will be to produce more food with limited supplies of water, and breeding for drought tolerance and water use efficiency are key to this. Globally, drought already results in greater yield loss than any other single biotic or abiotic factor and even in the UK drought losses are estimated to be 1-2 t/ha. However, Semenov (2008) considers that, in England and Wales, heat stress around flowering might represent a greater risk to wheat production in England than drought, because although the summer is predicted to be drier in the 2050s, winter is predicted to be wetter and water might still be available to the growing crop in late spring and summer.

This Handbook is specifically written keeping in mind of the students in particular and research scholars.

—Editor

Chapter 1

Introduction

Agricultural productivity is measured as the ratio of agricultural outputs to agricultural inputs. While individual products are usually measured by weight, their varying densities make measuring overall agricultural output difficult. Therefore, output is usually measured as the market value of final output, which excludes intermediate products such as corn feed used in the meat industry. This output value may be compared to many different types of inputs such as labour and land (yield). These are called partial measures of productivity. Agricultural productivity may also be measured by what is termed total factor productivity (TFP). This method of calculating agricultural productivity compares an index of agricultural inputs to an index of outputs. This measure of agricultural productivity was established to remedy the shortcomings of the partial measures of productivity; notably that it is often hard to identify the factors cause them to change. Changes in TFP are usually attributed to technological improvements.

Sources of Agricultural Productivity

Some sources of agricultural productivity are:

Mechanised Agriculture

Mechanized agriculture is the process of using agricultural machinery to mechanize the work of agriculture, massively increasing farm output and farm worker productivity. In modern times, powered machinery has replaced many jobs formerly carried out by men or animals such as oxen, horses and mules.

The history of agriculture contains many examples of tool use, but only in recent time has the high rate of machine use been at such a level.

The first pervasive mechanization of agriculture came with the introduction of the plough, usually powered by animals. It was invented in ancient Mesopotamia. Current mechanized agriculture includes the use of Tractors, trucks, combine harvesters, airplanes (crop dusters), helicopters, and other vehicles. Modern farms even sometimes use computers in conjunction with satellite imagery and GPS guidance to increase yields.

Mechanization was one of the factors responsible for urbanisation and industrial economies. Besides improving production efficiency, mechanization encourages large scale production and improves the quality of farm produce. On the other hand, it displaces unskilled farm labour, causes environmental pollution, deforestation and erosion.

History

Jethro Tull's seed drill (ca. 1701) was a mechanical seed spacing and depth placing device that increased crop yields and saved seed. It was an important factor in the British Agricultural Revolution.

Since the beginning of agriculture threshing was done by hand with a flail, requiring a great deal of labour. The threshing machine, which was invented in 1794 but not widely used for several more decades, simplified the operation and allowed the use of animal power.

Before the invention of the grain cradle (ca. 1790) an able bodied labourer could reap about one quarter acre of wheat in a day using a sickle. It was estimated that for each of Cyrus McCormick's horse pulled reapers (ca. 1830s) freed up five men for military service in the U.S. Civil War. Later innovations included raking and binding machines. By 1890 two men and two horses could cut, rake and bind 20 acres of wheat per day.

In the 1880s the reaper and threshing machine were combined into the combine harvester. These machines required large teams of horses or mules to pull.

Steam power was applied to threshing machines in the late 19th century. There were steam engines that moved around on wheels under their own power for supplying temporary power to stationary threshing machines. These were called *road engines,* and Henry Ford seeing one as a boy was inspired to build an automobile.

With internal combustion came the first modern tractors in the early 1900s, becoming more popular after the Fordson tractor (ca. 1917). At first reapers and combine harvesters were pulled by tractors, but in the 1930s self powered combines were developed.

The horse population in the U.S. began to decline in the 1920s after the conversion of agriculture and transportation to internal combustion. Peak tractor sales in the U.S. were around 1950. In addition to saving labour, this freed up much land previously used for supporting draft animals.

The greatest period of growth in agricultural productivity in the U.S. was from the 1940s to the 1970s, during which time agriculture was benefiting from internal combustion powered tractors and harvesting machines, chemical fertilizers and the green revolution.

High Yield Varieties, Which were the Basis of the Green Revolution

High-yielding varieties (HYVs) are any of a group of genetically enhanced cultivars of crops such as rice, maize and wheat that have an increased growth rate, an increased percentage of usable plant parts or an increased resistance against crop diseases. Those crops formed the basis for the Green Revolution.

Fertilizers: Primary plant nutrients: nitrogen, phosphorus and potassium and secondary nutrients nutrients such as sulfur, zinc, copper, manganese, calcium, magnesium and molybdenum on deficient soil.

Fertilizer (or fertiliser) is any organic or inorganic material of natural or synthetic origin (other than liming materials) that is added to a soil to supply one or more plant nutrients essential to the growth of plants. A recent assessment found that about 40 to 60% of crop yields are attributable to commercial fertilizer use.

Mined inorganic fertilizers have been used for many centuries, whereas chemically synthesized inorganic fertilizers were only widely developed during the industrial revolution. Increased understanding and use of fertilizers were important parts of the pre-industrial British Agricultural Revolution and the industrial Green Revolution of the 20th century.

Inorganic fertilizer use has also significantly supported global population growth — it has been estimated that almost half the people on the Earth are currently fed as a result of synthetic nitrogen fertilizer use.

Fertilizers typically provide, in varying proportions:

- six macronutrients: nitrogen (N), phosphorus (P), potassium (K), calcium (Ca), magnesium (Mg), and sulfur (S);
- seven micronutrients: boron (B), chlorine (Cl), copper (Cu), iron (Fe), manganese (Mn), molybdenum (Mo), and zinc (Zn).

The macronutrients are consumed in larger quantities and are present in plant tissue in quantities from 0.15% to 6.0% on a dry matter (0% moisture) basis (DM). Micronutrients are consumed in smaller quantities and are present in plant tissue on the order of parts per million (ppm), ranging from 0.15 to 400 ppm DM, or less than 0.04% DM.

Only three other macronutrients are required by all plants: carbon, hydrogen, and oxygen. These nutrients are supplied by water and carbon dioxide.

The nitrogen-rich fertilizer ammonium nitrate is also used as an oxidizing agent in improvised explosive devices, sometimes called *fertilizer bombs*, leading to sale regulations·

Labelling

The labelling of fertilizers varies by country in terms of analysis methodology and subsequent nutrient labeling. In most countries the macronutrients are labelled with an *NPK* analysis (in Australia, "N-P-K-S" adding sulfur).

The three numbers on the fertilizer label represent an analysis of the composition by weight. These three numbers correspond to nitrogen, phosphorus, and potassium (N-P-K) and always appear in that specific order. When a 4th number is included, it indicates the sulfur content (N-P-K-S).

While the number for "N" represents the percentage weight of nitrogen, the other two components are not for the analysis of the element, but rather, the analysis of the "available" or "soluble" form of the element. In traditional chemical analysis, the tests used treated the sample so as to measure the equivalent P_2O_5 and K_2O. For instance, some potassium-bearing rocks do not count as having available potassium.

The number for "P" is actually the weight of an equivalent quantity of P_2O_5 and not elemental phosphorus. In order to calculate the weight of P in the formulation, the weight of P_2O_5 can be multiplied by 0.44 to compensate for the weight of the oxygen in the molecule. For example, a bag of 10-10-10 has 10 pounds of nitrogen, 10 pounds of P_2O_5, but only 4.4 pounds of P.

Likewise, the number for "K" is actually the weight of an equivalent quantity of K_2O, and not elemental potassium. In order to calculate the weight of K in the formulation, the weight of K_2O can be multiplied by 0.83 to compensate for the weight of the oxygen in the molecule.

For example, a bag of 10-10-10 has 10 pounds of K_2O, but only 8.3 pounds of K.

As an example, the fertilizer *potash* (in modern times, potassium chloride) is composed of 52% potassium and 48% chlorine by weight; chemical analysis of 100g of potassium chloride (KCl), would show 63g of equivalent potassium oxide (K_2O) when done in the manner of fertilizer analysis. The *percentage yield* of K_2O from the original 100g of fertilizer is the number shown on the label. A potash fertilizer would thus be labelled 0-0-63, and *not* 0-0-52.

History

The modern understanding of plant nutrition dates to the 19th century and the work of Justus von Liebig, among others. Management of soil fertility, however, has been the pre-occupation of farmers for thousands of years.

Forms

Fertilizers come in various forms. The most typical form is granular fertilizer (powder form). The next most common form is liquid fertilizer; some advantages of liquid fertilizer are its immediate effect and wide coverage. There are also slow-release fertilizers (various forms including fertilizer spikes, tabs, etc.) which reduce the problem of "burning" the plants due to excess nitrogen.

More recently, organic fertilizer is on the rise as people are resorting to environmental friendly (or 'green') products. Although organic fertilizer usually contain less nutrients, some people still prefer organic due to natural ingredients·

Inorganic Fertilizer (Synthetic Fertilizer)

Fertilizers are broadly divided into *organic fertilizers* (composed of enriched organic matter—plant or animal), or *inorganic fertilizers* (composed of synthetic chemicals and/or minerals).

Inorganic fertilizer is often synthesized using the Haber-Bosch process, which produces ammonia as the end product. This ammonia is used as a feedstock for other nitrogen fertilizers, such as anhydrous ammonium nitrate and urea. These concentrated products may be diluted with water to form a concentrated liquid fertilizer (e.g. UAN). Ammonia can be combined with rock phosphate and potassium fertilizer in the Odda Process to produce *compound fertilizer*.

The use of synthetic nitrogen fertilizers has increased steadily in the last 50 years, rising almost 20-fold to the current rate of 100 million tonnes of nitrogen per year. The use of phosphate fertilizers has also increased from 9 million tonnes per year in 1960 to 40 million tonnes per year in 2000. A maize crop yielding 6-9 tonnes of grain per hectare requires 31–50 kg of phosphate fertilizer to be applied, soybean requires 20–25 kg per hectare. Yara International is the world's largest producer of nitrogen based fertilizers.

Controlled-release Types

Urea and formaldehyde, reacted together to produce sparingly soluble polymers of various molecular weights, is one of the oldest controlled-nitrogen-release technologies, having been first produced in 1936 and commercialised in 1955. The early product had 60 percent of the total nitrogen cold-water-insoluble, and the unreacted (quick release) less than 15%.

Methylene ureas were commercialised in the 1960s and 1970s, having 25 and 60% of the nitrogen cold-water-insoluble, and unreacted urea nitrogen in the range of 15 to 30%. Isobutylidene diurea, unlike the methylurea polymers, is a single crystalline solid of relatively uniform properties, with about 90% of the nitrogen water-insoluble.

In the 1960s the National Fertilizer Development Centre began developing Sulfur-coated urea; sulfur was used as the principle coating material because of its low cost and its value as a secondary nutrient. Usually there is another wax or polymer which seals the sulfur; the slow release properties depend on the degradation of the secondary sealant by soil microbes as well as mechanical imperfections (cracks, etc.) in the sulfur.

They typically provide 6 to 16 weeks of delayed release in turf applications. When a hard polymer is used as the secondary coating, the properties are a cross between diffusion-controlled particles and traditional sulfur-coated.

Other coated products use thermoplastics (and sometimes ethylene-vinyl acetate and surfactants, etc.) to produce diffusion-controlled release of urea or soluble inorganic fertilizers. "Reactive Layer Coating" can produce thinner, hence cheaper, membrane coatings by applying reactive monomers simultaneously to the soluble particles. "Multicote" is a process applying layers of low-cost fatty acid salts with a paraffin topcoat.

Besides being more efficient in the utilisation of the applied nutrients, slow-release technologies also reduce the impact on the environment and the contamination of the subsurface water.

Table 1: Top users of nitrogen-based fertilizer

Country	Total N use (Mt pa)	Amt. used (feed/pasture)
China	18.7	3.0
U.S.	9.1	4.7
France	2.5	1.3
Germany	2.0	1.2
Brazil	1.7	0.7
Canada	1.6	0.9
Turkey	1.5	0.3
U.K.	1.3	0.9
Mexico	1.3	0.3
Spain	1.2	0.5
Argentina	0.4	0.1

Application

Synthetic fertilizers are commonly used to treat fields used for growing maize, followed by barley, sorghum, rapeseed, soy and sunflower One study has shown that application of nitrogen fertilizer on off-season cover crops can increase the biomass (and subsequent green manure value) of these crops, while having a beneficial effect on soil nitrogen levels for the main crop planted during the summer season.

Nutrients in soil can be thrown out of balance with high concentrations of fertilizers. The interconnectedness and complexity of this soil 'food web' means any appraisal of soil function must necessarily take into account interactions with the living communities that exist within the soil. Stability of the system is reduced by the use of nitrogen-containing fertilizers, which cause soil acidification

Applying excessive amounts of fertilizer has negative environmental effects, and wastes the growers' time and money. To avoid over-application, the nutrient status of crops should be assessed. Nutrient deficiency can be detected by visually assessing the physical symptoms of the crop.

Nitrogen deficiency, for example has a distinctive presentation in some species. However, quantitative tests are more reliable for detecting nutrient deficiency before it has significantly affected the crop. Both soil tests and Plant Tissue Tests are used in agriculture to fine-tune nutrient management to the crops needs.

Problems with Inorganic Fertilizer

Trace Mineral Depletion

Many inorganic fertilizers may not replace trace mineral elements in the soil which become gradually depleted by crops. This depletion has been linked to studies which have shown a marked fall (up to 75%) in the quantities of such minerals present in fruit and vegetables.

In Western Australia deficiencies of zinc, copper, manganese, iron and molybdenum were identified as limiting the growth of broad-acre crops and pastures in the 1940s and 1950s Soils in Western Australia are very old, highly weathered and deficient in many of the major nutrients and trace elements Since this time these trace elements are routinely added to inorganic fertilizers used in agriculture in this state

Overfertilization

Over-fertilization of a vital nutrient can be as detrimental as underfertilization. "Fertilizer burn" can occur when too much fertilizer is applied, resulting in a drying out of the roots and damage or even death of the plant.

High Energy Consumption

In the USA in 2004, 317 billion cubic feet of natural gas was consumed in the industrial production of ammonia, less than 1.5% of total U.S. annual consumption of natural gas. A 2002 report suggested that the production of ammonia consumes about 5% of global natural gas consumption, which is somewhat under 2% of world energy production.

Ammonia is overwhelmingly produced from natural gas, but other energy sources, together with a hydrogen source, can be used for the production of nitrogen compounds suitable for fertilizers. The cost of natural gas makes up about 90% of the cost of producing ammonia. The increase in price of natural gases over the past decade, along with other factors such as increasing demand, have contributed to an increase in fertilizer price.

Long-Term Sustainability

Inorganic fertilizers are now produced in ways which theoretically cannot be continued indefinitely by definition as the resources used in their production are non-renewable. Potassium and phosphorus come from mines (or saline lakes such as the Dead Sea) and such resources are limited. However, more effective fertilizer utilisation practices may decrease present usage from mines. Improved knowledge of crop production practices can potentially decrease fertilizer usage of P and K without reducing the critical need to improve and increase crop yields. Atmospheric (*unfixed*) nitrogen is effectively unlimited (forming over 70% of the atmospheric gases), but this is not in a form useful to plants. To make nitrogen accessible to plants requires nitrogen fixation (conversion of atmospheric nitrogen to a plant-accessible form). Artificial nitrogen fertilizers are typically synthesized using fossil fuels such as natural gas and coal, which are limited resources. In lieu of converting natural gas to syngas for use in the Haber process, it is also possible to convert renewable biomass to syngas (or wood gas) to supply the necessary energy for the process, though the amount of land and resources (ironically often including fertilizer) necessary for such a project may be prohibitive.

Organic Fertilizer

Organic fertilizers include naturally occurring organic materials, (e.g. manure, worm castings, compost, seaweed, guano), or naturally occurring mineral deposits (e.g. saltpeter).

Benefits of Organic Fertilizer

Organic fertilizers have been known to improve biodiversity (soil life) and long-term productivity of soil, and may prove a large depository for excess carbon dioxide.

Organic nutrients increase the abundance of soil organisms by providing organic matter and micronutrients for organisms such as fungal mycorrhiza, (which aid plants in absorbing nutrients), and can drastically reduce external inputs of pesticides, energy and fertilizer, at the cost of decreased yield.

Disadvantages of Organic Fertilizers

- Organic fertilizers may contain pathogens and other disease causing organisms if not properly composted
- Nutrient contents are very variable and their release to available forms that the plant can use may not occur at the right plant growth stage
- Organic fertilizers are comparatively voluminous and can be too bulky to deploy the right amount of nutrients that will be beneficial to plants
- More expensive to produce
- not eco friendly.

Comparison with Inorganic Fertilizer

Organic fertilizer nutrient content, solubility, and nutrient release rates are typically all lower than inorganic fertilizers. One study found that over a 140-day period, after 7 leachings:

- Organic fertilizers had released between 25% and 60% of their nitrogen content
- Controlled release fertilizers (CRFs) had a relatively constant rate of release
- Soluble fertilizer released most of its nitrogen content at the first leaching

In general, the nutrients in organic fertilizer are both more dilute and also much less readily available to plants. According to the University of California's integrated pest management program, all *organic fertilizers* are classified as 'slow-release' fertilizers, and therefore cannot cause nitrogen burn.

Organic fertilizers from composts and other sources can be quite variable from one batch to the next. Without batch testing, amounts of applied nutrient cannot be precisely known. Nevertheless they are at least as effective as chemical fertilizers over longer periods of use.

Example of Organic Fertilizer

Chicken litter, which consists of chicken manure mixed with sawdust, is an organic fertilizer that has been shown to better condition soil for harvest than synthesized fertilizer. Researchers at the Agricultural Research Service (ARS) studied the effects of using chicken litter, an organic fertilizer, versus synthetic fertilizers on cotton fields, and found that fields fertilized with chicken litter had a 12% increase in cotton yields over fields fertilized with synthetic fertilizer. In addition to higher yields, researchers valued commercially sold chicken litter at a $17/ton premium (to a total valuation of $78/ton) over the traditional valuations of $61/ton due to value added as a soil conditioner.

Other ARS studies have found that algae used to capture nitrogen and phosphorus runoff from agricultural fields can not only prevent water contamination of these nutrients, but also can be used as an organic fertilizer. ARS scientists originally developed the "algal turf scrubber" to reduce nutrient runoff and increase quality of water flowing into streams, rivers, and lakes. They found that this nutrient-rich algae, once dried, can be applied to cucumber and corn seedlings and result in growth comparable to that seen using synthetic fertilizers.

Organic Fertilizer Sources

Animal

Animal-sourced and human urea are suitable for application organic agriculture, while pure synthetic forms of urea are not. The common thread that can be seen through these examples is that *organic* agriculture attempts to define itself through minimal processing (in contrast to the man-made Haber process), as well as being naturally occurring or via natural biological processes such as composting.

Besides immediate application of urea to the soil, urine can also be improved by converting it to struvite already done with human urine by a Dutch firm. The conversion is performed by adding magnesium to the urine. An added economical advantage of using urine as fertilizer is that it contains a large amount of phosphorus, a mineral whose production is rapidly decreasing (peak phosphorus) as the mines are running dry.

Sewage sludge (aka biosolids) use is only available to less than 1% of US ag land. USDA prohibits use of sewage sludge in organic agricultural operations in the U.S. has been extremely limited and rare due to of the practice (due to toxic metal accumulation, among other factors). The USDA now requires 3rd-party certification of high-nitrogen liquid organic fertilizers sold in the U.S.

Plant

Leguminous cover crops are also grown to enrich soil as a green manure through nitrogen fixation from the atmosphere; as well as phosphorus (through nutrient mobilisation) content of soils.

Mineral

Mined powdered limestone, rock phosphate and sodium nitrate, are inorganic (not of biologic origins) compounds which are energetically intensive to harvest and are approved for usage in organic agriculture in *minimal* amounts.

Negative Environmental Effects

Water Quality

Eutrophication: The nitrogen-rich compounds found in fertilizer runoff is the primary cause of a serious depletion of oxygen in many parts of the ocean, especially in coastal zones; the resulting lack of dissolved oxygen is greatly reducing the ability of these areas to sustain oceanic fauna. Visually, water may become cloudy and discoloured (green, yellow, brown, or red).

About half of all the lakes in the United States are now eutrophic, while the number of oceanic dead zones near inhabited coastlines are increasing. As of 2006, the application of nitrogen fertilizer is being increasingly controlled in Britain and the United States. If eutrophication *can* be reversed, it may take decades before the accumulated nitrates in groundwater can be broken down by natural processes.

Blue Baby Syndrome

High application rates of inorganic nitrogen fertilizers in order to maximise crop yields, combined with the high solubilities of these fertilizers leads to increased runoff into surface water as well as leaching into groundwater. The use of ammonium nitrate in *inorganic* fertilizers is particularly damaging, as plants absorb ammonium ions preferentially over nitrate ions, while excess nitrate ions which are not absorbed dissolve (by rain or irrigation) into runoff or groundwater.

Nitrate levels above 10 mg/L (10 ppm) in groundwater can cause 'blue baby syndrome' (acquired methemoglobinemia), leading to hypoxia (which can lead to coma and death if not treated).

Soil Acidification

Nitrogen-containing inorganic and organic fertilizers can cause soil acidification when added. This may lead to decreases in nutrient availability which may be offset by liming.

Persistent Organic Pollutants

Toxic persistent organic pollutants ("POPs"), such as Dioxins, polychlorinated dibenzo-p-dioxins (PCDDs), and polychlorinated dibenzofurans (PCDFs) have been detected in agricultural fertilizers and soil amendments

Heavy Metal Accumulation

The concentration of up to 100 mg/kg of cadmium in phosphate minerals (for example, minerals from Nauru and the Christmas islands) increases the contamination of soil with cadmium, for example in New Zealand.

Steel industry wastes, recycled into fertilizers for their high levels of zinc (essential to plant growth), wastes can include the following toxic metals: lead arsenic, cadmium, chromium, and nickel. The most common toxic elements in this type of fertilizer are mercury, lead, and arsenic. Concerns have been raised concerning fish meal mercury content by at least one source in Spain

Radioactive Element Accumulation

Uranium is another example of a contaminant often found in phosphate fertilizers (at levels from 7 to 100 pCi/g). Eventually these heavy metals can build up to unacceptable levels and build up in vegetable produce. Average annual intake of uranium by adults is estimated to be about 0.5 µg (500 mg) from ingestion of food and water and 0.6 µg from breathing air.

Also, highly radioactive Polonium-210 contained in phosphate fertilizers is absorbed by the roots of plants and stored in its tissues; tobacco derived from plants fertilized by rock phosphates contains Polonium-210 which emits alpha radiation estimated to cause about 11,700 lung cancer deaths each year worldwide.

For these reasons, it is recommended that nutrient budgeting, through careful observation and monitoring of crops, take place to mitigate the effects of excess fertilizer application.

Atmosphere

Methane emissions from crop fields (notably rice paddy fields) are increased by the application of ammonium-based fertilizers; these emissions contribute greatly to global climate change as methane is a potent greenhouse gas.

Through the increasing use of nitrogen fertilizer, which is added at a rate of 1 billion tons per year presently to the already existing amount of reactive nitrogen, nitrous oxide (N_2O) has become the third most important greenhouse gas after carbon dioxide and methane. It has a global warming potential 296 times larger than an equal mass of carbon dioxide and it also contributes to stratospheric ozone depletion. Storage and application of some nitrogen fertilizers in some weather or soil conditions can cause emissions of the potent greenhouse gas—nitrous oxide. Ammonia gas (NH_3) may be emitted following application of 'inorganic' fertilizers and/or manures and slurries.

The use of fertilizers on a global scale emits significant quantities of greenhouse gas into the atmosphere. Emissions come about through the use of:

- animal manures and urea, which release methane, nitrous oxide, ammonia, and carbon dioxide in varying quantities depending on their form (solid or liquid) and management (collection, storage, spreading)

- fertilizers that use nitric acid or ammonium bicarbonate, the production and application of which results in emissions of nitrogen oxides, nitrous oxide, ammonia and carbon dioxide into the atmosphere.

By changing processes and procedures, it is possible to mitigate some, but not all, of these effects on anthropogenic climate change.

Other Problems

Increased Pest Fitness

Excessive nitrogen fertilizer applications can also lead to pest problems by increasing the birth rate, longevity and overall fitness of certain agricultural pests, such as aphids (plant lice).

Fertility (Soil)

Fertile soil has the following properties:

- It is rich in nutrients necessary for basic plant nutrition, including nitrogen, phosphorus and potassium.
- It contains sufficient minerals (trace elements) for plant nutrition, including boron, chlorine, cobalt, copper, iron, manganese, magnesium, molybdenum, sulfur, and zinc.
- It contains soil organic matter that improves soil structure and soil moisture retention.
- Soil pH is in the range 6.0 to 6.8 for most plants but some prefer acid or alkaline conditions.
- Good soil structure, creating well drained soil, but some soils are wetter (as for producing rice) or drier (as for producing plants susceptible to fungi or rot) such as agave.
- A range of microorganisms that support plant growth.
- It often contains large amounts of topsoil.

In lands used for agriculture and other human activities, fertile soil typically arises from the use of soil conservation practices.

Soil Fertilization

Bioavailable nitrogen is the element in soil that is most often lacking. Phosphorus and potassium are also needed in substantial amounts. For this reason these three elements are always included in commercial fertilizers, and the content of each of these items is included on the bags of fertilizer. For example a 10-10-15 fertilizer has 10 percent nitrogen, 10 percent (P_2O_5) available phosphorus and 15 percent (K_2O) water soluble potassium.

Inorganic fertilizers are generally less expensive and have higher concentrations of nutrients than organic fertilizers. Some have criticized the use of inorganic fertilizers, claiming that the water-soluble nitrogen doesn't provide for the long-term needs of the plant and creates water pollution. Slow-release fertilizer, however, is less soluble and eliminates the biggest negative of fertilization, fertilizer burn. Additionally, most soluble fertilizers are coated, such as sulfur-coated urea.

In 2008 the cost of phosphorus as fertilizer more than doubled, while the price of rock phosphate as base commodity rose eight-fold. Recently the term peak phosphorus has been coined, due to the limited occurrence of rock phosphate in the world.

Soil can be revitalized through physical means such as soil steaming as well. Superheated steam is induced into the soil to kill pests and unblock nutrients.

Light and CO_2 Limitations

Photosynthesis This is the process whereby plants use light energy to drive chemical reactions which convert CO_2 into sugars. As such, all plants require access to both light and carbon dioxide to produce energy, grow and reproduce.

While typically limited by nitrogen, phosphorus and potassium, low levels of carbon dioxide can also act as a limiting factor on plant growth. Peer-reviewed and published scientific studies have shown that increasing CO_2 is highly effective at promoting plant growth up to levels over 300 ppm. Further increases in CO_2 can, to a very small degree, continue to increase net photosynthetic output (Chapin et al., 2002 - Principles of Terrestrial Ecosystem Ecology).

Since higher levels of CO_2 have only a minimal impact on photosynthetic output at present levels (presently around 380 ppm and increasing), we should not consider plant growth to be limited by carbon dioxide. Other biochemical limitations, such as soil organic content, nitrogen in the soil, phosphorus and potassium, are far more often in short supply. As such, neither commercial nor scientific communities look to air fertilization as an effective or economic method of increasing production in agriculture or natural ecosystems. Furthermore, since microbial decomposition occurs faster under warmer temperatures, higher levels of CO_2 (which is one of the causes of unusually fast climate change) should be expected to increase the rate at which nutrients are leached out of soils and may have a negative impact on soil fertility.

Soil Depletion

Soil depletion occurs when the components which contribute to fertility are removed and not replaced, and the conditions which support soil fertility are not maintained. This leads to poor crop yields. In agriculture, depletion can be due to excessively intense cultivation and inadequate soil management. One of the most widespread occurrences of soil depletion as of 2008 is in tropical zones where nutrient content of soils is low. The combined effects of growing population densities, large-scale industrial logging, slash-and-burn agriculture and ranching, and other factors, have in some places depleted soils through rapid and almost total nutrient removal.

Topsoil depletion occurs when the nutrient-rich organic topsoil, which takes hundreds to thousands of years to build up under natural conditions, is eroded or depleted of its original organic material. Historically, many past civilizations' collapses can be attributed to the depletion of the topsoil. Since the beginning of agricultural production in the Great Plains of North America in the 1880s, about one-half of its topsoil has disappeared.

Depletion may occur through a variety of other effects, including overtillage (which damages soil structure), overuse of inputs such as synthetic fertilizers and herbicides (which leave residues and buildups that inhibit microorganisms), and salinization of soil.

Importance of Agricultural Productivity

The productivity of a region's farms is important for many reasons. Aside from providing more food, increasing the productivity of farms affects the region's prospects for growth and competitiveness on the agricultural market, income distribution and savings, and labour migration. An increase in a region's agricultural productivity implies a more efficient distribution of scarce resources. As farmers adopt new techniques and differences in productivity arise, the more productive farmers benefit from an increase in their welfare while farmers who are not productive enough will exit the market to seek success elsewhere.

As a region's farms become more productive, its comparative advantage in agricultural products increases, which means that it can produce these products at a lower opportunity cost than can other regions. Therefore, the region becomes more competitive on the world market, which means that it can attract more consumers since they are able to buy more of the products offered for the same amount of money.

Increases in agricultural productivity lead also to agricultural growth and can help to alleviate poverty in poor and developing countries, where agriculture often employs the greatest portion of the population. As farms become more productive, the wages earned by those who work in agriculture increase. At the same time, food prices decrease and food supplies become more stable. Labourers therefore have more money to spend on food as well as other products. This also leads to agricultural growth. People see that there is a greater opportunity earn their living by farming and are attracted to agriculture either as owners of farms themselves or as labourers.

However, it is not only the people employed in agriculture who benefit from increases in agricultural productivity. Those employed in other sectors also enjoy lower food prices and a more stable food supply. Their wages may also increase.

Agricultural productivity is becoming increasingly important as the world population continues to grow. India, one of the world's most populous countries, has taken steps in the past decades to increase its land productivity. Forty years ago, North India produced only wheat, but with the advent of the earlier maturing high-yielding wheats and rices, the wheat could be harvested in time to plant rice. This wheat/rice combination is now widely used throughout the Punjab, Haryana, and parts of Uttar Pradesh. The wheat yield of three tons and rice yield of two tons combine for five tons of grain per hectare, helping to feed India's 1.1 billion people.

Agricultural Productivity and Sustainable Development

Increase in agricultural productivity are often linked with questions about sustainability and sustainable development. Changes in agricultural practices necessarily bring changes in demands on resources. This means that as regions implement measures to increase the productivity of their farm land, they must also find ways to ensure that future generations will also have the resources they will need to live and thrive.

U.S. Agriculture Productivity

Between 1950 and 2000, during the so called "second agricultural revolution of modern times", U.S. agricultural productivity rose fast, especially due to the development of new technologies. For example, the average amount of milk produced per cow increased from 5,314 pounds to 18,201 pounds per year (+242%), the average yield of corn rose from 39 bushels to 153 bushels per acre (+292%), and each farmer in 2000 produced on average 12 times as much farm output per hour worked as a farmer did in 1950.

Productive Farms

For many farmers (especially in non-industrial countries) agricultural productivity may mean much more. A productive farm is one that provides most of the resources necessary for the farmer's family to live, such as food, fuel, fibre, healing plants, etc. It is a farm which ensures food security as well as a way to sustain the well-being of a community. This implies that a productive farm is also one which is able to ensure proper management of natural resources, such as biodiversity, soil, water, etc. For most farmers, a productive farm would also produce more goods than required for the community in order to allow trade.

Diversity in agricultural production is one key to productivity, as it enables risk management and preserves potentials for adaptation and change. Monoculture is an example of such a nondiverse production system. In a monocultural system a farmer may produce only crops, but no livestock, or only livestock and no crop.

The benefits of raising livestock, among others, are that it provides multiple goods, such as food, wool, hides, and transportation. It also has an important value in term of social relationships (such as gifts in weddings). In case of famine, when crops are not sufficient to ensure food safety, livestock can be used as food. Livestock may also provide manure, which can be used to fertilize cultivated soils, which increases soil productivity. On the other hand, in an agricultural system based only on raising livestock, food has to be bought to other farmers, and wastes produced cannot be easily disposed of. Production has many functions, and diversity is the foundation of such production. To ignore the complex functions provided by a farm is thought by many to turn agricultural production into a commodity.

Abiotic Component

In biology and ecology, abiotic components are non-living chemical and physical factors in the environment which affect ecosystems. Abiotic phenomena underlie all of biology. Abiotic factors, while generally downplayed, can have enormous impact on ramonds evolution. Abiotic components are aspects of geodiversity. They can also be recognised as "abiotic pathogens"

From the viewpoint of biology, abiotic influences may be classified as light or more generally radiation, temperature, water, the chemical surrounding composed of the terrestrial atmospheric gases, as well as soil. The macroscopic climate often influences each of the above. Not to mention pressure and even sound waves if working with marine, or deep underground, biome.

Those underlying factors affect different plants, animals and fungi to different extents. Some plants are mostly water starved, so humidity plays a larger role in their biology. If there is little or no sunlight then plants may wither and die from not being able to get enough sunlight to do photosynthesis. Many archaebacteria require very high temperatures, or pressures, or unusual concentrations of all those who can figh chemical substances such as sulfur, because of their specialisation into extreme conditions. Certain fungi have evolved to survive mostly at the temperature, the humidity, and stability of their environment.

For example, there is a significant difference in access to water as well as humidity between temperate rainforests and deserts. This difference in water access causes a diversity in the types of plants and animals that grow in these areas.

Abiotic Stress

Abiotic stress is defined as the negative impact of non-living factors on the living organisms in a specific environment. The non-living variable must influence the environment beyond its normal range of variation to adversely affect the population performance or individual physiology of the organism in a significant way. Whereas a biotic stress would include such living disturbances as fungi or harmful insects, abiotic stress factors, or stressors, are naturally occurring, often intangible, factors such as intense sunlight or wind that may cause harm to the plants and animals in the area affected. Abiotic stress is essentially unavoidable.

Abiotic stress affects animals, but plants are especially dependent on environmental factors, so it is particularly constraining. Abiotic stress is the most harmful factor concerning the growth and productivity of crops worldwide. Research has also shown that abiotic stressors are at their most harmful when they occur together, in combinations of abiotic stress factors.

Examples of Abiotic Stress

Abiotic stress comes in many forms. The most common of the stressors are the easiest for people to identify, but there are many other, less recognisable abiotic stress factors which affect environments constantly. The most basic stressors include: high winds, extreme temperatures, drought, flood, and other natural disasters, such as tornados and wildfires. The lesser-known stressors generally occur on a smaller scale and so are less noticeable, but they include: poor edaphic conditions like rock content and pH, high radiation, compaction, contamination, and other, highly specific conditions like rapid rehydration during seed germination.

Effects of Abiotic Stress

Abiotic stress, as a natural part of every ecosystem, will affect organisms in a variety of ways. Although these effects may be either beneficial or detrimental, the location of the area is crucial in determining the extent of the impact that abiotic stress will have. The higher the latitude of the area affected, the greater the impact of abiotic stress will be on that area. So, a taiga or boreal forest is at the mercy of whatever abiotic stress factors may come along, while tropical zones are much less susceptible to such stressors.

Benefits

One example of a situation where abiotic stress plays a constructive role in an ecosystem is in natural wildfires. While they can be a human safety hazard, it is productive for these ecosystems to burn out every once in a while so that new organisms can begin to grow and thrive. Even though it is healthy for an ecosystem, a wildfire can still be considered an abiotic stressor, because it puts an obvious stress on individual organisms within the area. Every tree that is scorched and each bird nest that is devoured is a sign of the abiotic stress. On the larger scale, though, natural wildfires are positive manifestations of abiotic stress. What also needs to be taken into account when looking for benefits of abiotic stress, is that one phenomenon may not affect an entire ecosystem in the same way. While a flood will kill most plants living low on the ground in a certain area, if there is rice there, it will thrive in the wet conditions. Another example of this is in, phytoplankton and zooplankton. The same types of conditions are usually considered stressful for these two types of organisms. They act very similarly when exposed to ultraviolet light and most toxins, but at elevated temperatures the phytoplankton reacts negatively, while the thermophilic zooplankton reacts positively to the increase in temperature. The two may be living in the same environment, but an increase in temperature of the area would prove stressful only for one of the organisms. Lastly, abiotic stress has enabled species to grow, develop, and evolve, furthering natural selection as it picks out the weakest of a group of organisms. Both plants and animals have evolved mechanisms allowing them to survive extremes.

Detriments

The most obvious detriment concerning abiotic stress involves farming. It has been claimed by one study that abiotic stress causes the most crop loss of any other factor and that most major crops are reduced in their yield by more than 50% from their potential yield. It has also been speculated that this yield reduction will only worsen with the dramatic climate changes expected in the future. Because abiotic stress is widely considered a detrimental effect, the research on this branch of the issue is extensive.

In Plants

A plant's first line of defence against abiotic stress is in its roots. If the soil holding the plant is healthy and biologically diverse, the plant will have a higher chance of surviving stressful conditions.

Facilitation, or the positive interactions between different species of plants, is an intricate web of association in a natural environment. It is how plants work together. In areas of high stress, the level of facilitation is especially high as well. This could possibly be because the plants need a stronger network to survive in a harsher environment, so their interactions between species, such as cross-pollination or mutualistic actions, become more common to cope with the severity of their habitat. This facilitation will not go so far as to protect an entire species, however. For example, cold weather crops like rye, oats, wheat, and apples are expected to decline by about 15% in the next fifty years and strawberries will drop as much as 32% simply because of projected climate changes of a few degrees. Plants are extremely sensitive to such changes, and do not generally adapt quickly.

Plants also adapt very differently from one another, even from a plant living in the same area. When a group of different plant species was prompted by a variety of different stress signals, such as drought or cold, each plant responded uniquely. Hardly any of the responses were similar, even though the plants had become accustomed to exactly the same home environment. Rice (*Oryza sativa*) is a classic example. Rice is a staple food throughout the world, especially in China and India. Nowadays, due to climate change rice plants experience different types of abiotic stresses, like drought and high salinity. These stress conditions have a negative impact on rice production. Genetic diversity has been studied among several rice varieties with different genotypes using molecular markers. Several projects (DST/CSIR/UGC funded) related to abiotic stress tolerance in rice and its wild ancestor (*Oryza rufipogon*) were undertaken by many scientists in India, especially West Bengal.

In Animals

For animals, the most stressful of all the abiotic stressors is heat. This is because many species are unable to regulate their internal body temperature. Even in the species that are able to regulate their own temperature, it is not always a completely accurate system. Temperature determines metabolic rates, heart rates, and other very important factors within the bodies of animals, so an extreme temperature change can easily distress the animal's body. Animals can respond to extreme heat, for example, through natural heat acclimation or by burrowing into the ground to find a cooler space. It is also possible to see in animals that a high genetic diversity is beneficial in providing resiliency against harsh abiotic stressors. This acts as a sort of stock room when a species is plagued by the perils of natural selection. A variety of galling insects are among the most specialised and diverse herbivores on the planet, and their extensive protections against abiotic stress factors have helped the insect in gaining that position of honour.

In Endangered Species

Biodiversity is determined by many things, and one of them is abiotic stress. If an environment is highly stressful, biodiversity tends to be low. If abiotic stress does not have a strong presence in an area, the biodiversity will be much higher. This idea leads into the understanding of how abiotic stress and endangered species are related. It has been observed through a variety of environments that as the level of abiotic stress increases, the number of species decreases. This means that species are more likely to become population threatened, endangered, and even extinct, when and where abiotic stress is especially harsh.

Chapter 2

Biotic Stress

Biotic stress is stress that occurs as a result of damage done to plants by other living organisms, such as bacteria, viruses, fungi, parasites, beneficial and harmful insects, weeds, and cultivated or native plants.

Agriculture

It is a major focus of agricultural research, due to the vast economic losses caused by biotic stress to cash crops. The relationship between biotic stress and plant yield affects economic decisions as well as practical development. The impact of biotic injury on crop yield impacts population dynamics, plant-stressor coevolution, and ecosystem nutrient cycling. Biotic stress also impacts horticultural plant health and natural habitats ecology.

Biological Pest Control

Biological pest control herbivore agents feeding on invasive species results in biotic stress, part of reducing and controlling some noxious weeds in agricultural areas and natural ecosystems. The use of mottled water hyacinth weevil ("Neochetina eichhorniae") worldwide on water hyacinth (*Eichhornia crassipes*) is an example. More dominant cultivated or native plants in biological weed control can cause biotic stress from root, water, or nutrient competition below the surface, or from shading of sunlight above. This method can be part of habitat restoration projects.

Crop

A crop is a non-animal species or variety that is grown to be harvested as food, livestock fodder, fuel or for any other economic purpose. Major world crops include maize (corn), wheat, rice, soybeans, hay, potatoes and cotton. While the term "crop" most commonly refers to plants, it can also include species from other biological kingdoms.

For example, mushrooms like shiitake, which are in the fungi kingdom, can be referred to as crops. In addition, certain species of algae are also cultivated, although it is also harvested from the wild. In contrast, animal species that are raised by humans are called livestock, except those that are kept as pets. Microbial species, such as bacteria or viruses, are referred to as cultures. Microbes are not typically grown for food, but are rather used to alter food. For example, bacteria is used to ferment milk to produce yogurt.

Break Crop

Break crop is a term for the secondary crop within the practice of sustainable agriculture with intensive arable farming whereby as part of a crop rotation, a physiologically different crop is inserted into the main cropping plan in order to provide a "break" from the cycle of weeds, pests and diseases encountered with the latter. The aim is to optimise yields of the primary crops and therefore income while reducing the use, and cost, of pesticides. Nitrogen fixation can be another soil goal.

An example rotation would be winter oilseed rape as a break crop, followed by two crops of winter wheat, then winter barley or setaside. Another common example is maize (corn) that is typically rotated with cotton plantations.

Bumper Crop

In agriculture, a bumper crop refers to a particularly productive harvest yielded for a particular crop.

Example: "With all the rain we've had over the last few months, we are expecting a bumper crop this year."

The word "bumper" has a second definition meaning "something unusually large," which is where this term comes from

Cash Crop

In agriculture, a cash crop is a crop which is grown for profit.

The term is used to differentiate from subsistence crops, which are those fed to the producer's own livestock or grown as food for the producer's family. In earlier times cash crops were usually only a small (but vital) part of a farm's total yield, while today, especially in the developed countries, almost all crops are mainly grown for cash. In non-developed nations, cash crops are usually crops which attract demand in more developed nations, and hence have some export value. In many tropical and subtropical areas, jute, coffee, cocoa, sugar cane, bananas, oranges and cotton are common cash crops. In cooler areas, grain crops, oil-yielding crops and some vegetables and herbs are predominate; an example of this is the United States, where corn, wheat, soybean are the predominant cash crops. Coca, poppies and cannabis are other popular black-market cash crops, the prevalence of which varies. In the United States cannabis is considered by some to be the most valuable cash crop.

Prices for major cash crops are set in commodity markets with global scope, with some local variation (called basis) based on freight costs and local supply and demand balance. A consequence of this is that a nation, region, or individual producer relying on such a crop may suffer low prices should a bumper crop elsewhere lead to excess supply on the global markets. This system is criticized by traditional farmers. Coffee is a major part of this.

Issues involving subsidies and trade barriers on such crops have become controversial in discussions of globalisation. Many developing nations take the position that the current international trade system is unfair because it has caused tariffs to be lowered in industrial goods while allowing for low tariffs and agricultural subsidies for agricultural goods. This makes it difficult for a developing nation to export its goods overseas, and forces developing nations to compete with imported goods which are exported from developed nations at artificially low prices.

The practice of exporting at artificially low prices is known as dumping, and is illegal in most nations. Controversy over this issue led to the collapse of the Cancún trade talks in 2003, when the Group of 22 refused to consider agenda items proposed by the European Union unless the issue of agricultural subsidies were addressed.

Cash crop production in the United States is an often debated subject between traditional farmers and local food enthusiasts who support solely local farming practices. Cash cropping in the United States came to the forefront after the Baby Boom generation and the end of World War II as a way to feed the large population boom and continues to be the main factor in having an affordable food supply in the United States. According to the 1997 Ag Census, 90% of the farms in the United States are still owned by families, with an additional 6% owned by a partnership. Cash crop farmers are continually utilising cutting edge technology combined with time-tested practices to produce healthy, affordable food.

Catch Crop

In agriculture, a catch crop is a fast-growing crop that is grown simultaneously with, or between successive plantings of a main crop.

For example, radishes that mature from seed in 25–30 days can be grown between rows of most vegetables, and harvested long before the main crop matures. Or, a catch crop can be planted between the spring harvest and fall planting of some crops.

Catch cropping is a type of succession planting. It makes more efficient use of growing space.

Catch crops are also crops that are sown to prevent minerals being flushed away from the soil. By using catch crops, such as grain (millet,...) one can keep certain minerals not attached to the humous-clay connection (such as carbon (C) and other positively charged elements) in the soil for (many) years.

Cover Crop

Cover crops are crops planted primarily to manage soil fertility, soil quality, water, weeds, pests, diseases, biodiversity and wildlife in *agroecosystems* (Lu *et al.* 2000), ecological systems managed and largely shaped by humans across a range of intensities to produce food, feed, or fiber.

Cover crops are of interest in sustainable agriculture as many of them improve the sustainability of agroecosystem attributes and may also indirectly improve qualities of neighbouring natural ecosystems. Farmers choose to grow and manage specific cover crop types based on their own needs and goals, influenced by the biological, environmental, social, cultural, and economic factors of the food system within which farmers operate. (Snapp *et al.* 2005)

Soil Fertility Management

One of the primary uses of cover crops is to increase soil fertility. These types of cover crops are referred to as "green manure." They are used to manage a range of soil macronutrients and micronutrients. Of the various nutrients, the impact that cover crops have on nitrogen management has received the most attention from researchers and farmers, because nitrogen is often the most limiting nutrient in crop production.

Often, green manure crops are grown for a specific period, and then plowed under before reaching full maturity in order to improve soil fertility and quality.

Green manure crops are commonly leguminous, meaning they are part of the Fabaceae (pea) family. This family is unique in that all of the species in it set pods, such as bean, lentil, lupins and alfalfa. Leguminous cover crops are typically high in nitrogen and can often provide the required quantity of nitrogen for crop production. In conventional farming, this nitrogen is typically applied in chemical fertilizer form. This quality of cover crops is called fertilizer replacement value (Thiessen-Martens *et al.* 2005).

Another quality unique to leguminous cover crops is that they form symbiotic relationships with the rhizobial bacteria that reside in legume root nodules. Lupins is nodulated by the soil microorganism *Bradyrhizobium* sp. (Lupinus). Bradyrhizobia are encountered as microsymbionts in other leguminous crops (Argyrolobium, Lotus, Ornithopus, Acacia, Lupinus) of Mediterranean origin. These bacteria convert biologically unavailable atmospheric nitrogen gas (N2) to biologically available mineral nitrogen (NH4+) through the process of biological nitrogen fixation.

Prior to the advent of the Haber-Bosch process, an energy-intensive method developed to carry out industrial nitrogen fixation and create chemical nitrogen fertilizer, most nitrogen introduced to ecosystems arose through biological nitrogen fixation (Galloway *et al.* 1995). Some scientists believe that widespread biological nitrogen fixation, achieved mainly through the use of cover crops, is the only alternative to industrial nitrogen fixation in the effort to maintain or increase future food production levels (Bohlool *et al.* 1992, Peoples and Craswell 1992, Giller and Cadisch 1995). Industrial nitrogen fixation has been criticized as an unsustainable source of nitrogen for food production due to its reliance on fossil fuel energy and the environmental impacts associated with chemical nitrogen fertilizer use in agriculture (Jensen and Hauggaard-Nielsen 2003). Such widespread environmental impacts include nitrogen fertilizer losses into waterways, which can lead to eutrophication (nutrient loading) and ensuing hypoxia (oxygen depletion) of large bodies of water.

An example of this lies in the Mississippi Valley Basin, where years of fertilizer nitrogen loading into the watershed from agricultural production have resulted in a hypoxic "dead zone" off the Gulf of Mexico the size of New Jersey (Rabalais *et al.* 2002). The ecological complexity of marine life in this zone has been diminishing as a consequence (CENR 2000). As well as bringing nitrogen into agroecosystems through biological nitrogen fixation, types of cover crops known as "catch crops" are used to retain and recycle soil nitrogen already present. The catch crops take up surplus nitrogen remaining from fertilization of the previous crop, preventing it from being lost through leaching (Morgan *et al.* 1942), or gaseous denitrification or volatilization (Thorup-Kristensen *et al.* 2003).

Catch crops are typically fast-growing annual cereal species adapted to scavenge available nitrogen efficiently from the soil (Ditsch and Alley 1991). The nitrogen tied up in catch crop biomass is released back into the soil once the catch crop is incorporated as a green manure or otherwise begins to decompose.

An example of green manure use comes from Nigeria, where the cover crop *Mucuna pruriens* (velvet bean) has been found to increase the availability of phosphorus in soil after a farmer applies rock phosphate (Vanlauwe *et al.* 2000).

Soil Quality Management

Cover crops can also improve soil quality by increasing soil organic matter levels through the input of cover crop biomass over time. Increased soil organic matter enhances soil structure, as well as the water and nutrient holding and buffering capacity of soil (Patrick *et al.* 1957). It can also lead to increased soil carbon sequestration, which has been promoted as a strategy to help offset the rise in atmospheric carbon dioxide levels (Kuo *et al.* 1997, Sainju *et al.* 2002, Lal 2003).

Although cover crops can perform multiple functions in an agroecosystem simultaneously, they are often grown for the sole purpose of preventing soil erosion. Soil erosion is a process that can irreparably reduce the productive capacity of an agroecosystem. Dense cover crop stands physically slow down the velocity of rainfall before it contacts the soil surface, preventing soil splashing and erosive surface runoff (Romkens *et al.* 1990). Additionally, vast cover crop root networks help anchor the soil in place and increase soil porosity, creating suitable habitat networks for soil macrofauna (Tomlin *et al.* 1995).

Soil quality is managed to produce optimum circumstances for crops to flourish. The principal factors of soil quality are soil salination, pH, microorganism balance and the prevention of soil contamination.

Water Management

By reducing soil erosion, cover crops often also reduce both the rate and quantity of water that drains off the field, which would normally pose environmental risks to waterways and ecosystems downstream (Dabney *et al.* 2001). Cover crop biomass acts as a physical barrier between rainfall and the soil surface, allowing raindrops to steadily trickle down through the soil profile. Also, as stated above, cover crop root growth results in the formation of soil pores, which in addition to enhancing soil macrofauna habitat provides pathways for water to filter through the soil profile rather than draining off the field as surface flow. With increased water infiltration, the potential for soil water storage and the recharging of aquifers can be improved (Joyce *et al.* 2002).

Just before cover crops are killed (by such practices including mowing, tilling, discing, rolling, or herbicide application) they contain a large amount of moisture. When the cover crop is incorporated into the soil, or left on the soil surface, it often increases soil moisture. In agroecosystems where water for crop production is in short supply, cover crops can be used as a mulch to conserve water by shading and cooling the soil surface. This reduces evaporation of soil moisture. In other situations farmers try to dry the soil out as quickly as possible going into the planting season. Here prolonged soil moisture conservation can be problematic.

While cover crops can help to conserve water, in temperate regions (particularly in years with below average precipitation) they can draw down soil water supply in the spring, particularly if climatic growing conditions are good. In these cases, just before crop planting, farmers often face a tradeoff between the benefits of increased cover crop growth and the drawbacks of reduced soil moisture for cash crop production that season.

Weed Management

Thick cover crop stands often compete well with weeds during the cover crop growth period, and can prevent most germinated weed seeds from completing their life cycle and reproducing. If the cover crop is left on the soil surface rather than incorporated into the soil as a green manure after its growth is terminated, it can form a nearly impenetrable mat. This drastically reduces light transmittance to weed seeds, which in many cases reduces weed seed germination rates (Teasdale 1993). Furthermore, even when weed seeds germinate, they often run out of stored energy for growth before building the necessary structural capacity to break through the cover crop mulch layer. This is often termed the *cover crop smother effect* (Kobayashi *et al.* 2003). Some cover crops suppress weeds both during growth and after death

(Blackshaw *et al.* 2001). During growth these cover crops compete vigorously with weeds for available space, light, and nutrients, and after death they smother the next flush of weeds by forming a mulch layer on the soil surface.

For example, Blackshaw *et al.* (2001) found that when using *Melilotus officinalis* (yellow sweetclover) as a cover crop in an improved fallow system (where a fallow period is intentionally improved by any number of different management practices, including the planting of cover crops), weed biomass only constituted between 1-12% of total standing biomass at the end of the cover crop growing season. Furthermore, after cover crop termination, the yellow sweetclover residues suppressed weeds to levels 75-97% lower than in fallow (no yellow sweetclover) systems.

In addition to competition-based or physical weed suppression, certain cover crops are known to suppress weeds through allelopathy (Creamer *et al.* 1996, Singh *et al.* 2003).

This occurs when certain biochemical cover crop compounds are degraded that happen to be toxic to, or inhibit seed germination of, other plant species. Some well known examples of allelopathic cover crops are *Secale cereale* (rye), *Vicia villosa* (hairy vetch), *Trifolium pratense* (red clover), *Sorghum bicolour* (sorghum-sudangrass), and species in the brassicaceae family, particularly mustards (Haramoto and Gallandt 2004). In one study, rye cover crop residues were found to have provided between 80% and 95% control of early season broadleaf weeds when used as a mulch during the production of different cash crops such as soybean, tobacco, corn, and sunflower (Nagabhushana *et al.* 2001).

In a recent study released by the Agricultural Research Service (ARS) scientists examined how rye seeding rates and planting patterns affected cover crop production. The results show that planting more pounds per acre of rye increased the cover crop's production as well as decreased the amount of weeds. The same was true when scientists tested seeding rates on legumes and oats; a higher density of seeds planted per acre decreased the amount of weeds and increased the yield of legume and oat production. The planting patterns, which consisted of either traditional rows or grid patterns, did not seem to make a significant impact on the cover crop's production or on the weed production in either cover crop. The ARS scientists concluded that increased seeding rates could be an effective method of weed control.

Disease Management

In the same way that allelopathic properties of cover crops can suppress weeds, they can also break disease cycles and reduce populations of bacterial and fungal diseases (Everts 2002), and parasitic nematodes (Potter *et al.* 1998, Vargas-Ayala *et al.* 2000). Species in the brassicaceae family, such as mustards, have been widely shown to suppress fungal disease populations through the release of naturally occurring toxic chemicals during the degradation of glucosinolade compounds in their plant cell tissues (Lazzeri and Manici 2001).

Pest Management

Some cover crops are used as so-called "trap crops", to attract pests away from the crop of value and toward what the pest sees as a more favourable habitat (Shelton and Badenes-Perez 2006). Trap crop areas can be established within crops, within farms, or within landscapes. In many cases the trap crop is grown during the same season as the food crop being produced. The limited area occupied by these trap crops can be treated with a pesticide once pests are drawn to the trap in large enough numbers to reduce the pest populations. In some organic systems, farmers drive over the trap crop with a large vacuum-based implement to physically pull the pests off the plants and out of the field (Kuepper and Thomas 2002). This system has been recommended for use to help control the lygus bugs in organic strawberry production (Zalom *et al.* 2001).

Other cover crops are used to attract natural predators of pests by providing elements of their habitat. This is a form of biological control known as habitat augmentation, but achieved with the use of cover crops (Bugg and Waddington 1994). Findings on the relationship between cover crop presence and predator/pest population dynamics have been mixed, pointing toward the need for detailed information on specific cover crop types and management practices to best complement a given integrated pest management strategy. For example, the predator mite Euseius tularensis (Congdon) is known to help control the pest citrus thrips in Central California citrus orchards. Researchers found that the planting of several different leguminous cover crops (such as bell bean, woollypod vetch, New Zealand white clover, and Austrian winter pea) provided sufficient pollen as a feeding source to cause a seasonal increase in Congdon populations, which with good timing could potentially introduce enough predatory pressure to reduce pest populations of citrus thrips (Grafton-Cardwell *et al.* 1999).

Diversity and Wildlife

Although cover crops are normally used to serve one of the above discussed purposes, they often simultaneously improve farm habitat for wildlife. The use of cover crops adds at least one more dimension of plant diversity to a cash crop rotation. Since the cover crop is typically not a crop of value, its management is usually less intensive, providing a window of "soft" human influence on the farm. This relatively "hands-off" management, combined with the increased on-farm heterogeneity created by the establishment of cover crops, increases the likelihood that a more complex trophic structure will develop to support a higher level of wildlife diversity (Freemark and Kirk 2001).

In one study, researchers compared arthropod and songbird species composition and field use between conventionally and cover cropped cotton fields in the Southern United States. The cover cropped cotton fields were planted to clover, which was left to grow in between cotton rows throughout the early cotton growing season (stripcover cropping). During the migration and breeding season, they found that songbird densities were 7–20 times higher in the cotton fields with integrated clover cover crop than in the conventional cotton fields. Arthropod abundance and biomass was also higher in the clover cover cropped fields throughout much of the songbird breeding season, which was attributed to an increased supply of flower nectar from the clover. The clover cover crop enhanced songbird habitat by providing cover and nesting sites, and an increased food source from higher arthropod populations (Cederbaum *et al.* 2004).

Crop Diversity

Crop diversity is the variance in genetic and phenotypic characteristics of plants used in agriculture. Crops may vary in seed size, branching pattern, in height, flower colour, fruiting time, or flavour. They may also vary in less obvious characteristics such as their response to heat, cold or drought, or their ability to resist specific diseases and pests. It is possible to discover variation in almost every conceivable trait, including nutritional qualities, preparation and cooking techniques, and of course how a crop tastes. And if a trait cannot be found in the crop itself, it can often be found in a wild relative of the crop; a plant that has similar species that have not been farmed or used in agriculture, but exist in the wild. Diversity in a crop can also result from different growing conditions: a crop growing in nutrient poor soil is likely to be shorter than a crop growing in more fertile soil. In addition, and perhaps most importantly, diversity of a harvested plant can be the result of genetic differences: a crop may have genes conferring early maturity or disease resistance. It is these heritable traits that are of special interest as they are passed on from generation to generation and collectively determine a crop's overall characteristics and future potential. Through combining genes for different traits in desired combinations, plant breeders are able to develop new crop varieties to meet specific conditions. A new variety might, for example, be higher yielding, more disease resistant and have a longer shelf life than the varieties from which it was bred. The practical use of crop diversity goes back to early agricultural methods of crop rotation and fallow fields, planting and harvesting one type of crop on a plot of land one year, and using a different crop the next based on differences in a plant's nutrient needs. Both farmers and scientists must continually draw on the irreplaceable resource of genetic diversity to ensure productive harvests, as genetic variability provides farmers resilience to pests and diseases and allows scientists access to a more diverse genetic bank. Diversification of harvests and maintaining wild biodiversity in crop relatives influence many aspects of human and global interaction, being important for environmental and species sustainability.

Benefits to the Environment

The loss of biodiversity is considered one of today's most serious environmental concerns by the Food and Agriculture Organisation of the United Nations. According to some estimates, if current trends persist, as many as half of all plant species could face extinction. Among the many threatened species are wild relatives of our crops – species that could contribute invaluable traits to future crop varieties. It has been estimated that 6% of wild relatives of cereal crops (wheat, maize, rice, sorghum etc.) are under threat as are 18% of legume species (the wild relatives of beans, peas and lentils) and 13% of species within the botanical family that includes potato, tomato, eggplant, and pepper. The wise use of crop genetic diversity in plant breeding can contribute significantly to protecting the environment. Crop varieties that are resistant to pests and diseases can reduce the need for application of harmful pesticides more vigorous varieties can better compete with weeds; reducing the need for applying herbicides as in the case study at Aarhus University in Denmark using more robust maize; drought resistant plants can help save water through reducing the need for irrigation; deeper rooting varieties can help stabilise soils; and varieties that are more efficient in their use of nutrients require less fertilizer. Most importantly, perhaps, productive agricultural systems reduce or eliminate the need to cut down forest or clear fragile lands to create more farmland for food production.

Crop Diversity and the Economy

Agriculture is the economic foundation of most countries, and for developing countries the most likely source of economic growth. Growth is most rapid where agricultural productivity has risen the most, and the reverse is also true. Growth in agriculture, although beneficial for the wider economy, benefits the poor most, and by providing affordable food these benefits extend beyond the 70% of the world's poorest people who live in rural areas and for whose livelihoods agriculture remains central.

Ensuring agriculture is able to play this fundamental role requires a range of improvements including: the growing of higher value crops, promoting value-adding activities through, for example, improved processing, expanding access to markets, and lowering food prices through increasing production, processing and marketing efficiency, particularly for subsistence and very low income farming families. Fundamental to all of these potential solutions is crop diversity – the diversity that enables farmers and plant breeders to develop higher yielding, more productive varieties having improved quality characteristics required by farmers and desired by consumers. They can breed varieties better suited to particular processing methods or that store longer or that can be transported with less loss. They can produce varieties that resist pests and diseases and are drought tolerant, providing more protection against crop failure and better insulating poor farmers from risk. Agriculture's part in fighting poverty is complex, but without the genetic diversity found within crops, it cannot fulfill its potential.

Disease Threats to Crops with Low Genetic Diversity

One particular threat to mass producing plants for harvest is their susceptibility to diseases. Generally speaking, a species has a range of genetic variability that allows for individuals and/or populations within that species to survive should a stressor or disturbance occur. In the case of agriculture, this is a tricky business to ensure, as seeds are planted under uniform conditions. For example, monocultural agriculture potentially elicits low crop diversity (especially if the seeds were mass produced or cloned). It is possible that a single pest or disease could wipe out entire areas of a crop due to this uniformity. One of the more historically known examples of harvests that suffered from low crop diversity was the Irish Potato Famine of 1845-1847.

One growing danger to present day agriculture is something called wheat rust: the name given from the reddish spores, it is a fungus that attaches to plants and breaks them down for food. A new form of the wheat disease - stem rust, strain Ug99 - has spread from Africa across to the Arabian Peninsula. This development was summarised on the January 16th 2007 by the international research centres Borlaug Global Rust Initiative and the Agricultural Research Service of the United States Department of Agriculture over 2 years of observation after its initial outbreak. The Ug99 stem rust has recently proven to be even more virulent than other forms. Observations from field trials in Kenya showed that more than 85% of wheat samples, including cultivars from the major wheat producing regions in the world, have succumbed to the pest. This is a serious pest alert for one of the major food crops of the world. The key to overcome the threat is genetic resistance found in certain wheat varieties. As Nobel laureate Norman Borlaug puts it: "We know what to do and how to do it. All we need are the financial resources, scientific cooperation and political will to contain this threat to world food security."

Reports from Burundi and Angola warn of another looming food crisis partly caused by outbreaks of the African Cassava Mosaic Virus (ACMD). Creating a "mosaic" of decay on the plants leaves, ACMD is responsible for the loss of a million tons of food each year. The Famine Early Warning Network of USAID reports from Angola that pockets of food insecurity exist in a number of districts partly due to the impacts of mosaic virus on the cassava crop. Likewise FAO (Food and Agriculture Organisation of the United Nations) has warned about food insecurity in north and east central Burundi and one of the factors causing the precarious situation is declining yam harvests and the losses of cassava crops to the mosaic virus. CMD also affects people already exposed to malnutrition and with limited coping mechanisms. CMD continues to be prevalent in all the main cassava-growing areas in the Great Lakes region of east Africa, causing between 20 and 90 percent crop losses in the Congo. Breeders and relief agencies work together to fight the disease, and the FAO emergency relief and rehabilitation program is engaged in a project to assist vulnerable returnee populations in the African Great Lakes Region through mass propagation and distribution of CMD resistant or highly tolerant cassava planting materials.

A well known occurrence of disease susceptibility in crops lacking diversity concerns the Gros Michel, a seedless banana that saw world marketing in the 1940s. As the market demand became high for this particular species, growers and farmers of the Gros Michel banana began to use this species almost exclusively.

Genetically, these bananas are duplicates of every other in their species due to its self-pollinating reproductive style, and because of this lack of genetic diversity, are now virtually extinct due to a single fungus; Panama Disease. This fungus (also known as Fusarium wilt), which infected Gros Michel banana crops in the 1950s, completely wiped out the Gros Michel as the predecessor to the current, and most popular, banana on the market: the Cavendish.

Organisations, Technology and Solutions

The implications of crop diversity are at both the local and world level, and numerous organisations are emerging with great global backing in response to this ideology. International Plant Genetic Resources Institute (IPGRI – now known as Bioversity International), the International Institute of Tropical Agriculture (ITTA), the Borlaug Global Rust Initiative, and the International Network for Improvement of Banana and Plantain (INIBAP) are a few of the most prominent. Members of the United Nations, at the World Summit on Sustainable Development 2002 at Johannesburg, said that crop diversity is in danger of being lost if measures are not taken.

One such step taken in the action against the loss of biodiversity among crops is called gene banking. There are a number of organisations that enlist teams of local farmers to grow native varieties, particularly those that are threatened by extinction due to lack of modern-day use.

There are also local, national and international efforts to preserve agricultural genetic resources through ex situ (off-site) methods such as seed and sperm banks for further research and/or crop breeding. Some of the major germplasm storage efforts include:

- The Global Crop Diversity Trust is an independent international organisation which exists to ensure the conservation and availability of crop diversity for food security worldwide. It was established through a partnership between the United Nations Food and Agriculture Organisation (FAO) and the Consultative Group on International Agricultural Research (CGIAR) acting through Bioversity International.

- The Consultative Group on International Agricultural Research (CGIAR) is a consortium of International Agriculture Research Centres (IARC) and others that each conduct research on and preserve germplasm from a particular crop or animal species. The CGIAR holds one of the world's largest off site collections of plant genetic resources in trust for the world community. It contains over 500,000 accessions of more than 3,000 crop, forage, and agro-forestry species. The collection includes farmers' varieties and improved varieties and, in substantial measure, the wild species from which those varieties were created.

- National germplasm storage centres including the U.S. Department of Agriculture's National Centre for Genetic Resources Preservation, India's National Bureau of Animal Genetic Resources (NBAGR), the Taiwan Livestock Research Institute, and the Australian Network of Plant Genetic Resource Centres.

- Organisations such as the World Resources Institute (WRI) and the World Conservation Union (IUCN) are non-profit organisations that provide funding and other support to off site and on site conservation efforts. The wise use of crop genetic diversity in plant breeding and genetic modification can also contribute significantly to protecting the biodiversity in crops. Crop varieties with specifically modified genes grow resistances to pests and diseases. One successful example of this is the insertion of the gene from the soil bacterium Bacillus thuringiensis.

- Bacillus thuringiensis (Bt) is a soil bacterium that produces a natural insecticide toxin.

- Genes from Bt can be inserted into crop plants to make them capable of producing an insecticidal toxin and therefore a resistance to certain pests.

- There are no known adverse human health effects associated with Bt corn.

- Bt corn can adversely affect non-target insects if they are closely related to the target pest, as is the case with Monarch butterfly. These adverse effects are considered minor, relative to those associated with the alternative of blanket insecticide applications.

Crop Residue

Field residues are materials left in an agricultural field or orchard after the crop has been harvested. These residues include stalks and stubble (stems), leaves, and seed pods. Good management of *field residues* can increase efficiency of irrigation and control of erosion.

Process residues are those materials left after the processing of the crop into a usable resource. These residues include husks, seeds, bagasse, and roots. They can be used as animal fodder and soil amendment, fertilizers and in manufacturing.

Crop Rotation

Crop rotation is the practice of growing a series of dissimilar types of crops in the same area in sequential seasons.

Crop rotation confers various benefits to the soil. A traditional element of crop rotation is the replenishment of nitrogen through the use of green manure in sequence with cereals and other crops. Crop rotation also mitigates the build-up of pathogens and pests that often occurs when one species is continuously cropped, and can also improve soil structure and fertility by alternating deep-rooted and shallow-rooted plants. Crop rotation is one component of polyculture.

History

Historic crop rotation methods are mentioned in Roman literature, and referred to by several civilizations in Asia and on three major elements: sophisticated systems of crop rotation, highly developed irrigation techniques and the introduction of a large variety of crops which were studied and cataloged according to the season, type of land and amount of water they require. Numerous farming encyclopedias were produced.

In Europe, since the times of Charlemagne, there was a transition from a two-field crop rotation to a three-field crop rotation. Under a two-field rotation, half the land was planted in a year while the other half lay fallow. Then, in the next year, the two fields were reversed. Under three-field rotation, the land was divided into three parts. One section was planted in the Autumn with winter wheat or rye. The next Spring, the second field was planted with other crops such as peas, lentils, or beans and the third field was left fallow. The three fields were rotated in this manner so that every three years, a field would rest and be unplanted. Under the two field system, if one has a total of 600 acres (2.4 km²) of fertile land, one would only plant 300 acres.

Under the new three-field rotation system, one would plant (and thereby harvest) 400 acres. But, the additional crops had a more significant effect than mere productivity. Since the Spring crops were mostly legumes, they increased the overall nutrition of the people of Northern Europe. From the end of the Middle Ages until the 20th century, the three-year rotation was practiced by farmers in Europe with a rotation of rye or winter wheat, followed by spring oats or barley, then letting the soil rest (leaving it fallow) during the third stage. That suitable rotations made it possible to restore or to maintain a productive soil has long been recognised by planting spring crops for livestock in place of grains for human consumption.

A four-field rotation was pioneered by farmers, namely in the region Waasland in the early 16th century and popularised by the British agriculturist Charles Townshend in the 18th century. The system (wheat, turnips, barley and clover), opened up a fodder crop and grazing crop allowing livestock to be bred year-round. The four-field crop rotation was a key development in the British Agricultural Revolution. George Washington Carver pioneered crop rotation methods in the United States by teaching southern farmers to rotate soil depleting crops like cotton with soil enriching crops like peanuts and peas.

In the Green Revolution, the traditional practice of crop rotation gave way in some parts of the world to the practice of supplementing the chemical inputs to the soil through top dressing with fertilizers, e.g., adding ammonium nitrate or urea and restoring soil pH with lime in the search for increased yields, preparing soil for specialist crops, and seeking to reduce waste and inefficiency by simplifying planting and harvesting.

Rationale

Growing the same crop in the same place for many years in a row disproportionately depletes the soil of certain nutrients. With rotation, a crop that leaches the soil of one kind of nutrient is followed during the next growing season by a dissimilar crop that returns that nutrient to the soil or draws a different ratio of nutrients: for example, rice followed by cotton.

Implementation

Choice of Crops

The choice & sequence of rotation crops depends on the nature of the soil, the climate, & precipitation which together determine the type of plants that may be cultivated. Other important aspects of farming such as crop marketing & economic variables must also be considered when deciding crop rotations.

Crop rotations may include 2 to 6 or more crop rotations over numerous seasons. A two crop rotation such as corn and soybean in cash grains or corn and alfalfa in forage systems use legumes to help fix nitrogen in the soil for utilisation over the long term. Multiple cropping systems, such as intercropping or companion planting, offer more diversity and complexity within the same season or rotation. Carrots can be shaded by tomatoes and loosen soil below them. Double cropping is common where two crops, typically of different species, are grown sequentially in the same growing season. Winter rye and barley can be sown after oats or rice and harvested before the next crop goes in of oats or rice. These systems can maximise benefits of the rotation as well as available land resources.

More complex rotations commonly utilise animal resources for greater use of on-farm nutrient management and additional farm products. A soil-feeding crop of clover could be replaced or aided by an application of manure to set up a field for a double crop of winter grains after potatoes. Soil building and pest population management benefits can be further utilised with different complexities of crop rotation. In general the complexity of a field's rotation is limited by what soil, climate, and other environmental conditions permit. This also includes the current or desired management tools and goals of the farmer.

Incorporation of Animals

In Sub-Saharan Africa, as animal husbandry becomes less of a nomadic practice many herders have begun integrating crop production into their practice. This is known as mixed farming, or the practice of crop cultivation with the incorporation of raising cattle, sheep and/ or goats by the same economic entity, is increasing in commonality. This interaction between the animal, the land and the crops are being done on a small scale all across this region.

Crop residues provide animal feed, while the animals provide manure for replenishing crop nutrients and draft power. Both processes are extremely important in this region of the world as it is expensive and logistically unfeasible to transport in synthetic fertilizers and large-scale machinery. As an additional benefit, the cattle, sheep and/ or goat provide milk and can act as a cash crop in the times of economic hardship.

Benefits

Using crop rotation farmers can keep their fields under continuous production, instead of letting them lie fallow, as well as reducing the need for artificial fertilizers, both of which can be expensive.

A general effect of crop rotation is that there is a geographic mixing of crops, which can slow the spread of pests and diseases during the growing season. The different crops can also reduce the effects of adverse weather for the individual farmer and, by requiring planting and harvest at different times, allow more land to be farmed with the same amount of machinery and labour.

Agronomists describe the benefits to yield in rotated crops as "The Rotation Effect". There are many found benefits of rotation systems: however, there is no specific scientific basis for the sometimes 10-25% yield increase in a crop grown in rotation versus monoculture. The factors related to the increase are simply described as alleviation of the negative factors of monoculture cropping systems. Explanations due to improved nutrition; pest, pathogen, and weed stress reduction; and improved soil structure have been found in some cases to be correlated, but causation has not been determined for the majority of cropping systems.

Other benefits of rotation cropping systems include production costs advantages. Overall financial risks are more widely distributed over more diverse production of crops and/or livestock. Less reliance is placed on purchased inputs and overtime crops can maintain production goals with fewer inputs. This in tandem with greater short and long term yields makes rotation a powerful tool for improving agricultural systems.

Nutrients

Rotating crops adds nutrients to the soil. Legumes, plants of the family Fabaceae, for instance, have nodules on their roots which contain nitrogen-fixing bacteria. It therefore makes good sense agriculturally to alternate them with cereals (family Poaceae) and other plants that require nitrates. An extremely common modern crop rotation is alternating soybeans and maize (corn). In subsistence farming, it also makes good nutritional sense to grow beans & grain at the same time in different fields.

Pest Control

Crop rotation is also used to control pests & diseases that can become established in the soil over time. The changing of crops in a sequence tends to decrease the population level of pests. Plants within the same taxonomic family tend to have similar pests and pathogens. By regularly changing the planting location, the pest cycles can be broken or limited. For example, root-knot nematode is a serious problem for some plants in warm climates & sandy soils, where it slowly builds up to high levels in the soil, & can severely damage plant productivity by cutting off circulation from the plant roots. Growing a crop that is not a host for root-knot nematode for one season greatly reduces the level of the nematode in the soil, thus making it possible to grow a susceptible crop the following season without needing soil fumigation.

It is also difficult to control weeds similar to the crop which may contaminate the final produce. For instance, ergot in weed grasses is difficult to separate from harvested grain. A different crop allows the weeds to be eliminated, breaking the ergot cycle.

This principle is of particular use in organic farming, where pest control may be achieved without synthetic pesticides.

Soil Erosion

Crop rotation can greatly affect the amount of soil lost from erosion by water. In areas that are highly susceptible to erosion, farm management practices such as zero and reduced tillage can be supplemented with specific crop rotation methods to reduce raindrop impact, sediment detachment, sediment transport, surface runoff, and soil loss.

Protection against soil loss is maximised with rotation methods that leave the greatest mass of crop stubble (plant residue left after harvest) on top of the soil. Stubble cover in contact with the soil minimises erosion from water by reducing overland flow velocity, stream power, and thus the ability of the water to detach and transport sediment. Soil Erosion and Cill prevent the disruption and detachment of soil aggregates that cause macrospores to block, infiltration to decline, and runoff to increase. This significantly improves the resilience of soils when subjected to periods of erosion and stress.

The effect of crop rotation on erosion control varies by climate. In regions under relatively consistent climate conditions, where annual rainfall and temperature levels are assumed, rigid crop rotations can produce sufficient plant growth and soil cover. In regions where climate conditions are less predictable, and unexpected periods of rain and drought may occur, a more flexible approach for soil cover by crop rotation is necessary. An opportunity cropping system promotes adequate soil cover under these erratic climate conditions. In an opportunity cropping system, crops are grown when soil water is adequate and there is a reliable sowing window. This form of cropping system is likely to produce better soil cover than a rigid crop rotation because crops are only sown under optimal conditions, whereas rigid systems are sown in the best conditions available.

Crop rotations also affect the timing and length of when a field is subject to fallow. This is very important because depending on a particular region's climate, a field could be the most vulnerable to erosion when it is under fallow. Efficient fallow management is an essential part of reducing erosion in a crop rotation system. Zero tillage is a fundamental management practice that promotes crop stubble retention under longer unplanned fallows when crops cannot be planted. Such management practices that succeed in retaining suitable soil cover in areas under fallow will ultimately reduce soil loss.

Additional Soil Improvements

The use of different species in rotation allows for increased soil organic matter (SOM), greater soil structure, and improvement of the chemical and biological soil environment for crops. With more SOM, water infiltration and retention improves, providing increased drought tolerance and decreased erosion. Soil aggregation allows greater nutrient retention and utilisation, decreasing the need for added nutrients. Soil microorganisms also improve nutrient availability and decrease pathogen and pest activity through competition. In addition, plants produce root exudates and other chemicals which manipulate their soil environment as well as their weed environment. Thus rotation allows increased yields from nutrient availability but also alleviation of allelopathy and competitive weed environments.

Risks

Balancing the commitment to new crops or livestock with increased yield potentials and long term sustainability is the task of many farmers and agricultural scientists. With this research many new rotations have been developed and become widely accepted.

Risks of crop rotation include less overall profitability due to decreased acreage of the most valuable crop. Greater investment and lower relative efficiency in machinery used for different crops is also a possible outcome. More complex rotations require more crop species and livestock. This means the farmer must have additional skills and make more time and equipment investments initially. Also the more complex the system, the less flexible it becomes in terms of long term land management. Starting a rotation of a new crop may add profitability and farm resilience over time, but benefits are initially subject to being over-shadowed by volatile markets or high startup investments which can take time to overcome. Overall many farmers and agronomists agree finding a suitable rotation can benefit the overall productivity and sustainability of the farm.

Crop Weed

Crop weeds are weeds that grow amongst crops. Examples of crop weeds include chickweed, barnyard grass and dandelion. The dandelion, while sometimes used as a food source, can also prove harmful to crops. Crop weeds can inhibit the growth of crops, contaminate harvested crops and often spread rapidly. They can also host crop pests such as aphids, fungal rots and viruses.

Crop Wild Relative

A crop wild relative (CWR) is a wild plant closely related to a domesticated plant. It may be a wild ancestor of the domesticated plant, or another closely related taxon.

Overview

The wild relatives of crop plants constitute an increasingly important resource for improving agricultural production and for maintaining sustainable agro-ecosystems. With the advent of climate change and greater ecosystem instability CWRs are likely to prove a critical resource in ensuring food security for the new millennium. It was Nikolai Vavilov, a Russian Botanist who first realised the importance of crop wild relatives in the early 20th century. Genetic material from CWRs has been utilised by humans for thousands of years to improve the quality and yield of crops. Farmers have used traditional breeding methods for millennia, wild maize (*Zea mexicana*) is routinely grown alongside maize to promote natural crossing and improve yields. More recently, plant breeders have utilised CWR genes to improve a wide range of crops like rice (*Oryza sativa*), tomato (*Solanum lycopersicum*) and grain legumes. CWRs have contributed many useful genes to crop plants, and modern varieties of most major crops now contain genes from their wild relatives. Therefore CWRs are wild plants related to socio-economically important species including food, fodder and forage crops, medicinal plants, condiments, ornamental, and forestry species, as well as plants used for industrial purposes, such as oils and fibres, and to which they can contribute beneficial traits. A CWR can be defined as "... a wild plant taxon that has an indirect use derived from its relatively close genetic relationship to a crop..."

Conservation of Crop Wild Relatives

CWRs are essential components of natural and agricultural ecosystems and hence are indispensable for maintaining ecosystem health. Their conservation and sustainable use is very important for improving agricultural production, increasing food security, and maintaining a healthy environment. The natural populations of many CWRs are increasingly at risk. They are threatened by habitat loss through the destruction and degradation of natural environment or their conversion to other uses. Deforestation is leading to the loss of many populations of important wild relatives of fruit, nut, and industrial crops. Populations of wild relatives of cereal crops that occur in arid or semi-arid lands are being severely reduced by over grazing and resulting desertification. The growing industrialisation of agriculture is drastically reducing the occurrence of CWRs within the traditional agro-ecosystems. The wise conservation and use of CWRs are essential elements for increasing food security, eliminating poverty, and maintaining the environment.

Examples of Wild Relatives

Grains

- Wheat - Einkorn wheat
- Barley - Wild Barley
- Oats - A. byzantina.

Vegetables

Note: Many different vegetables share one common ancestor, particularly in the Brassica family and plants. Many vegetables are also hybrids of different species, again this is particularly true of Brassicas.

- Lettuce - Lactuca sativa
- Cabbage - Brassica oleracea ("Wild Cabbage")
- Leek - Allium ampeloprasum ("Wild Leek")
- Onion - Allium cepa

- Turnip - Brassica rapa
- Brocolli - Brassica oleracea
- Rutabaga - Brassica napus
- Carrot - Daucus Carota ("Wild Carrot").

Fruits

- Strawberry - Fragaria × ananassa
- Alpine Strawberry - Fragaria vesca
- Cherry - Prunus avium
- Orange - a hybrid between Pomelo and Mandarins
- Banana - A hybrid between Musa acuminata and Musa balbisiana
- Apple - mostly Malus sieversii, but with some cultivars perhaps belonging to Malus sylvestris or being a hybrid of the two.
- Pear - Pyrus communis
- Plum - A hybrid between Prunus spinosa and Prunus cerasifera

Crop-lien System

The crop-lien system is a credit system that became widely used by farmers in the United States in the South from the 1860s to the 1920s.

After the American Civil War, farmers in the South had little cash. The crop-lien system was a way for farmers to get credit before the planting season by borrowing against the value for anticipated harvests. Local merchants provided food and supplies all year long on credit; when the cotton crop was harvested farmers turned it over to the merchant to pay back their loan. Sometimes there was cash left over; when cotton prices were low, the crop did not cover the debt and the farmer started the next year in the red. The credit system was used by land owners, sharecroppers and tenant farmers.

The merchants had to borrow the money to buy supplies, and in turn charged the farmer interest as well as a higher price for merchandise bought on such credit. The merchant insisted that more cotton (or some other cash crop) be grown—nothing else paid well—and thus came to dictate the crops that a farmer grew. When farmers suddenly left the area, the bills went unpaid and the merchant had to absorb the loss, as well as the risk that cotton prices would fall so the raw cotton he was given at harvest time was worth less than the amount he loaned during the year.

In the early 20th century, because of automobiles, higher cotton prices and growing consumerism, city department stores gradually played a more important role than isolated country stores in Southern economic life. Women shopped in increasing numbers, paying with cash or in monthly installments. With the increasing importance of advertising, Southern economic life became more modernised.

Energy Crop

An energy crop is a plant grown as a low cost and low maintenance harvest used to make biofuels, or combusted for its energy content to generate electricity or heat. Energy crops are generally categorized as woody or herbaceous (grassy).

Commercial energy crops are typically densely planted, high yielding crop species where the energy crops will be burnt to generate power. Woody crops such as Willow or Poplar are widely utilised, as well as temperate grasses such as Miscanthus and Pennisetum purpureum (both known as elephant grass). If carbohydrate content is desired for the production of biogas, whole-crops such as maize, Sudan grass, millet, white sweet clover and many others, can be made into silage and then converted into biogas.

Through genetic modification and application of biotechnology plants can be manipulated to create greater yields, reduce associated costs and require less water. However, high energy yield can be realised with existing cultivars.

Types of Energy Crops

Solid Biomass

Energy generated by burning plants grown for the purpose, often after the dry matter is pelletized. Energy crops are used for firing power plants, either alone or co-fired with other fuels. Alternatively they may be used for heat or combined heat and power (CHP) production.

Gas Biomass (Methane)

Anaerobic digesters or biogas plants can be directly supplemented with energy crops once they have been ensiled into silage. The fastest growing sector of German biofarming has been in the area of "Renewable Energy Crops" on nearly 500,000 ha land (2006). Energy crops can also be grown to boost gas yields where feedstocks have a low energy content, such as manures and spoiled grain. It is estimated that the energy yield presently of bioenergy crops converted via silage to methane is about 2 GWh/km². Small mixed cropping enterprises with animals can use a portion of their acreage to grow and convert energy crops and sustain the entire farms energy requirements with about 1/5 the acreage.

In Europe and especially Germany, however, this rapid growth has occurred only with substantial government support, as in the German bonus system for renewable energy. Similar developments of integrating crop farming and bioenergy production via silage-methane have been almost entirely overlooked in N. America, where political and structural issues and a huge continued push to centralize energy production has overshadowed positive developments.

Liquid Biomass

Biodiesel

European production of biodiesel from energy crops has grown steadily in the last decade, principally focused on rapeseed used for oil and energy. Production of oil/biodiesel from rape covers more than 12,000 km² in Germany alone, and has doubled in the past 15 years. Typical yield of oil as pure biodiesel may be is 100,000 L/km² or more, making biodiesel crops economically attractive, provided sustainable crop rotations exist that are nutrient-balanced and preventative of the spread of disease such as clubroot. Biodiesel yield of soybeans is significantly lower than that of rape.

Table : Typical oil extractable by weight

Crop	Oil %
copra	62
castor seed	50
sesame	50
groundnut kernel	42
jatropha	40
rapeseed	37
palm kernel	36
mustard seed	35
sunflower	32
palm fruit	20
soybean	14
cotton seed	13

Bioethanol

Energy crops for biobutanol are grasses. Two leading non-food crops for the production of cellulosic bioethanol are switchgrass and giant miscanthus. There has been a preoccupation with cellulosic bioethanol in America as the agricultural structure supporting biomethane is absent in many regions, with no credits or bonus system in place. Consequently a lot of private money and investor hopes are being pinned on marketable and patentable innovations in enzyme hydrolysis and the like and therefore America is viewed by some technology planners as falling further behind Europe in real bioenenergy gains.

Bioethanol also refers to the technology of using animal and human grains, principally corn (maize seed) to make ethanol directly through fermentation, a process that is widely reputed to consume as much energy as it produces, therefore being non-sustainable. New developments in converting grain stillage (referred to as distillers grain stillage or DGS) into biogas energy looks promising as a means to improve the poor energy ratio of this type of bioethanol process. 2007 saw a setback in the economics of building grain refineries in the USA while the shipment of grains and ethanol by rail car has prompted the train industries largest growth phase since 50 years.

By Dedication

Dedicated energy crops are non-food energy crops as giant miscanthus, switchgrass, jatropha, fungi, and algae.

Also byproducts (green waste) of food and non-food energy crops can be used to produce biofuels.

Fiber Crop

Fiber crops are field crops grown for their fibres, which are traditionally used to make paper, cloth, or rope. The fibers may be chemically modified, like in viscose or cellophane. In recent years materials scientists have begun exploring further use of these fibers in composite materials.

Fiber crops are generally harvestable after a single growing season, as distinct from trees, which are typically grown for many years before being harvested for wood pulp fiber. In specific circumstances, fiber crops can be superior to wood pulp fiber in terms of technical performance, environmental impact or cost. There are a number of issues regarding the use of fiber crops to make pulp. One of these is seasonal availability. While trees can be harvested continuously, many field crops are harvested once during the year and must be stored such that the crop doesn't rot over a period of many months. Considering that many pulp mills require several thousand tonnes of fiber source per day, storage of the fiber source can be a major issue.

Botanically, the fibers harvested from many of these plants are bast fibers; the fibers come from the phloem tissue of the plant. The other fiber crop fibers are seed padding, leaf fiber, or other parts of the plant.

Fiber Crops

- Bast fibers (Stem-skin fibers)
  - o Jute (widely used, cheapest fiber after cotton)
  - o Flax (produces linen)
  - o Indian hemp (The Dogbane used by native Americans.)
  - o Hemp (A soft, strong fiber, edible seeds.)
  - o Hoopvine (Also used for barrel hoops and baskets, edible leaves, medicine.)
  - o Kenaf (The interior of the plant stem is also used for fiber. Edible leaves.)
  - o Nettles
  - o Ramie (A nettle, stronger than cotton or flax, makes "China grass cloth")
- Other fibers (Leaf, fruit, and other fibers)
  - o Abacá (A banana, producing "manila" rope from leaves)
  - o Bamboo fiber
  - o Bowstring Hemp, (An old use of a common decorative agave, also Sansevieria roxburghiana, Sansevieria hyacinthoides)
  - o Cotton
  - o Coir (fiber from the coconut shell)
  - o Esparto
  - o Henequen (An agave, useful fiber, but not as high quality as sisal)
  - o Kapok
  - o Milkweed
  - o Papaya

o   Phormium ("New Zealand Flax", an agave)

o   Sisal (Often termed agave)

o   Umbrella plant

o   Yucca (An agave).

Fiber Dimensions

Source of Pulp	Fiber Length, mm	Fiber Diametre, μm
Softwood	3.0	30
Hardwood	1.0	16
Wheat straw	1.5	13
Rice straw	1.5	9
Esparto grass	1.1	10
Reed	1.5	13
Bagasse	1.7	20
Bamboo	2.7	14
Cotton	25.0	20

Industrial Crop

An industrial crop is a crop grown to produce goods to be used in the production sector, rather than food for consumption. Industrial crops impact the economy by providing a product which lessens the need for imports.

Purpose of Industrial Crops

Industrial crops is a designation given to an enterprise that attempts to raise farm sector income, and provide economic development activities for rural areas. Industrial crops also attempt to provide products that can be used as substitutes for imports from other nations.

Examples of Industrial Crops

For example, to produce fibre for clothing. Some examples include flax, hemp, cotton, tobacco or silk. Fiber crops are amongst the most common industrial crops. They are different from cash crops as they do not supply to an industry

Dangers of Industrial Crops Contamination

The danger of industrial crops is not the products being grown, but rather the items used to stimulate the growth of the these crops.

Intercropping

Intercropping is the practice of growing two or more crops in proximity. The most common goal of intercropping is to produce a greater yield on a given piece of land by making use of resources that would otherwise not be utilised by a single crop. Careful planning is required, taking into account the soil, climate, crops, and varieties. It is particularly important not to have crops competing with each other for physical space, nutrients, water, or sunlight. Examples of intercropping strategies are planting a deep-rooted crop with a shallow-rooted crop, or planting a tall crop with a shorter crop that requires partial shade. Inga alley cropping has been proposed as an alternative to the ecological destruction of Slash-and-burn farming.

When crops are carefully selected, other agronomic benefits are also achieved. Lodging-prone plants, those that are prone to tip over in wind or heavy rain, may be given structural support by their companion crop (Trenbath 1976). Delicate or light sensitive plants may be given shade or protection, or otherwise wasted space can be utilised. An example is the tropical multi-tier system where coconut occupies the upper tier, banana the middle tier, and pineapple, ginger, or leguminous fodder, medicinal or aromatic plants occupy the lowest tier.

Intercropping of compatible plants also encourages biodiversity, by providing a habitat for a variety of insects and soil organisms that would not be present in a single-crop environment. This biodiversity can in turn help to limit outbreaks of crop pests (Altieri 1994) by increasing the diversity or abundance of natural enemies, such as spiders or parasitic wasps. Increasing the complexity of the crop environment through intercropping also limits the places where pests can find optimal foraging or reproductive conditions.

The degree of spatial and temporal overlap in the two crops can vary somewhat, but both requirements must be met for a cropping system to be an intercrop. Numerous types of intercropping, all of which vary the temporal and spatial mixture to some degree, have been identified (Andrews & Kassam 1976). These are some of the more significant types:

- Mixed intercropping, as the name implies, is the most basic form in which the component crops are totally mixed in the available space.
- Row cropping involves the component crops arranged in alternate rows. Variations include alley cropping, where crops

are grown in between rows of trees, and strip cropping, where multiple rows, or a strip, of one crop are alternated with multiple rows of another crop.

- Intercropping also uses the practice of sowing a fast growing crop with a slow growing crop, so that the fast growing crop is harvested before the slow growing crop starts to mature. This obviously involves some temporal separation of the two crops.

- Further temporal separation is found in relay cropping, where the second crop is sown during the growth, often near the onset of reproductive development or fruiting, of the first crop, so that the first crop is harvested to make room for the full development of the second.

Multiple Cropping

In agriculture, multiple cropping is the practice of growing two or more crops in the same space during a single growing season. It is a form of polyculture. It can take the form of double-cropping, in which a second crop is planted after the first has been harvested, or relay cropping, in which the second crop is started amidst the first crop before it has been harvested. A related practice, companion planting, is sometimes used in gardening and intensive cultivation of vegetables and fruits. One example of multi-cropping is tomatoes + onions + marigold; the marigolds repel some tomato pests.

Multiple cropping is found in many agricultural traditions. In the Garhwal Himalaya of India, a practice called *baranaja* involves sowing 12 or more crops on the same plot, including various types of beans, grams, and millets, and harvesting them at different times.

In the cultivation of rice, multiple cropping requires effective irrigation, especially in areas with a dry season. Rain that falls during the wet season permits the cultivation of rice during that period, but during the other half of the year, water cannot be channelled into the rice fields without an irrigation system. The Green Revolution in Asia led to the development of high-yield varieties of rice, which required a substantially shorter growing season of 100 days, as opposed to traditional varieties, which needed 150 to 180 days. Due to this, multiple cropping became more prevalent in Asian countries.

According to the Government-owned North Korean news outlet KCNA, the late Kim Il Sung had a "profound knowledge of double cropping". Kim allegedly "tested double cropping in the experimental fields belonging to his residence for years in his lifetime". One kind of multiple cropping is intercropping, where an additional crop is planted in the spaces available between the main crop.

Nurse Crop

In agriculture, a nurse crop is an annual crop used to assist in establishment of a perennial crop. The widest use of nurse crops is in the establishment of legumaceous plants such as alfalfa, clover, and trefoil. Occasionally nurse crops are used for establishment of perennial grasses.

Nurse crops reduce the incidence of weeds, prevent erosion, and prevent excessive sunlight from reaching tender seedlings. Often the nurse crop can be harvested for grain, straw, hay, or pasture. Oats are the most common nurse crop, though other annual grains are also used.

Permanent Crop

A permanent crop is one produced from plants which last for many seasons, rather than being replanted after each harvest.

As used in The World Factbook land use statistics the term comprises land cultivated for crops like citrus, olives, coffee, and rubber; it includes land under flowering shrubs, fruit trees, nut trees, and vines, but excludes land under trees grown for wood or timber.

Protein Crop

Protein crops are crops that provide substantial protein, a large class of naturally occurring complex combinations of amino acids. Such crops, including various oilseeds and grains, are important in meeting the nutrient requirements of farm animals. EU Common Agricultural Policy designates certain protein crops as eligible for support, such as peas, field beans, and sweet lupins.

Sharecropping

Sharecropping is a system of agriculture in which a landowner allows a tenant to use the land in return for a share of the crop produced on the land (e.g. 50% of the crop). This should not be confused with a crop fixed rent contract, in which a landowner allows a tenant to use the land in return for a fixed amount of crop per unit of land (e.g. 1 Tonne per hectare). Sharecropping has a long history and there are a wide range of different situations and types of agreements that have encompassed the system. Some are governed by tradition, others by law. Legal contract systems such as the Italian mezzadria, the French métayage, and Spanish Mediero occur widely. Islamic law contains a traditional "musaqat" sharecropping agreement for the cultivation of orchards.

Overview

Sharecropping has benefits and costs for both the owners and the croppers. It encourages the cropper to remain on the land throughout the harvest season to work the land, solving the harvest rush problem. At the same time, since the cropper pays in shares of his harvest, owners and croppers share the risk of harvests being large or small and prices being high or low. Because tenants benefit from larger harvests, they have an incentive to work harder and invest in better methods than in a slave plantation system. However, by dividing the working force into many individual workers, large farms no longer benefit from economies of scale. On the whole, sharecropping was not as economically efficient as the gang agriculture of slave plantations.

Sharecropping occurred extensively in colonial Africa, Scotland, and Ireland and came into wide use in the Southern United States during the Reconstruction era (1865–1877). The South had been devastated by war; planters had ample land but little money for wages or taxes. At the same time, most of the former slaves had labour but no money and no land; they rejected the kind of gang labour that typified slavery. The solution was the sharecropping system focused on cotton, which was the only crop that could generate cash for the croppers, landowners, merchants and the tax collector. Poor white farmers, who previously had done little cotton farming, needed cash as well and became sharecroppers.

Jeffery Paige made a distinction between centralised sharecropping found on cotton plantations and the decentralised sharecropping with other crops. The former is characterised by political conservatism and long lasting tenure. Tenants are tied to the landlord through the plantation store. Their work is heavily supervised as slave plantations were. This form of tenure tends to be replaced by wage labour as markets penetrate. Decentralised sharecropping involves virtually no role for the landlord: plots are scattered, peasants manage their own labour and the landowners do not manufacture the crops. Leases are very short which leads to peasant radicalism. This form of tenure becomes more common when markets penetrate.

Use of the sharecropper system has also been identified in England (as the practice of "farming to halves"). It is still used in many rural poor areas today, notably in Pakistan and India. Although there is a perception that sharecropping was exploitative, "evidence from around the world suggests that sharecropping is often a way for differently endowed enterprises to pool resources to mutual benefit, overcoming credit restraints and helping to manage risk."

It can have more than a passing similarity to serfdom or indenture, and has therefore been seen as an issue of land reform in contexts such as the Mexican Revolution. However, Nyambara states that Eurocentric historiographical devices such as 'feudalism' or 'slavery' often qualified by weak prefixes like 'semi-' or 'quasi-' are not helpful in understanding the antecedents and functions of sharecropping in Africa. Sharecropping agreements can however be made fairly, as a form of tenant farming or sharefarming that has a variable rental payment, paid in arrears. There are three different types of contracts.

1.   Workers can rent plots of land from the owner for a certain sum and keep the whole crop.
2.   Workers work on the land and earn a fixed wage from the land owner but keep some of the crop.
3.   No money changes hands but the worker and land owner each keep a share of the crop.

Advantages and Disadvantages

The advantages of sharecropping in other situations include enabling access for women to arable land where ownership rights are vested only in men. Paige pointed out that sharecropping was economically inefficient in a free market. However, many outside factors make it efficient. One factor is slave emancipation: sharecropping provided the freed slaves of the USA, Brazil and the late Roman Empire with land access. It is efficient also as a way of escaping inflation, hence its rise in sixteenth century France and Italy. Landlords opt for sharecropping to avoid the administrative costs and shirking that occurs on plantations and haciendas. It is preferred to cash tenancy because cash tenants take all the risks, and any harvest failure will hurt them and not the landlord. Therefore, they tend to demand lower rents than sharecroppers.

Regions

Africa

In settler colonies of colonial Africa, sharecropping was a feature of the agricultural life. White farmers, who owned most of the land, were frequently unable to work the whole of their farm for lack of capital. They therefore allowed black farmers to work the excess on a sharecropping basis. In South Africa the 1913 Natives' Land Act outlawed the ownership of land by blacks in areas designated for white ownership and effectively reduced the status of most sharecroppers to tenant farmers and then to farm labourers. In the 1960s, generous subsidies to white farmers meant that most farmers could afford to work their entire farms, and sharecropping faded out.

The arrangement has reappeared in other African countries in modern times, including Ghana and Zimbabwe.

United States

Sharecropping became widespread as a response to economic upheaval caused by the emancipation of slaves and disenfranchisement of poor whites in the agricultural South during Reconstruction. Plantations had first relied on slaves for cheap labour. Prior to emancipation, sharecropping was limited to poor landless whites, usually working marginal lands for absentee landlords. Following emancipation, sharecropping came to be an economic arrangement that largely maintained the status quo between black and white through legal means. One area that has attracted scholars interested in the rise or origins of Sharecropping is the Natchez District, roughly centred in Adams County Mississippi and the County seat, Natchez. The location of the city, with access to the Mississippi river, but high on a bluff and safe from flooding, meant that the records of the cotton trading Gentry survived the natural disasters. The Civil War largely bypassed the city, saving the records from man made disasters as well. The mass influx of immigrants in the 1900s brought an increase in sharecropping during the World War I era. Sharecroppers worked a section of the plantation independently, usually growing cotton, tobacco, rice, and other cash crops and received a small portion of the parcel's output.

Although the sharecropping system was primarily a post-Civil War development, it did exist in antebellum Mississippi, especially in the northeastern part of the state, an area with few slaves or plantations, and most probably also existed in Tennessee. Sharecropping, along with tenant farming, was a dominant form in the cotton South from the 1870s to the 1950s, among both blacks and whites, but it has largely disappeared.

After the American Civil War, plantation owners had to borrow money to produce crops. Interest rates on these loans were around

15%. The indebtedness of cotton planters increased through the early 1940s, and the average plantation fell into bankruptcy about every twenty years. It is against this backdrop that the wealthiest owners maintained their concentrated ownership of the land.

In Reconstruction-era United States, sharecropping was one of few options for penniless freedmen to conduct subsistence farming and support themselves and their families. Other solutions included the crop-lien system (where the farmer was extended credit for seed and other supplies by the merchant), a rent labour system (where the former slave rents their land but keeps their entire crop), and the wage system (worker earns a fixed wage, but keeps none of their crop). Sharecropping was by far the most economically efficient, as it provided incentives for workers to produce a bigger harvest. It was a stage beyond simple hired labour, because the sharecropper had an annual contract. During Reconstruction, the Freedman's Bureau wrote and enforced the contracts.

However, sharecropping was an easy way for white former slave owners to take advantage of uneducated freedmen. Former slaves had little to no education, so the landowner could draw up a 70-30 contract instead of half.

Croppers were assigned a plot of land to work, and in exchange owed the owner a share of the crop at the end of the season, usually one-half. The owner provided the tools and farm animals. Farmers who owned their own mule and plow were at a higher stage and are called tenant farmers; they paid the landowner less, usually only a third of each crop. In both cases the farmer kept the produce of gardens.

The sharecropper purchased seed, tools and fertilizer, as well as food and clothing, on credit from a local merchant, or sometimes from a plantation store. When the harvest came, the cropper would harvest the whole crop and sell it to the merchant who had extended credit. Purchases and the landowner's share were deducted and the cropper kept the difference—or added to his debt.

Though the arrangement protected sharecroppers from the negative effects of a bad crop, many sharecroppers (both black and white) were economically confined to serf-like conditions of poverty. To work the land, sharecroppers had to buy seed and implements, sometimes from the plantation owner who often charged exorbitant prices against the sharecropper's next season. Arrangements also typically gave half or less of the crop to the sharecropper, and the sale price in some cases was set by the landowner. Lacking the resources to market their crops independently, the sharecropper was sometimes compensated in scrip redeemable only at the plantation.

Thus the cost of production and price of sale were both largely controlled by the land owner, with the sharecropper having little, if any, margin for profit. These factors made sharecroppers dependent on the plantation owners in a way that perpetuated some of the aspects of slavery, and in the late 19th century maintained a stable, low-cost work force that replaced slave labour; it was the bottom rung in the Southern tenancy ladder.

By the early 1930s there were 5.5 million white tenants, sharecroppers, and mixed cropping/labourers in the United States, and 3 million blacks. In Tennessee whites made up two thirds or more of the sharecroppers. In Mississippi, by 1900, 36% of all white farmers were tenants or sharecroppers, while 85 percent of black farmers were. Sharecropping continued to be a significant institution in Tennessee agriculture for more than sixty years after the Civil War, peaking in importance in the early 1930s, when sharecroppers operated approximately one-third of all farm units in the state.

The situation of landless farmers who challenged the system in the rural south as late as 1941 has been described thus: "he is at once a target subject of ridicule and vitriolic denunciation; he may even be waylaid by hooded or unhooded leaders of the community, some of whom may be public officials. If a white man persists in 'causing trouble', the night riders may pay him a visit, or the officials may haul him into court; if he is a Negro, a mob may hunt him down."

Sharecroppers formed unions in the 1930s, beginning in Tallapoosa County, Alabama in 1931, and Arkansas in 1934. Membership in the Southern Tenant Farmers Union included both blacks and poor whites. As leadership strengthened, meetings became more successful, and protest became more vigorous, landlords responded with a wave of terror.

Sharecroppers' strikes in Arkansas and the Bootheel of Missouri, the 1939 Missouri Sharecroppers' Strike, were documented in the film *Oh Freedom After While.*

In the 1930s and 1940s, increasing mechanization virtually brought the institution of sharecropping to an end in the United States. The sharecropping system in the U.S. increased during the Great Depression with the creation of tenant farmers following the failure of many small farms throughout the Dustbowl. Traditional sharecropping declined after mechanization of farm work became economical in the mid-20th century As a result, many sharecroppers were forced off the farms, and migrated to the industrialised North to work in factories, or become migrant workers in the Western United States during World War II.

Sharecropping Agreements

Typically, a sharecropping agreement would specify which party was expected to cover certain expenses, like seed, fertilizer, weed control, irrigation district assessments, and fuel. Sometimes the sharecropper covered those costs, but they expected a larger share of the crop in return. The agreement would also indicate whether the sharecropper would use his own equipment to raise the crops, or use the landlord's equipment. The agreement would also indicate whether the landlord would pick up his or her share of the crop in the field or whether the sharecropper would deliver it (and where it would be delivered.)

For example, a landowner may have a sharecropper farming an irrigated hayfield. The sharecropper uses his own equipment, and covers all the costs of fuel and fertilizer. The landowner pays the irrigation district assessments and does the irrigating himself. The sharecropper cuts and bales the hay, and delivers one-third of the baled hay to the landlord's feedlot. The sharecropper might also leave the landlord's share of the baled hay in the field, where the landlord would fetch it when he wanted hay.

Another arrangement could have the sharecropper delivering the landlord's share of the product to market, in which case the landlord would get his share in the form of the sale proceeds. In that case, the agreement should indicate the timing of the delivery to market, which can have a significant effect on the ultimate price of some crops. The market timing decision should probably be decided shortly before harvest, so that the landlord has more complete information about the area's harvest, to determine whether the crop will earn more money immediately after harvest, or whether it should be stored until the price rises. Market timing can entail storage costs and losses to spoilage as well, for some crops.

Farmer's Cooperatives

Cooperative farming exists in many forms throughout the United States, Canada, and the rest of the world. Various arrangements can be made through collective bargaining or purchasing to get the best deals on seeds, supplies, and equipment. For example, members of a farmers' cooperative who cannot afford heavy equipment of their own can lease them for nominal fees from the cooperative. Farmers' cooperatives can also allow groups of small farmers and dairymen to manage pricing and prevent undercutting by competitors.

Economic Theories of Share Tenancy

The theory of share tenancy was long dominated by Alfred Marshall's famous footnote 5, wherein he illustrated the inefficiency of agricultural share-contracting. Steven N.S. Cheung (1969), challenged this view, showing that with sufficient competition and in the absence of transaction costs, share tenancy will be equivalent to competitive labour markets and therefore efficient. He also showed that in the presence of transaction costs, share-contracting may be preferred to either wage contracts or rent contracts—due to the mitigation of labour shirking and the provision of risk sharing. Joseph Stiglitz (1974, 1988), suggested that if share tenancy is only a labour contract, then it is only pairwise-efficient and that land-to-the-tiller reform would improve social efficiency by removing the necessity for labour contracts in the first place. Reid (1973), Murrel (1983), Roumasset (1995) and Allen and Lueck (2004) provided transaction cost theories of share-contracting, wherein tenancy is more of a partnership than a labour contract and both landlord and tenant provide multiple inputs. It has been also argued that the sharecropping institution can be explained by factors such as informational asymmetry, moral hazard or limited liability.

Underutilised Crop

Neglected and underutilised crops are domesticated plant species that have been used for centuries or even millennia for their food, fibre, fodder, oil or medicinal properties, but have been reduced in importance over time owing to particular supply and use constraints. These can inter alia include poor shelf life, unrecognised nutritional value, poor consumer awareness and reputational problems ("poor people's food"). As the demand for plant and crop attributes changes (re-appraisal or discovery of nutritional traits, culinary value, adaptation to climate change, etc.), neglected crops can overcome the constraints to the wider production and use. As a matter of fact, many formerly neglected crops are now globally significant crops (oilpalm, soybean, kiwi fruit). Although the options for scaling-up neglected crops for large-scale agriculture appear to be increasingly exhausted, many species have the potential to contribute to food security, nutrition, dietary and culinary diversification, health and income generation. They also provide environmental services. It is impossible to define what would constitute "proper" or "correct" levels of utilisation, however, it is evident that many neglected species are under-utilised relative to their nutritional value and productivity.

Overview

Just three crops - maize, wheat and rice - account for about 40% of the world's consumption of calories and protein. 95% of the world's food needs are provided for by just 30 species of plants. In stark contrast, at least 12,650 species names have been compiled as edible (Kunkel 1984). Neglected and underutilised plants are those that could be - and, in many cases, historically have been - used for food and other uses on a larger scale. Such crop species have also been described as "minor", "orphan", "promising" and "little-used". They continue to play an important role in the subsistence and economy of poor people throughout the developing world, particularly in the agrobiodiversity-rich tropics. Despite their potential for dietary diversification and the provision of micronutrient such as vitamins and minerals, they continue to attract little research and development attention. Alongside their commercial potential, many of the underutilised crops also provide important environmental services, as they are adapted to marginal soil and climate conditions.

It is difficult to precisely define which attributes makes a crop "underutilised", but often they display the following features:

- Linkage with the cultural heritage of their places of origin
- Local and traditional crops whose distribution, biology, cultivation and uses are poorly documented
- Adaptation to specific agro-ecological niches and marginal land
- Weak or no formal seed supply systems
- Traditional uses in localized areas
- Produced in traditional production systems with little or no external inputs
- Receive little attention from research, extension services, policy and decision makers, donors, technology providers and consumers
- May be highly nutritious and/or have medicinal properties or other multiple uses.

Agriculture

Agriculture (also called farming or husbandry) is the cultivation of animals, plants, fungi and other life forms for food, fiber, and other products used to sustain life. Agriculture was the key implement in the rise of sedentary human civilization, whereby farming of domesticated species created food surpluses that nurtured the development of civilization. The study of agriculture is known as agricultural science. Agriculture is also observed in certain species of ant and termite, but generally speaking refers to human activities.

The history of agriculture dates back thousands of years, and its development has been driven and defined by greatly different climates, cultures, and technologies. However, all farming generally relies on techniques to expand and maintain the lands suitable for raising domesticated species. For plants, this usually requires some form of irrigation, although there are methods of dryland farming; pastoral herding on rangeland is still the most common means of raising livestock. In the developed world, industrial agriculture based on large-scale monoculture has become the dominant system of modern farming, although there is growing support for sustainable agriculture (e.g. permaculture or organic agriculture).

Modern agronomy, plant breeding, pesticides and fertilizers, and technological improvements have sharply increased yields from cultivation, but at the same time have caused widespread ecological damage and negative human health effects. Selective breeding and modern practices in animal husbandry such as intensive pig farming have similarly increased the output of meat, but have raised concerns about animal cruelty and the health effects of the antibiotics, growth hormones, and other chemicals commonly used in industrial meat production.

The major agricultural products can be broadly grouped into foods, fibers, fuels, and raw materials. In the 21st century, plants have been used to grow biofuels, biopharmaceuticals, bioplastics, and pharmaceuticals. Specific foods include cereals, vegetables, fruits, and meat. Fibers include cotton, wool, hemp, silk and flax. Raw materials include lumber and bamboo. Other useful materials are produced by plants, such as resins. Biofuels include methane from biomass, ethanol, and biodiesel. Cut flowers, nursery plants, tropical fish and birds for the pet trade are some of the ornamental products. Regarding food production, the World Bank targets agricultural food production and water management as an increasingly global issue that is fostering an important and growing debate. In 2007, one third of the world's workers were employed in agriculture. The services sector has overtaken agriculture as the economic sector employing the most people worldwide. Despite the size of its workforce, agricultural production accounts for less than five percent of the gross world product (an aggregate of all gross domestic products).

Etymology

The word *agriculture* is the English adaptation of Latin *agricultûra*, from *ager*, "a field", and *cultûra*, "cultivation" in the strict sense of "tillage of the soil". Thus, a literal reading of the word yields "tillage of a field/of fields".

Overview

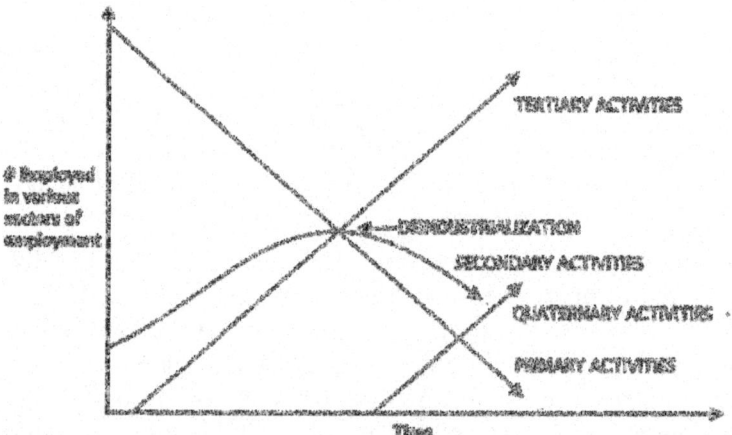

Figure: Clark's Sector Model (1950): The percent of the human population working in primary sector activities such as agriculture has decreased over time.

Agriculture has played a key role in the development of human civilization. Until the Industrial Revolution, the vast majority of the human population laboured in agriculture. The type of agriculture they developed was typically subsistence agriculture in which farmers raised most of their crops for consumption on farm, and there was only a small portion left over for the payment of taxes, dues, or trade. In subsistence agriculture cropping decisions are made with an eye to what the family needs for food, and to make clothing, and not the world marketplace. Development of agricultural techniques has steadily increased agricultural productivity, and the widespread diffusion of these techniques during a time period is often called an agricultural revolution. A remarkable shift in agricultural practices has occurred over the past century in response to new technologies, and the development of world markets. This also led to technological improvements in agricultural techniques, such as the Haber-Bosch method for synthesizing ammonium nitrate which made the traditional practice of recycling nutrients with crop rotation and animal manure less necessary.

Synthetic nitrogen, along with mined rock phosphate, pesticides and mechanization, have greatly increased crop yields in the early 20th century. Increased supply of grains has led to cheaper livestock as well. Further, global yield increases were experienced later in the 20th century when high-yield varieties of common staple grains such as rice, wheat, and corn (maize) were introduced as a part of the Green Revolution. The Green Revolution exported the technologies (including pesticides and synthetic nitrogen) of the developed world to the developing world. Thomas Malthus famously predicted that the Earth would not be able to support its growing population, but technologies such as the Green Revolution have allowed the world to produce a surplus of food.

Many governments have subsidized agriculture to ensure an adequate food supply. These agricultural subsidies are often linked to the production of certain commodities such as wheat, corn (maize), rice, soybeans, and milk. These subsidies, especially when instituted by developed countries have been noted as protectionist, inefficient, and environmentally damaging.

In the past century agriculture has been characterised by enhanced productivity, the use of synthetic fertilizers and pesticides, selective breeding, mechanization, water contamination, and farm subsidies. Proponents of organic farming such as Sir Albert Howard argued in the early 20th century that the overuse of pesticides and synthetic fertilizers damages the long-term fertility of the soil. While this feeling lay dormant for decades, as environmental awareness has increased in the 21st century there has been a movement towards sustainable agriculture by some farmers, consumers, and policymakers.

In recent years there has been a backlash against perceived external environmental effects of mainstream agriculture, particularly regarding water pollution, resulting in the organic movement. One of the major forces behind this movement has been the European Union, which first certified organic food in 1991 and began reform of its Common Agricultural Policy (CAP) in 2005 to phase out commodity-linked farm subsidies, also known as decoupling. The growth of organic farming has renewed research in alternative technologies such as integrated pest management and selective breeding. Recent mainstream technological developments include genetically modified food.

In late 2007, several factors pushed up the price of grains consumed by humans as well as used to feed poultry and dairy cows and other cattle, causing higher prices of wheat (up 58%), soybean (up 32%), and maize (up 11%) over the year. Food riots took place in several countries across the world. Contributing factors included drought in Australia and elsewhere, increasing demand for grain-fed animal products from the growing middle classes of countries such as China and India, diversion of foodgrain to biofuel production and trade restrictions imposed by several countries.

An epidemic of stem rust on wheat caused by race Ug99 is currently spreading across Africa and into Asia and is causing major concern. Approximately 40% of the world's agricultural land is seriously degraded. In Africa, if current trends of soil degradation continue, the continent might be able to feed just 25% of its population by 2025, according to UNU's Ghana-based Institute for Natural Resources in Africa.

History

Agricultural practices such as irrigation, crop rotation, fertilizers, and pesticides were developed long ago, but have made great strides in the past century. The history of agriculture has played a major role in human history, as agricultural progress has been a crucial factor in worldwide socio-economic change. Division of labour in agricultural societies made commonplace specialisations rarely seen in hunter-gatherer cultures. So, too, are arts such as epic literature and monumental architecture, as well as codified legal systems. When farmers became capable of producing food beyond the needs of their own families, others in their society were freed to devote themselves to projects other than food acquisition. Historians and anthropologists have long argued that the development of agriculture made civilization possible. The total world population probably never exceeded 15 million inhabitants before the invention of agriculture.

Ancient Origins

The Fertile Crescent of Western Asia, Egypt, and India were sites of the earliest planned sowing and harvesting of plants that had previously been gathered in the wild. Independent development of agriculture occurred in northern and southern China, Africa's Sahel, New Guinea and several regions of the Americas. The eight so-called Neolithic founder crops of agriculture appear: first emmer wheat and einkorn wheat, then hulled barley, peas, lentils, bitter vetch, chick peas and flax.

By 7000 BC, small-scale agriculture reached Egypt. From at least 7000 BC the Indian subcontinent saw farming of wheat and barley, as attested by archaeological excavation at Mehrgarh in Balochistan in what is present day Pakistan. By 6000 BC, mid-scale farming was entrenched on the banks of the Nile. This, as irrigation had not yet matured sufficiently. About this time, agriculture was developed independently in the Far East, with rice, rather than wheat, as the primary crop. Chinese and Indonesian farmers went on to domesticate taro and beans including mung, soy and azuki. To complement these new sources of carbohydrates, highly organised net fishing of rivers, lakes and ocean shores in these areas brought in great volumes of essential protein. Collectively, these new methods of farming and fishing inaugurated a human population boom that dwarfed all previous expansions and continues today.

By 5000 BC, the Sumerians had developed core agricultural techniques including large-scale intensive cultivation of land, monocropping, organised irrigation, and the use of a specialised labour force, particularly along the waterway now known as the Shatt al-Arab, from its Persian Gulf delta to the confluence of the Tigris and Euphrates. Domestication of wild aurochs and mouflon into cattle and sheep, respectively, ushered in the large-scale use of animals for food/fiber and as beasts of burden. The shepherd joined the farmer as an essential provider for sedentary and seminomadic societies. Maize, manioc, and arrowroot were first domesticated in the Americas as far back as 5200 BC.

The potato, tomato, pepper, squash, several varieties of bean, tobacco, and several other plants were also developed in the Americas, as was extensive terracing of steep hillsides in much of Andean South America. The Greeks and Romans built on techniques pioneered by the Sumerians, but made few fundamentally new advances. Southern Greeks struggled with very poor soils, yet managed to become a dominant society for years. The Romans were noted for an emphasis on the cultivation of crops for trade. In the same region, a parallel agricultural revolution occurred, resulting in some of the most important crops grown today. In Mesoamerica wild teosinte was transformed through human selection into the ancestor of modern maize, more than 6000 years ago. It gradually spread across North America and was the major crop of Native Americans at the time of European exploration. Other Mesoamerican crops include hundreds of varieties of squash and beans. Cocoa was also a major crop in domesticated Mexico and Central America. The turkey, one of the most important meat birds, was probably domesticated in Mexico or the U.S. Southwest. In the Andes region of South America the major domesticated crop was potatoes, domesticated perhaps 5000 years ago. Large varieties of beans were domesticated, in South America, as well as animals, including llamas, alpacas, and guinea pigs. Coca, still a major crop, was also domesticated in the Andes.

A minor centre of domestication, the indigenous people of the Eastern U.S. appear to have domesticated numerous crops. Sunflowers, tobacco, varieties of squash and Chenopodium, as well as crops no longer grown, including marshelder and little barley were domesticated. Other wild foods may have undergone some selective cultivation, including wild rice and maple sugar. The most common varieties of strawberry were domesticated from Eastern North America.

By 3500 BC, the simplest form of the plough was developed, called the ard. Before this period, simple digging sticks or hoes were used. These tools would have also been easier to transport, which was a benefit as people only stayed until the soil's nutrients were depleted. However, through excavations in Mexico it has been found that the continuous cultivating of smaller pieces of land would also have been a sustaining practice. Additional research in central Europe later revealed that agriculture was indeed practiced at this method. For this method, ards were thus much more efficient than digging sticks.

Middle Ages

During the Middle Ages, farmers in North Africa, the Near East, and Europe began making use of agricultural technologies including irrigation systems based on hydraulic and hydrostatic principles, machines such as norias, water-raising machines, dams, and reservoirs. This combined with the invention of a three-field system of crop rotation and the moldboard plow greatly improved agricultural efficiency. In the European medieval period, agriculture was considered part of the set of *seven mechanical arts.* "Between 1413 and 1635, the top 5 percent of villagers tripled the amount of arable land they held, (Duplessis). The wealth getting wealthier.

Modern Era

After 1492, a global exchange of previously local crops and livestock breeds occurred. Key crops involved in this exchange included the tomato, maize, potato, manioc, cocoa bean and tobacco going from the New World to the Old, and several varieties of wheat, spices, coffee, and sugar cane going from the Old World to the New. The most important animal exportation from the Old World to the New were those of the horse and dog (dogs were already present in the pre-Columbian Americas but not in the numbers and breeds suited to farm work). Although not usually food animals, the horse (including donkeys and ponies) and dog quickly filled essential production roles on western-hemisphere farms.

The potato became an important staple crop in northern Europe. Since being introduced by Portuguese in the 16th century, maize and manioc have replaced traditional African crops as the continent's most important staple food crops.

By the early 19th century, agricultural techniques, implements, seed stocks and cultivar had so improved that yield per land unit was many times that seen in the Middle Ages. Although there is a vast and interesting history of crop cultivation before the dawn of the 20th century, there is little question that the work of Charles Darwin and Gregor Mendel created the scientific foundation for plant breeding that led to its explosive impact over the past 150 years.

With the rapid rise of mechanization in the late 19th century and the 20th century, particularly in the form of the tractor, farming tasks could be done with a speed and on a scale previously impossible. These advances have led to efficiencies enabling certain modern farms in the United States, Argentina, Israel, the United Kingdom Germany, and a few other nations to output volumes of high-quality produce per land unit at what may be the practical limit.

The Haber-Bosch method for synthesizing ammonium nitrate represented a major breakthrough and allowed crop yields to overcome previous constraints. In the past century agriculture has been characterised by enhanced productivity, the substitution of synthetic fertilizers and pesticides for labour, water pollution, and farm subsidies. In recent years there has been a backlash against the external environmental effects of conventional agriculture, resulting in the organic movement.

The cereals rice, corn, and wheat provide 60% of human food supply. Between 1700 and 1980, "the total area of cultivated land worldwide increased 466%" and yields increased dramatically, particularly because of selectively bred high-yielding varieties, fertilizers, pesticides, irrigation, and machinery. For example, irrigation increased corn yields in eastern Colorado by 400 to 500% from 1940 to 1997.

However, concerns have been raised over the sustainability of intensive agriculture. Intensive agriculture has become associated with decreased soil quality in India and Asia, and there has been increased concern over the effects of fertilizers and pesticides on the environment, particularly as population increases and food demand expands. The monocultures typically used in intensive agriculture increase the number of pests, which are controlled through pesticides. Integrated pest management (IPM), which "has been promoted for decades and has had some notable successes" has not significantly affected the use of pesticides because policies encourage the use of pesticides and IPM is knowledge-intensive.

Although the "Green Revolution" significantly increased rice yields in Asia, yield increases have not occurred in the past 15–20 years. The genetic "yield potential" has increased for wheat, but the yield potential for rice has not increased since 1966, and the yield potential for maize has "barely increased in 35 years". It takes a decade or two for herbicide-resistant weeds to emerge, and insects become resistant to insecticides within about a decade. Crop rotation helps to prevent resistances.

Agricultural exploration expeditions, since the late 19th century, have been mounted to find new species and new agricultural practices in different areas of the world. Two early examples of expeditions include Frank N. Meyer's fruit- and nut-collecting trip to China and Japan from 1916-1918 and the Dorsett-Morse Oriental Agricultural Exploration Expedition to China, Japan, and Korea from 1929-1931 to collect soybean germplasm to support the rise in soybean agriculture in the United States.

In 2009, the agricultural output of China was the largest in the world, followed by the European Union, India and the United States, according to the International Monetary Fund. Economists measure the total factor productivity of agriculture and by this measure agriculture in the United States is roughly 2.6 times more productive than it was in 1948.

Six countries - the US, Canada, France, Australia, Argentina and Thailand - supply 90% of grain exports. Water deficits, which are already spurring heavy grain imports in numerous middle-sized countries, including Algeria, Iran, Egypt, and Mexico, may soon do the same in larger countries, such as China or India.

Crop Production Systems

Cropping systems vary among farms depending on the available resources and constraints; geography and climate of the farm; government policy; economic, social and political pressures; and the philosophy and culture of the farmer. Shifting cultivation (or slash and burn) is a system in which forests are burnt, releasing nutrients to support cultivation of annual and then perennial crops for a period of several years.

Then the plot is left fallow to regrow forest, and the farmer moves to a new plot, returning after many more years (10-20). This fallow period is shortened if population density grows, requiring the input of nutrients (fertilizer or manure) and some manual pest control. Annual cultivation is the next phase of intensity in which there is no fallow period. This requires even greater nutrient and pest control inputs.

Further industrialisation lead to the use of monocultures, when one cultivar is planted on a large acreage. Because of the low biodiversity, nutrient use is uniform and pests tend to build up, necessitating the greater use of pesticides and fertilizers. Multiple cropping, in which several crops are grown sequentially in one year, and intercropping, when several crops are grown at the same time are other kinds of annual cropping systems known as polycultures.

In tropical environments, all of these cropping systems are practiced. In subtropical and arid environments, the timing and extent of agriculture may be limited by rainfall, either not allowing multiple annual crops in a year, or requiring irrigation. In all of these environments perennial crops are grown (coffee, chocolate) and systems are practiced such as agroforestry. In temperate environments, where ecosystems were predominantly grassland or prairie, highly productive annual cropping is the dominant farming system.

The last century has seen the intensification, concentration and specialisation of agriculture, relying upon new technologies of agricultural chemicals (fertilizers and pesticides), mechanization, and plant breeding (hybrids and GMO's). In the past few decades, a move towards sustainability in agriculture has also developed, integrating ideas of socio-economic justice and conservation of resources and the environment within a farming system. This has led to the development of many responses to the conventional agriculture approach, including organic agriculture, urban agriculture, community supported agriculture, ecological or biological agriculture, integrated farming and holistic management, as well as an increased trend towards agricultural diversification.

Crop Statistics

Important categories of crops include grains and pseudograins, pulses (legumes), forage, and fruits and vegetables. Specific crops are cultivated in distinct growing regions throughout the world. In millions of metric tons, based on FAO estimate.

Table : Top agricultural products, by crop types(million tonnes) 2004 data

Cereals	2,263
Vegetables and melons	866
Roots and Tubers	715
Milk	619
Fruit	503
Meat	259
Oilcrops	133
Fish (2001 estimate)	130
Eggs	63
Pulses	60
Vegetable Fiber	30

Source:Food and Agriculture Organisation (FAO)

Table : Top agricultural products, by individual crops(million tonnes) 2004 data

Sugar Cane	1,324
Maize	721
Wheat	627
Rice	605
Potatoes	328
Sugar Beet	249
Soybean	204
Oil Palm Fruit	162
Barley	154
Tomato	120

Source:Food and Agriculture Organisation (FAO)

Livestock Production Systems

Animals, including horses, mules, oxen, camels, llamas, alpacas, and dogs, are often used to help cultivate fields, harvest crops, wrangle other animals, and transport farm products to buyers. Animal husbandry not only refers to the breeding and raising of animals for meat or to harvest animal products (like milk, eggs, or wool) on a continual basis, but also to the breeding and care of species for work and companionship. Livestock production systems can be defined based on feed source, as grassland - based, mixed, and landless.

Grassland based livestock production relies upon plant material such as shrubland, rangeland, and pastures for feeding ruminant animals. Outside nutrient inputs may be used, however manure is returned directly to the grassland as a major nutrient source. This system is particularly important in areas where crop production is not feasible because of climate or soil, representing 30-40 million pastoralists. Mixed production systems use grassland, fodder crops and grain feed crops as feed for ruminant and monogastic (one stomach; mainly chickens and pigs) livestock. Manure is typically recycled in mixed systems as a fertilizer for crops. Approximately 68% of all agricultural land is permanent pastures used in the production of livestock.

Landless systems rely upon feed from outside the farm, representing the de-linking of crop and livestock production found more prevalently in OECD member countries. In the U.S., 70% of the grain grown is fed to animals on feedlots. Synthetic fertilizers are more heavily relied upon for crop production and manure utilisation becomes a challenge as well as a source for pollution.

Production Practices

Tillage is the practice of plowing soil to prepare for planting or for nutrient incorporation or for pest control. Tillage varies in intensity from conventional to no-till. It may improve productivity by warming the soil, incorporating fertilizer and controlling weeds, but also renders soil more prone to erosion, triggers the decomposition of organic matter releasing CO_2, and reduces the abundance and diversity of soil organisms.

Pest control includes the management of weeds, insects/mites, and diseases. Chemical (pesticides), biological (biocontrol), mechanical (tillage), and cultural practices are used. Cultural practices include crop rotation, culling, cover crops, intercropping, composting, avoidance, and resistance. Integrated pest management attempts to use all of these methods to keep pest populations below the number which would cause economic loss, and recommends pesticides as a last resort.

Nutrient management includes both the source of nutrient inputs for crop and livestock production, and the method of utilisation of manure produced by livestock. Nutrient inputs can be chemical inorganic fertilizers, manure, green manure, compost and mined minerals. Crop nutrient use may also be managed using cultural techniques such as crop rotation or a fallow period. Manure is used either by holding livestock where the feed crop is growing, such as in managed intensive rotational grazing, or by spreading either dry or liquid formulations of manure on cropland or pastures.

Water management is where rainfall is insufficient or variable, which occurs to some degree in most regions of the world. Some farmers use irrigation to supplement rainfall. In other areas such as the Great Plains in the U.S. and Canada, farmers use a fallow year to conserve soil moisture to use for growing a crop in the following year. Agriculture represents 70% of freshwater use worldwide.

Processing, Distribution, and Marketing

In the United States, food costs attributed to processing, distribution, and marketing have risen while the costs attributed to farming have declined. This is related to the greater efficiency of farming, combined with the increased level of value addition (e.g. more highly processed products) provided by the supply chain. From 1960 to 1980 the farm share was around 40%, but by 1990 it had declined to 30% and by 1998, 22.2%. Market concentration has increased in the sector as well, with the top 20 food manufacturers accounting for half the food-processing value in 1995, over double that produced in 1954. As of 2000 the top six US supermarket groups had 50% of sales compared to 32% in 1992. Although the total effect of the increased market concentration is likely increased efficiency, the changes redistribute economic surplus from producers (farmers) and consumers, and may have negative implications for rural communities.

Crop Alteration and Biotechnology

Crop alteration has been practiced by humankind for thousands of years, since the beginning of civilization. Altering crops through breeding practices changes the genetic make-up of a plant to develop crops with more beneficial characteristics for humans, for example, larger fruits or seeds, drought-tolerance, or resistance to pests. Significant advances in plant breeding ensued after the work of geneticist Gregor Mendel. His work on dominant and recessive alleles gave plant breeders a better understanding of genetics and brought great insights to the techniques utilised by plant breeders. Crop breeding includes techniques such as plant selection with desirable traits, self-pollination and cross-pollination, and molecular techniques that genetically modify the organism.

Domestication of plants has, over the centuries increased yield, improved disease resistance and drought tolerance, eased harvest and improved the taste and nutritional value of crop plants. Careful selection and breeding have had enormous effects on the characteristics of crop plants. Plant selection and breeding in the 1920s and 1930s improved pasture (grasses and clover) in New Zealand. Extensive X-ray and ultraviolet induced mutagenesis efforts (i.e. primitive genetic engineering) during the 1950s produced the modern commercial varieties of grains such as wheat, corn (maize) and barley.

The Green Revolution popularized the use of conventional hybridization to increase yield many folds by creating "high-yielding varieties". For example, average yields of corn (maize) in the USA have increased from around 2.5 tons per hectare (t/ha) (40 bushels per acre) in 1900 to about 9.4 t/ha (150 bushels per acre) in 2001. Similarly, worldwide average wheat yields have increased from less than 1 t/ha in 1900 to more than 2.5 t/ha in 1990. South American average wheat yields are around 2 t/ha, African under 1 t/ha, Egypt and Arabia up to 3.5 to 4 t/ha with irrigation. In contrast, the average wheat yield in countries such as France is over 8 t/ha. Variations in yields are due mainly to variation in climate, genetics, and the level of intensive farming techniques (use of fertilizers, chemical pest control, growth control to avoid lodging).

Genetic Engineering

Genetically Modified Organisms (GMO) are organisms whose genetic material has been altered by genetic engineering techniques generally known as recombinant DNA technology. Genetic engineering has expanded the genes available to breeders to utilise in creating desired germlines for new crops. After mechanical tomato-harvesters were developed in the early 1960s, agricultural scientists genetically modified tomatoes to be more resistant to mechanical handling. More recently, genetic engineering is being employed in various parts of the world, to create crops with other beneficial traits. New research on woodland strawberry genome was found to be short and easy to manipulate. Researchers now have tools to improve strawberry flavours and aromas of cultivated strawberries as stated in a publication by Nature Genetics.

Herbicide-Tolerant GMO Crops

Roundup Ready seed has a herbicide resistant gene implanted into its genome that allows the plants to tolerate exposure to glyphosate. *Roundup* is a trade name for a glyphosate-based product, which is a systemic, nonselective herbicide used to kill weeds. *Roundup Ready* seeds allow the farmer to grow a crop that can be sprayed with glyphosate to control weeds without harming the resistant crop. Herbicide-tolerant crops are used by farmers worldwide. Today, 92% of soybean acreage in the US is planted with genetically modified herbicide-tolerant plants. With the increasing use of herbicide-tolerant crops, comes an increase in the use of glyphosate-based herbicide sprays. In some areas glyphosate resistant weeds have developed, causing farmers to switch to other herbicides. Some studies also link widespread glyphosate usage to iron deficiencies in some crops, which is both a crop production and a nutritional quality concern, with potential economic and health implications.

Insect-Resistant GMO Crops

Other GMO crops used by growers include insect-resistant crops, which have a gene from the soil bacterium *Bacillus thuringiensis* (Bt), which produces a toxin specific to insects. These crops protect plants from damage by insects; one such crop is Starlink. Another is cotton, which accounts for 63% of US cotton acreage.

Some believe that similar or better pest-resistance traits can be acquired through traditional breeding practices, and resistance to various pests can be gained through hybridization or cross-pollination with wild species. In some cases, wild species are the primary source of resistance traits; some tomato cultivars that have gained resistance to at least 19 diseases did so through crossing with wild populations of tomatoes.

Costs and Benefits of GMOs

Genetic engineers may someday develop transgenic plants which would allow for irrigation, drainage, conservation, sanitary engineering, and maintaining or increasing yields while requiring fewer fossil fuel derived inputs than conventional crops. Such developments would be particularly important in areas which are normally arid and rely upon constant irrigation, and on large scale farms. However, genetic engineering of plants has proven to be controversial. Many issues surrounding food security and environmental impacts have risen regarding GMO practices. For example, GMOs are questioned by some ecologists and economists concerned with GMO practices such as terminator seeds, which is a genetic modification that creates sterile seeds. Terminator seeds are currently under strong international opposition and face continual efforts of global bans.

Another controversial issue is the patent protection given to companies that develop new types of seed using genetic engineering. Since companies have intellectual ownership of their seeds, they have the power to dictate terms and conditions of their patented product. Currently, ten seed companies control over two-thirds of the global seed sales. Vandana Shiva argues that these companies are guilty of biopiracy by patenting life and exploiting organisms for profit Farmers using patented seed are restricted from saving seed for subsequent plantings, which forces farmers to buy new seed every year. Since seed saving is a traditional practice for many farmers in both developing and developed countries, GMO seeds legally bind farmers to change their seed saving practices to buying new seed every year.

Locally adapted seeds are an essential heritage that has the potential to be lost with current hybridized crops and GMOs. Locally adapted seeds, also called land races or crop eco-types, are important because they have adapted over time to the specific micro-climates, soils, other environmental conditions, field designs, and ethnic preference indigenous to the exact area of cultivation. Introducing GMOs and hybridized commercial seed to an area brings the risk of cross-pollination with local land races Therefore, GMOs pose a threat to the sustainability of land races and the ethnic heritage of cultures. Once seed contains transgenic material, it becomes subject to the conditions of the seed company that owns the patent of the transgenic material.

Modern Agriculture

Modern agriculture is a term used to describe the wide majority of production practices employed by America's farmers. The term depicts the push for innovation, stewardship and advancements continually made by growers to sustainability produce higher-quality products with a reduced environmental impact. Intensive scientific research and robust investment in modern agriculture during the past 50 years has helped farmers double food production.

Safety

The agriculture industry works with government agencies and other organisations to ensure that farmers have access to the technologies required to support modern agriculture practices. Farmers are supported by education and certification programs that ensure they apply agricultural practices with care and only when required.

Sustainability

Technological advancements help provide farmers with tools and resources to make farming more sustainable. New technologies have given rise to innovations like conservation tillage, a farming process which helps prevent land loss to erosion, water pollution and enhances carbon sequestration.

Affordability

The goal of modern agriculture practices is to help farmers provide an affordable supply of food to meet the demands of a growing population. With modern agriculture, more crops can be grown on less land allowing farmers to provide an increased supply of food at an affordable price.

Food Safety, Labeling and Regulation

Food security issues also coincide with food safety and food labeling concerns. Currently a global treaty, the BioSafety Protocol, regulates the trade of GMOs. The EU currently requires all GMO foods to be labelled, whereas the US does not require transparent labeling of GMO foods. Since there are still questions regarding the safety and risks associated with GMO foods, some believe the public should have the freedom to choose and know what they are eating and require all GMO products to be labelled. The Food and Agriculture Organisation of the United Nations (FAO) leads international efforts to defeat hunger and provides a neutral forum where nations meet as equals to negotiate agreements and debate food policy and the regulation of agriculture. According to Dr. Samuel Jutzi, director of FAO's animal production and health division, lobbying by "powerful" big food corporations has stopped reforms that would improve human health and the environment. The "real, true issues are not being addressed by the political process because of the influence of lobbyists, of the true powerful entities," he said, speaking at the Compassion in World Farming annual forum. For example, recent proposals for a voluntary code of conduct for the livestock industry that would have provided incentives for improving standards for health, and environmental regulations, such as the number of animals an area of land can support without long-term damage, were successfully defeated due to large food company pressure.

Environmental Impact

Agriculture imposes external costs upon society through pesticides, nutrient runoff, excessive water usage, and assorted other problems. A 2000 assessment of agriculture in the UK determined total external costs for 1996 of £2,343 million, or £208 per hectare. A 2005 analysis of these costs in the USA concluded that cropland imposes approximately $5 to 16 billion ($30 to $96 per hectare), while livestock production imposes $714 million. Both studies concluded that more should be done to internalize external costs, and neither included subsidies in their analysis, but noted that subsidies also influence the cost of agriculture to society. Both focused on purely fiscal impacts. The 2000 review included reported pesticide poisonings but did not include speculative chronic effects of pesticides, and the 2004 review relied on a 1992 estimate of the total impact of pesticides.

In 2010, the International Resource Panel of the United Nations Environment Programme published a report assessing the environmental impacts of consumption and production. The study found that agriculture and food consumption are two of the most important drivers of environmental pressures, particularly habitat change, climate change, water use and toxic emissions.

Agriculture accounts for 70 per cent of withdrawals of freshwater resources. However, increasing pressure being placed on water resources by industry, cities and the involving biofuels industry means that water scarcity is increasing and agriculture is facing the challenge of producing more food for the world's growing population with fewer water resources. Scientists are also realising that water resources need to be allocated to maintain natural environmental services, such as protecting towns from flooding, cleaning ecosystems and supporting fish stocks. In the book *Out of Water: From abundance to scarcity and how to solve the world's water problems*, authors Colin Chartres and Samyukta Varma of the International Water Management Institute lay down a six-point plan of action for addressing the global challenge of producing sufficient food for the world with dwindling water resources. One of the actions they say is required is to ensure all water systems, such as lakes and rivers, have water allocated to environmental flow. A key player who is credited to saving billions of lives because of his revolutionary work in developing new agricultural techniques is Norman Borlaug. His transformative work brought high-yield crop varieties to developing countries and earned him an unofficial title as the father of the Green Revolution.

Livestock Issues

A senior UN official and co-author of a UN report detailing this problem, Henning Steinfeld, said "Livestock are one of the most significant contributors to today's most serious environmental problems". Livestock production occupies 70% of all land used for agriculture, or 30% of the land surface of the planet. It is one of the largest sources of greenhouse gases, responsible for 18% of the world's greenhouse gas emissions as measured in CO_2 equivalents. By comparison, all transportation emits 13.5% of the CO_2. It produces 65% of human-related nitrous oxide (which has 296 times the global warming potential of CO_2) and 37% of all human-induced methane (which is 23 times as warming as CO_2. It also generates 64% of the ammonia emission. Livestock expansion is cited as a key factor driving deforestation, in the Amazon basin 70% of previously forested area is now occupied by pastures and the remainder used for feedcrops. Through deforestation and land degradation, livestock is also driving reductions in biodiversity.

Land Transformation and Degradation

Land transformation, the use of land to yield goods and services, is the most substantial way humans alter the Earth's ecosystems, and is considered the driving force in the loss of biodiversity. Estimates of the amount of land transformed by humans vary from 39–50%. Land degradation, the long-term decline in ecosystem function and productivity, is estimated to be occurring on 24% of land worldwide, with cropland over-represented. The UN-FAO report cites land management as the driving factor behind degradation and reports that 1.5 billion people rely upon the degrading land. Degradation can be deforestation, desertification, soil erosion, mineral depletion, or chemical degradation (acidification and salinization).

Eutrophication

Eutrophication, excessive nutrients in aquatic ecosystems resulting in algal blooms and anoxia, leads to fish kills, loss of biodiversity, and renders water unfit for drinking and other industrial uses. Excessive fertilization and manure application to cropland, as well as high livestock stocking densities cause nutrient (mainly nitrogen and phosphorus) runoff and leaching from agricultural land. These nutrients are major nonpoint pollutants contributing to eutrophication of aquatic ecosystems.

Pesticides

Pesticide use has increased since 1950 to 2.5 million tons annually worldwide, yet crop loss from pests has remained relatively constant. The World Health Organisation estimated in 1992 that 3 million pesticide poisonings occur annually, causing 220,000 deaths. Pesticides select for pesticide resistance in the pest population, leading to a condition termed the 'pesticide treadmill' in which pest resistance warrants the development of a new pesticide.

An alternative argument is that the way to 'save the environment' and prevent famine is by using pesticides and intensive high yield farming, a view exemplified by a quote heading the Centre for Global Food Issues website: 'Growing more per acre leaves more land for nature'. However, critics argue that a trade-off between the environment and a need for food is not inevitable, and that pesticides simply replace good agronomic practices such as crop rotation.

Climate Change

Climate change has the potential to affect agriculture through changes in temperature, rainfall (timing and quantity), CO_2, solar radiation and the interaction of these elements. Agriculture can both mitigate or worsen global warming. Some of the increase in CO_2 in the atmosphere comes from the decomposition of organic matter in the soil, and much of the methane emitted into the atmosphere is caused by the decomposition of organic matter in wet soils such as rice paddies. Further, wet or anaerobic soils also lose nitrogen through denitrification, releasing the greenhouse gases nitric oxide and nitrous oxide. Changes in management can reduce the release of these greenhouse gases, and soil can further be used to sequester some of the CO_2 in the atmosphere.

International Economics and Market Reports

Differences in economic development, population density and culture mean that the farmers of the world operate under very different conditions.

A US cotton farmer may receive US$230 in government subsidies per acre planted (in 2003), while farmers in Mali and other third-world countries do without. When prices decline, the heavily subsidized US farmer is not forced to reduce his output, making it difficult for cotton prices to rebound, but his Mali counterpart may go broke in the meantime.

A livestock farmer in South Korea can calculate with a (highly subsidized) sales price of US$1300 for a calf produced. A South American Mercosur country rancher calculates with a calf's sales price of US$120–200 (both 2008 figures). With the former, scarcity and high cost of land is compensated with public subsidies, the latter compensates absence of subsidies with economics of scale and low cost of land.

In the Peoples Republic of China, a rural household's productive asset may be one hectare of farmland. In Brazil, Paraguay and other countries where local legislature allows such purchases, international investors buy thousands of hectares of farmland or raw land at prices of a few hundred US$ per hectare.

To promote exports of agricultural products, many government agencies publish on the web economic studies and reports categorized by product and country. Among these agencies include four of the largest exporters of agricultural products, such as the FAS of the United States Department of Agriculture, Agriculture and Agri-Food Canada (AAFC), Austrade, and NZTE. The Federation of International Trade Associations publishes studies and reports by FAS and AAFC, as well as other non-governmental organisations.

List of Countries by Agricultural Output

Below is a list of countries by agricultural output in 2010.

Table : Agricultural output in 2010 (Nominal)

Rank	Country	Output in billions of US$
—	World	3,585.829
1	China	599.582
—	European Union	293.080
2	India	284.524
3	United States	161.236
4	Brazil	142.141
5	Indonesia	108.130
6	Japan	76.424
7	Turkey	71.218
8	Nigeria	65.041
9	Russia	58.603
10	France	51.651

Table : Agricultural output in 2010 (PPP)

Rank	Country	Output in billions of US$
—	World	4,233.098
1	China	1,028.742
2	India	751.173
—	European Union	273.068
3	United States	161.236
4	Indonesia	157.572
5	Brazil	147.700
6	Nigeria	113.385
7	Pakistan	101.348
8	Turkey	92.209
9	Iran	90.052
10	Russia	88.918

Energy and Agriculture

Since the 1940s, agricultural productivity has increased dramatically, due largely to the increased use of energy-intensive mechanization, fertilizers and pesticides. The vast majority of this energy input comes from fossil fuel sources. Between 1950 and 1984, the Green Revolution transformed agriculture around the globe, with world grain production increasing by 250% as world population doubled. Modern agriculture's heavy reliance on petrochemicals and mechanization has raised concerns that oil shortages could increase costs and reduce agricultural output, causing food shortages.

Table : Agriculture and food system share (%) of total energy consumption by three industrialised nations

Country	Year	Agriculture (direct & indirect)	Food System
United Kingdom	2005	1.9	11
United States of America	1996	2.1	10
United States of America	2002	2.0	14
Sweden	2000	2.5	13

Modern or industrialised agriculture is dependent on fossil fuels in two fundamental ways: 1) direct consumption on the farm and 2) indirect consumption to manufacture inputs used on the farm. Direct consumption includes the use of lubricants and fuels to operate farm vehicles and machinery; and use of gas, liquid propane, and electricity to power dryers, pumps, lights, heaters, and coolers. American farms directly consumed about 1.2 exajoules (1.1 quadrillion BTU) in 2002, or just over 1 percent of the nation's total energy.

Indirect consumption is mainly oil and natural gas used to manufacture fertilizers and pesticides, which accounted for 0.6 exajoules (0.6 quadrillion BTU) in 2002. The energy used to manufacture farm machinery is also a form of indirect agricultural energy consumption, but it is not included in USDA estimates of U.S. agricultural energy use. Together, direct and indirect consumption by U.S. farms accounts for about 2 percent of the nation's energy use. Direct and indirect energy consumption by U.S. farms peaked in 1979, and has gradually declined over the past 30 years.

Food systems encompass not just agricultural production, but also off-farm processing, packaging, transporting, marketing, consumption, and disposal of food and food-related items. Agriculture accounts for less than one-fifth of food system energy use in the United States.

In 2007, higher incentives for farmers to grow non-food biofuel crops combined with other factors (such as over-development of former farm lands, rising transportation costs, climate change, growing consumer demand in China and India, and population growth) to cause food shortages in Asia, the Middle East, Africa, and Mexico, as well as rising food prices around the globe. As of December 2007, 37 countries faced food crises, and 20 had imposed some sort of food-price controls. Some of these shortages resulted in food riots and even deadly stampedes.

The biggest fossil fuel input to agriculture is the use of natural gas as a hydrogen source for the Haber-Bosch fertilizer-creation process. Natural gas is used because it is the cheapest currently available source of hydrogen. When oil production becomes so scarce that natural gas is used as a partial stopgap replacement, and hydrogen use in transportation increases, natural gas will become much more expensive. If the Haber Process is unable to be commercialised using renewable energy (such as by electrolysis) or if other sources of hydrogen are not available to replace the Haber Process, in amounts sufficient to supply transportation and agricultural needs, this major source of fertilizer would either become extremely expensive or unavailable. This would either cause food shortages or dramatic rises in food prices.

Mitigation of Effects of Petroleum Shortages

In the event of a petroleum shortage, organic agriculture can be more attractive than conventional practices that use petroleum-based pesticides, herbicides, or fertilizers. Some farmers using modern organic-farming methods have reported yields as high as those available from conventional farming. Organic farming may however be more labour-intensive and would require a shift of the workforce from urban to rural areas. The reconditioning of soil to restore nutrients lost during the use of monoculture agriculture techniques also takes time.

It has been suggested that rural communities might obtain fuel from the biochar and synfuel process, which uses agricultural *waste* to provide charcoal fertilizer, some fuel *and* food, instead of the normal food vs fuel debate. As the synfuel would be used on-site, the process would be more efficient and might just provide enough fuel for a new organic-agriculture fusion.

It has been suggested that some transgenic plants may some day be developed which would allow for maintaining or increasing yields while requiring fewer fossil-fuel-derived inputs than conventional crops. The possibility of success of these programs is questioned by ecologists and economists concerned with unsustainable GMO practices such as terminator seeds.

While there has been some research on sustainability using GMO crops, at least one prominent multi-year attempt by Monsanto Company has been unsuccessful, though during the same period traditional breeding techniques yielded a more sustainable variety of the same crop.

Electrical Energy Efficiency on Farms

Policy

Agricultural policy focuses on the goals and methods of agricultural production. At the policy level, common goals of agriculture include:

- Conservation
- Economic stability
- Environmental sustainability
- Food quality: Ensuring that the food supply is of a consistent and known quality.
- Food safety: Ensuring that the food supply is free of contamination.

- Food security: Ensuring that the food supply meets the population's needs.
- Poverty reduction.

Agronomy

Agronomy is the science and technology of producing and using plants for food, fuel, feed, fiber, and reclamation. Agronomy encompasses work in the areas of plant genetics, plant physiology, meteorology, and soil science. Agronomy is the application of a combination of sciences like biology, chemistry, economics, ecology, earth science, and genetics. Agronomists today are involved with many issues including producing food, creating healthier food, managing environmental impact of agriculture, and creating energy from plants. Agronomists often specialise in areas such as crop rotation, irrigation and drainage, plant breeding, plant physiology, soil classification, soil fertility, weed control, insect and pest control.

Plant Breeding

This area of agronomy involves selective breeding of plants to produce the best crops under various conditions. Plant breeding has increased crop yields and has improved the nutritional value of numerous crops, including corn, soybeans, and wheat. It has also led to the development of new types of plants. For example, a hybrid grain called triticale was produced by crossbreeding rye and wheat. Triticale contains more usable protein than does either rye or wheat. Agronomy has also been instrumental in fruit and vegetable production research. It is understood that the role of agronomist includes seeing whether produce from a field of 'x' meets the following conditions: 1. Land and water access, 2. Commercialisation (market), 3. Quality and quantity of inputs, 4. Risk protection (insurance), 5. Agricultural credit.

Biotechnology

Agronomists use biotechnology to extend and expedite the development of desired characteristics listed in the Plant Breeding section. Biotechnology is often a lab activity requiring field testing of the new crop varieties that are developed.

In addition to increasing crop yields agronomic biotechnology is increasingly being applied for novel uses other than food. For example, oilseed is at present used mainly for margarine and other food oils, but it can be modified to produce fatty acids for detergents, substitute fuels and petrochemicals.

Soil Science

Agronomists study sustainable ways to make soils more productive and profitable. They classify soils and reproduce them to determine whether they contain substances vital to plant growth such as compounds of nitrogen, phosphorus, and potassium.

If a certain soil is deficient in these substances, fertilizers may provide them. Soil science also involves investigation of the movement of nutrients through the soil, the amount of nutrients absorbed by a plant's roots, and the development of roots and their relation to the soil.

History

Prior to the development of pedology in the 19th century, agricultural soil science (or edaphology) was the only branch of soil science. The bias of early soil science toward viewing soils only in terms of their agricultural potential continues to define the soil science profession in both academic and popular settings as of 2006. (Baveye, 2006)

Current Status

Agricultural soil science studies the chemical, physical, biological, and mineralogical composition of soils as they relate to agriculture. Agricultural soil scientists develop methods that will improve the use of soil and increase the production of food and fiber crops.

Emphasis continues to grow on the importance of soil sustainability. Soil degradation such as erosion, compaction, lowered fertility, and contamination continue to be serious concerns. They conduct research in irrigation and drainage, tillage, soil classification, plant nutrition, soil fertility, and other areas.

Soil Fertility

Agricultural soil scientists study ways to make soils more productive. They classify soils and test them to determine whether they contain nutrients vital to plant growth. Such nutritional substances include compounds of nitrogen, phosphorus, and potassium. If a certain soil is deficient in these substances, fertilizers may provide them.

Agricultural soil scientists investigate the movement of nutrients through the soil, and the amount of nutrients absorbed by a plant's roots. Agricultural soil scientists also examine the development of roots and their relation to the soil. Some agricultural soil scientists try to understand the structure and function of soils in relation to soil fertility. They grasp the structure of soil as porous solid. The solid frames of soil consist of mineral derived from the rocks and organic matter originated from the dead bodies of various organisms. The pore space of the soil is essential for the soil to become productive.

Small pores serve as water reservoir supplying water to plants and other organisms in the soil during the rain-less period. The water in the small pores of soils is not pure water; they call it soil solution. In soil solution, various plant nutrients derived from minerals and organic matters in the soil are there. Indeed the soil solution is the milk to plants.

Large pores serve as water drainage pipe to allow the excessive water pass through the soil, during the heavy rains. They also serve as air tank to supply oxygen to plant roots and other living beings in the soil. In short, agricultural soil scientists see the soil as a vessel, the most precious one for us, containing all of the substances needed by the plants and other living beings on earth.

Soil Preservation

In addition, agricultural soil scientists develop methods to preserve the agricultural productivity of soil and to decrease the effects on productivity of erosion by wind and water. For example, a technique called contour plowing may be used to prevent soil erosion and conserve rainfall. Researchers in agricultural soil science also seek ways to use the soil more effectively in addressing associated challenges. Such challenges include the beneficial reuse of human and animal wastes using agricultural crops; agricultural soil management aspects of preventing water pollution and the build-up in agricultural soil of chemical pesticides.

Employment of Agricultural Soil Scientists

Most agricultural soil scientists are consultants, researchers, or teachers. Many work in the developed world as farm advisors, agricultural experiment stations, federal, state or local government agencies, industrial firms, or universities. Within the USA they may be trained through the USDA's Cooperative Extension Service offices, although other countries may use universities, research institutes or research agencies. Elsewhere, agricultural soil scientists may serve in international organisations such as the Agency for International Development and the Food and Agriculture Organisation of the United Nations.

Soil Conservation

In addition, agronomists develop methods to preserve the soil and to decrease the effects of erosion by wind and water. For example, a technique called contour plowing may be used to prevent soil erosion and conserve rainfall. Researchers in agronomy also seek ways to use the soil more effectively in solving other problems. Such problems include the disposal of human and animal wastes; water pollution; and the build-up in the soil of pesticides.

No-tilling crops is a technique now used to help prevent erosion. Planting of soil binding grasses along contours can be tried in steep slopes. For better effect, contour drains of depths up to 1 metre may help retain the soil and prevent permanent wash off.

Agroecology

Agroecology is the management of agricultural systems with an emphasis on ecological and environmental perspectives. This area is closely associated with work in the areas of sustainable agriculture, organic farming, alternative food systems and the development of alternative cropping systems.

Theoretical Modelling

Theoretical production ecology tries to quantitatively study the growth of crops. The plant is treated as a kind of biological factory, which processes light, carbon dioxide, water and nutrients into harvestable parts. Main parameters kept into consideration are temperature, sunlight, standing crop biomass, plant production distribution, nutrient and water supply.

Modelling

Modelling is essential in theoretical production ecology. Unit of modelling usually is the crop, the assembly of plants per standard surface unit. Analysis results for an individual plant are generalised to the standard surface, e.g. the Leaf Area Index is the generalised surface of all crop leaves per surface unit.

Processes

The usual system of describing plant production divides the plant production process into at least five separate processes, which are influenced by several external parameters.

Two cycles of biochemical reactions constitute the basis of plant production, the light reaction and the dark reaction.

- In the light reaction, sunlight photons are absorbed by chloroplasts which split water into an electron, proton and oxygen radical which is recombined with another radical and released as molecular oxygen. The recombination of the electron with the proton yields the energy carriers NADH and ATP. The rate of this reaction often depends on sunlight intensity, leaf area index, leaf angle and amount of chloroplasts per leaf surface unit. The maximum theoretical gross production rate under optimum growth conditions is approximately 250 kg per hectare per day.

- The dark reaction or Calvin cycle ties atmospheric carbon dioxide and uses NADH and ATP to convert it into sucrose. The available NADH and ATP, as well as temperature and carbon dioxide levels determine the rate of this reaction. Together those two reactions are termed photosynthesis. The rate of photosynthesis is determined by the interaction of a number of factors including temperature, light intensity and carbon dioxide.

- The produced carbohydrates are transported to other plant parts, such as storage organs and converted into secondary products, such as amino acids, lipids, cellulose and other chemicals needed by the plant or used for respiration. Lipids, sugars, cellulose and starch can be produced without extra elements. The conversion of carbohydrates into amino acids and nucleic acids requires nitrogen, phosphorus and sulfur. Chlorophyll production requires magnesium, while several enzymes and coenzymes require trace elements. This means, nutrient supply influences this part of the production chain. Water supply is essential for transport, hence limits this too.

- The production centres, i.e. the leaves, are sources, the storage organs, growth tips or other destinations for the photosynthetic production are sinks. The lack of sinks can be a limiting factor for production too, as happens e.g. in apple orchards where insects or night frost have destroyed the blossoms and the produced assimilates cannot be converted into apples. Biennial and perennial plants employ the stored starch and fats in their storage organs to produce new leafs and shoots the next year.

- The amount of crop biomass and the relative distribution of biomass over leafs, stems, roots and storage organs determines the respiration rate. The amount of biomass in leafs determines the leaf area index, which is important in calculating the gross photosynthetic production.

- extensions to this basic model can include insect and pest damage, intercropping, climatical changes, etc.

Parameters

Important parameters in theoretical production models thus are:

Climate

- Temperature - The temperature determines the speed of respiration and the dark reaction. A high temperature combined with a low intensity of sunlight means a high loss by respiration. A low temperature combined with a high intensity of sunlight means that NADH and ATP heap up but cannot be converted into glucose because the dark reaction cannot process them swiftly enough.

- Light - Light, also called photosynthetic Active Radiation (PAR) is the energy source for green plant growth. PAR powers the light reaction, which provides ATP and NADPH for the conversion of carbon dioxide and water into carbohydrates and molecular oxygen. When temperature, moisture, carbon dioxide and nutrient levels are optimal, light intensity determines maximum production level.

- Carbon dioxide levels - Atmospheric carbon dioxide is the sole carbon source for plants. About half of all proteins in green leaves have the sole purpose of capturing carbon dioxide.

Although CO_2 levels are constant under natural circumstances, CO_2 fertilization is common in greenhouses and is known to increase yields by on average 24%.

C_4 plants like maize and sorghum can achieve a higher yield at high solar radiation intensities, because they prevent the leaking of captured carbon dioxide due of the spatial separation of carbon dioxide capture and carbon dioxide use in the dark reaction. This means that their photorespiration is almost zero. This advantage is sometimes offset by a higher rate of maintenance respiration. In most models for natural crops, carbon dioxide levels are assumed to be constant.

Crop

- Standing crop biomass - Unlimited growth is an exponential process, which means that the amount of biomass determines the production. Because an increased biomass implies higher respiration per surface unit and a limited increase in intercepted light, crop growth is a sigmoid function of crop biomass.

- Plant production distribution - Usually only a fraction of the total plant biomass consists of useful products, e.g. the seeds in pulses and cereals, the tubers in potato and cassava, the leafs in sisal and spinach etc. The yield of usable plant portions will increase when the plant allocates more nutrients to this parts, e.g. the high-yielding varieties of wheat and rice allocate 40% of their biomass into wheat and rice grains, while the traditional varieties achieve only 20%, thus doubling the effective yield.

Different plant organs have a different respiration rate, e.g. a young leaf has a much higher respiration rate than roots, storage tissues or stems do. There is a distinction between "growth respiration" and "maintenance respiration".

Sinks, such as developing fruits, need to be present. They are usually represented by a discrete switch, which is turned on after a certain condition, e.g. critical daylength has been met.

Care

- Water supply - Because plants use passive transport to transfer water and nutrients from their roots to the leafs, water supply is essential to growth, even so that water efficiency rates are known for different crops, e.g. 5000 for sugar cane, meaning that each kilogram of produced sugar requires up to 5000 liters of water.

- Nutrient supply - Nutrient supply has a twofold effect on plant growth. A limitation in nutrient supply will limit biomass production as per Liebig's Law of the Minimum. With some crops, several nutrients influence the distribution of plant products in the plants. A nitrogen gift is known to stimulate leaf growth and therefore can work adversely on the yield of crops which are accumulating photosynthesis products in storage organs, such as ripening cereals or fruit-bearing fruit trees.

Phases in Crop Growth

Theoretical production ecology assumes that the growth of common agricultural crops, such as cereals and tubers, usually consists of four (or five) phases:

- Germination - Agronomical research has indicated a temperature dependence of germination time (GT, in days). Each crop has a unique critical temperature (CT, dimension temperature) and temperature sum (dimensions temperature times time), which are related as follows.

$$GT = \frac{TS}{\sum_{k=1}^{N}(T - T_{crit})}$$

When a crop has a temperature sum of e.g. 150 °C ·d and a critical temperature of 10 °C, it will germinate in 15 days when temperature is 20 °C, but in 10 days when temperature is 25 °C. When the temperature sum exceeds the threshold value, the germination process is complete.

- Initial spread - In this phase, the crop does not cover the field yet. The growth of the crop is linearly dependent on leaf area index, which in its turn is linearly dependent on crop biomass. As a result, crop growth in this phase is exponential.

- Total coverage of field - in this phase, growth is assumed to be linearly dependent on incident light and respiration rate, as nearly 100% of all incident light is intercepted. Typically, LAI is above two to three in this phase. This phase of vegetative growth ends when the plant gets a certain environmental or internal signal and starts generative growth (as in cereals and pulses) or the storage phase (as in tubers).

- Allocation to storage organs - in this phase, up to 100% of all production is directed to the storage organs. Generally, the leafs are still intact and as a result, gross primary production stays the same. Prolonging this phase, e.g. by careful fertilization, water and pest management results directly in a higher harvest.

- Ripening - in this phase, leafs and other production structures slowly die off. Their carbohydrates and proteins are transported to the storage organs. As a result, the LAI and, hence, the primary production decreases.

Existing Plant Production Models

Plant production models exist in varying levels of scope (cell, physiological, individual plant, crop, geographical region, global) and of generality: the model can be crop-specific or be more generally applicable. In this section the emphasis will be on crop-level based models as the crop is the main area of interest from an agronomical point of view.

As of 2005, several crop production models are in use. The crop growth model SUCROS has been developed during more than 20 years and is based on earlier models. Its latest revision known dates from 1997.

The IRRI and Wageningen University more recently developed the rice growth model ORYZA2000. This model is used for modelling rice growth. Both crop growth models are open source. Other more crop-specific plant growth models exist as well.

Sucros

SUCROS is programmed in the Fortran computer programming language. The model can and has been applied to a variety of weather regimes and crops. Because the source code of Sucros is open source, the model is open to modifications of users with FORTRAN programming experience.

The official maintained version of SUCROS comes into two flavours: SUCROS I, which has non-inhibited unlimited crop growth (which means that only solar radiation and temperature determine growth) and SUCROS II, in which crop growth is limited only by water shortage.

ORYZA2000

The ORYZA2000 rice growth model has been developed at the IRRI in cooperation with Wageningen University. This model, too, is programmed in FORTRAN. The scope of this model is limited to rice, which is the main food crop for Asia.

Other Models

The United States Department of Agriculture has sponsored a number of applicable crop growth models for various major US crops, such as cotton, soy bean, wheat and rice. Other widely-used models are the precursor of SUCROS (SWATR), CERES, several incarnations of PLANTGRO, SUBSTOR, the FAO-sponsored CROPWAT, AGWATER and the erosion-specific model EPIC.

A less mechanistic growth and competition model, called the Conductance Model, has been developed, mainly at Warwick-HRI, Wellesbourne, UK. This model simulates light interception and growth of individual plants based on the lateral expansion of their crown zone areas. Competition between plants is simulated by a set algorithms related to competition for space and resultant light intercept as the canopy closes.

Some versions of the model assume overtopping of some species by others. Although the model cannot take account of water or mineral nutrients, it can simulate individual plant growth, variability in growth within plant communities and inter-species competition. This model was written in Matlab.

Agronomy Schools

Agronomy programs are offered at colleges, universities, and specialised agricultural schools. Agronomy programs often involve classes across a range of departments including agriculture, biology, chemistry, and physiology. They can usually take from four to twelve years. Many companies will pay an agronomist-in-training's way through college if they agree to work for them when they graduate.

Career Outlook

Due to the continued growth of the global population—and the consequent expanding need for study of food crops and agriculture in general—the outlook for agronomy and agronomists is excellent Past agricultural research has created higher yielding crops, crops with better resistance to pests and plant pathogens, and more effective fertilizers and pesticides. Research is still necessary, however, particularly as insects and diseases continue to adapt to pesticides and as soil fertility and water quality continue to need improvement.

Emerging biotechnologies will play an ever larger role in agricultural research. Scientists will be needed to apply these technologies to the creation of new food products and other advances. Moreover, increasing demand is expected for biofuels and other agricultural products used in industrial processes. Agricultural scientists will be needed to find ways to increase the output of crops used in these products.

Agronomists will also be needed to balance increased agricultural output with protection and preservation of soil, water, and ecosystems. They increasingly encourage the practice of sustainable agriculture by developing and implementing plans to manage pests, crops, soil fertility and erosion, and animal waste in ways that reduce the use of harmful chemicals and do little damage to farms and the natural environment.

Most agronomists are consultants, researchers, or teachers. Many work for agricultural experiment stations, federal or state government agencies, industrial firms, or universities. Agronomists also serve in such international organisations as the Agency for International Development, The United States Department of Agriculture, and the Food and Agriculture Organisation of the United Nations.

Agronomists career options are expanding rapidly with possible ties with golf landscaping including topsoil analysis and drainage conditions. They often work in conjunction with landscape architects and engineers to determine the best soil qualities/conditions to suit the site specifications.

Chapter 3

Mechanical Stress Induces Biotic and Abiotic Stress

Plants are continuously exposed to a myriad of abiotic and biotic stresses. However, the molecular mechanisms by which these stress signals are perceived and transduced are poorly understood. To begin to identify primary stress signal transduction components, we have focused on genes that respond rapidly (within 5 min) to stress signals. Because it has been hypothesized that detection of physical stress is a mechanism common to mounting a response against a broad range of environmental stresses, we have utilised mechanical wounding as the stress stimulus and performed whole genome microarray analysis of Arabidopsis thaliana leaf tissue. This led to the identification of a number of rapid wound responsive (RWR) genes. Comparison of RWR genes with published abiotic and biotic stress microarray datasets demonstrates a large overlap across a wide range of environmental stresses. Interestingly, RWR genes also exhibit a striking level and pattern of circadian regulation, with induced and repressed genes displaying antiphasic rhythms.

Using bioinformatic analysis, we identified a novel motif over-represented in the promoters of RWR genes, herein designated as the Rapid Stress Response Element (RSRE). We demonstrate in transgenic plants that multimerized RSREs are sufficient to confer a rapid response to both biotic and abiotic stresses in vivo, thereby establishing the functional involvement of this motif in primary transcriptional stress responses. Collectively, our data provide evidence for a novel cis-element that is distributed across the promoters of an array of diverse stress-responsive genes, poised to respond immediately and coordinately to stress signals. This structure suggests that plants may have a transcriptional network resembling the general stress signalling pathway in yeast and that the RSRE element may provide the key to this coordinate regulation.

Plants are persistently challenged with numerous biotic and abiotic environmental stresses. To cope with environmental stresses plants have evolved phytohormones such as jasmonic acid, salicylic acid, ethylene, and abscisic acid, which are utilised to regulate plant responses to both abiotic and biotic stresses with considerable signaling crosstalk. While these phytohormone pathways have been well studied, knowledge of stress perception and initial signaling events, aside from plant pathogen interactions, are less defined. It is known that application of insect oral secretions containing protein fragments of chloroplastic ATP synthase or application of purified oligouronides (OGAs) derived from the plant cell wall are capable of inducing plant defence responses, although a receptor has not yet been identified. Additionally, a cellulose synthase (*CESA3*) mutant *cev1* shows enhanced resistance to powdery mildew as a result of constitutive increase in jasmonic acid levels in these plants. This has led to the hypothesis that mechanical disruption of the cell wall may result in stress signaling. The perception of cold stress has been hypothesized to be mediated through the detection of changes in membrane fluidity and protein conformation. Finally, secondary messengers such as Ca^{2+}, reactive oxygen species (ROS), and phosphatidic acid have been implicated in initial signaling cascades in response to both abiotic and biotic stresses.

One mechanism of response to stress that has been studied extensively in yeast and animals is the general stress response (GSR) (also referred to as the cellular stress response). The GSR acts in a transient manner in response to a diverse array of stresses. The GSR is initiated in response to strain imposed by environmental forces on macromolecules such as membrane lipids, proteins, and/or DNA. A critical aspect of the GSR, downstream of perception of macromolecular damage, is generation of ROS. Furthermore, key molecular components of the GSR are evolutionarily conserved in all organisms.

To better understand plant stress responses, transcript profiling experiments have been successfully employed for many different abiotic and biotic stresses. One common emerging theme from these experiments is that abiotic and biotic stresses regulate different but overlapping sets of genes. For example, cDNA–amplified fragment length polymorphism analysis of the Avr9- and Cf-9-mediated defence response in tobacco cell culture revealed overlap between race-specific resistance and response to wounding. Additionally, partial genome microarray analysis of the *Arabidopsis* wound response revealed that a number of wound-responsive genes encode proteins known to be involved in pathogen defence. Examination of the AtGenExpress abiotic datasets demonstrates that the initial transcriptional abiotic stress response may comprise a core set of multi-stress-responsive genes. The abiotic stress response then becomes stress specific at later time points. Finally, recent analysis of the AtGenExpress abiotic and biotic datasets has uncovered ~200 genes that are expressed in response to a broad range of stresses, which may represent the GSR of *Arabidopsis*.

Recently, a shift in stress tolerance engineering has been proposed that transfers the focus from pathway endpoints to factors governing upstream reactions. Focusing on upstream signaling components may enable the engineering of multi-stress tolerance. Identification of *cis*-regulatory elements for use in synthetic promoters to confer stress tolerance has also recently been proposed. Towards this aim, we have utilised mechanical wounding, as it uniquely confers an instantaneous and synchronous stimulus, to identify primary stress-responsive transcripts. Comparison of the 5 min rapid wound response (RWR) genes we identified with published transcript profiles demonstrated a large overlap with previously identified abiotic and biotic stress-responsive genes. Notably, RWR genes also exhibit a striking level and pattern of circadian regulation. Further investigation via real-time quantitative RT-PCR (RT-qPCR) of a wounding time course revealed genes that are expressed rapidly and transiently as well as rapidly and stably. Two rapidly and transiently expressed genes, *ETHYLENE RESPONSE FACTOR #018* (*ERF#018; AT1G74930*) and *CCR4-ASSOCIATED FACTOR 1* (*CAF1-like; AT3G44260*), were confirmed as wound and biotic stress inducible in vivo using stable transgenic lines expressing transcriptional luciferase fusions. Detailed analysis of the RWR promoters identified a novel *cis*-regulatory element we term the rapid stress response element (RSRE), which is sufficient to confer reporter gene induction in response to abiotic and biotic stress. RWR genes identified in this study may represent initial components of the GSR and be useful in engineering multi-stress tolerance.

Results/Discussion

Transcript Profiling Identifies Rapid Wound Response Genes

To identify primary stress-responsive transcripts we utilised Agilent microarrays to monitor gene expression changes 5 min after mechanical wounding of *Arabidopsis* rosette leaves. Because of the short duration of our stress treatment we hypothesized that expression changes would be low. In order to accurately detect these changes we utilised three biological replicates of pooled plants per treatment. In addition, two technical replicates, with dye swap of each technical replicate, were performed on each biological replicate.

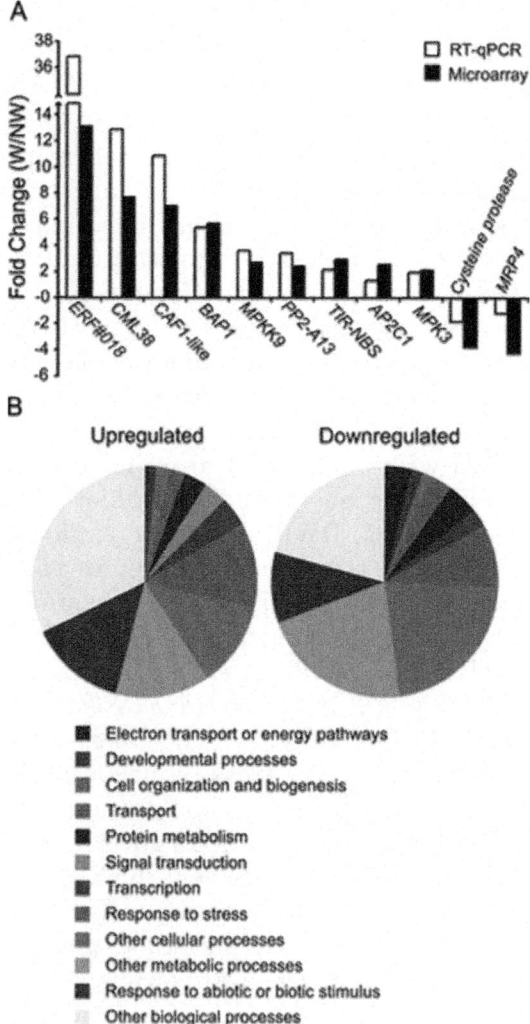

Figure: Verification and Functional Classification of the RWR Genes

(A) RT-qPCR expression analysis of selected genes normalised to the 60S ribosomal protein L14 (At4g27090) measured in the same samples. The resulting relative expression was then used to calculate fold change upon wounding. Data are means of $n = 3$.

(B) Functional classification of RWR genes using GO annotations.

Using this approach, we found that the expression of 162 genes was upregulated and the expression of 44 genes was downregulated at least 2-fold and had a p-value d" 0.01 five min after mechanical wounding. The expression level of selected RWR genes representing a range of high-to-low-fold change was then validated using RT-qPCR. The expression changes determined by RT-qPCR data are in good agreement with the fold change observed by microarray with a spearman rank order correlation coefficient of 0.927 (p-value = 0.000). RWR genes were then classified according to gene ontology (GO) terms in order to provide insight into their biological function. The two largest defined classes of GO terms involve response to stress or abiotic/biotic stimuli. It is also of interest to note that genes classified for an involvement in signal transduction were observed in upregulated but not downregulated RWR genes. These data indicate that 5 min of mechanical wounding was sufficient for induction of known stress-responsive genes as well as unknown genes that may play a role in multi-stress responses.

In Vivo Validation of Rapid Wound Response Genes

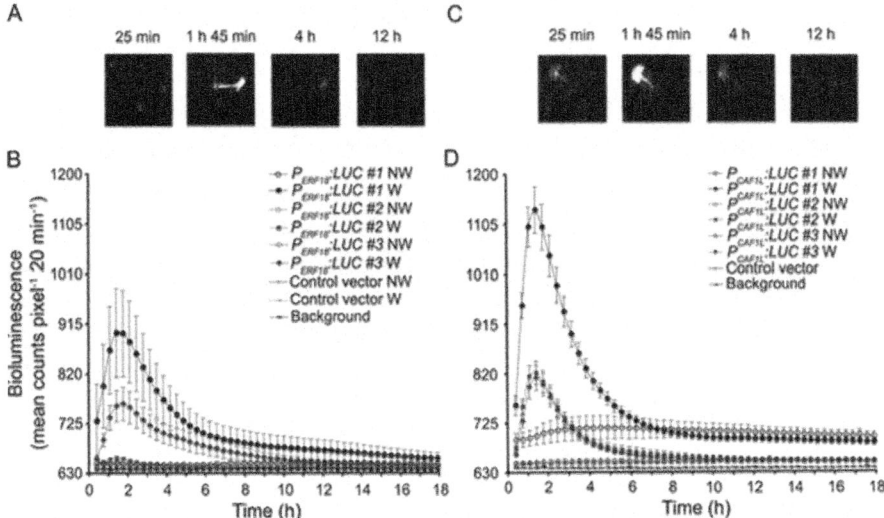

Figure: In Vivo Monitoring of RWR Gene Induction Following Wounding

(A) Image of an individual P_{ERF18}:LUC #3 transgenic plant over time.

(B) Luciferase activity of three independent transgenic lines expressing transcriptional P_{ERF18}:LUC fusions. Luciferase activity was calculated in wounded leaves (single leaf per plant) and NW leaves (on NW plants). Data are means of n = 12 ± SEM.

(C) Image of an individual P_{CAFIL}:LUC transgenic plant over time.

(D) Luciferase activity of three independent transgenic lines expressing transcriptional P_{CAFIL}:LUC fusions. Luciferase activity was calculated in wounded leaves (single leaf per plant) and NW leaves (on NW plants). Data are means of n = 18 ± SEM.

We next created stable transgenic lines expressing transcriptional fusions of the *ERF#018* and *CAF1-like* promoters to luciferase to validate in vivo RWR genes and to investigate their temporal expression pattern.

For each construct, three independent T2 lines were imaged to control for positional effects of the transgene insertion site. Luciferase activity was then monitored following the wounding of a single leaf per plant to enable the observation of whether the induced activity occurred only locally or also systemically.

The wound-induced expression of P_{ERF18}:*LUC* occurs rapidly and peaks ~1 h 45 min after the wound stimulus. Additionally, expression of P_{ERF18}:*LUC* was observed in the petiole and shoot apex. The expression of $P_{CAF1-like}$:*LUC* was detected rapidly and peaked ~1 h 25 min surrounding the wound site.

These data provide in vivo confirmation that RWR genes do respond rapidly to mechanical wounding.

Expression of RWR Genes over Time

To gain further insight into how the RWR genes may be acting, we performed a RT-qPCR time-course on selected genes. We classified genes as rapidly and stably expressed if 60 min post wounding they remained greater than 2-fold induced.

In contrast, we classified genes that had decreased in expression to less than half of maximal expression by 60 min post wounding as rapidly and transiently expressed. Among the rapidly and stably expressed transcripts are genes with either a known or predicted role in stress signal transduction events.

CML38 is a calmodulin-like gene that is predicted to be a sensor of Ca^{2+}, a known secondary messenger of stress responses. *MPKK9*, a MAPK signal transduction component, was also identified as rapidly and stably expressed.

Additionally, the transcription factor *WRKY40*, which is known to be involved in pathogen defence, was identified. Finally, *BAP1*, a negative regulator of defence responses whose binding of phospholipids is enhanced by calcium, was shown to respond rapidly and stably to wounding.

The upregulation of RWR genes in a rapid and stable manner may indicate that these genes play a more prolonged role in response to stress.

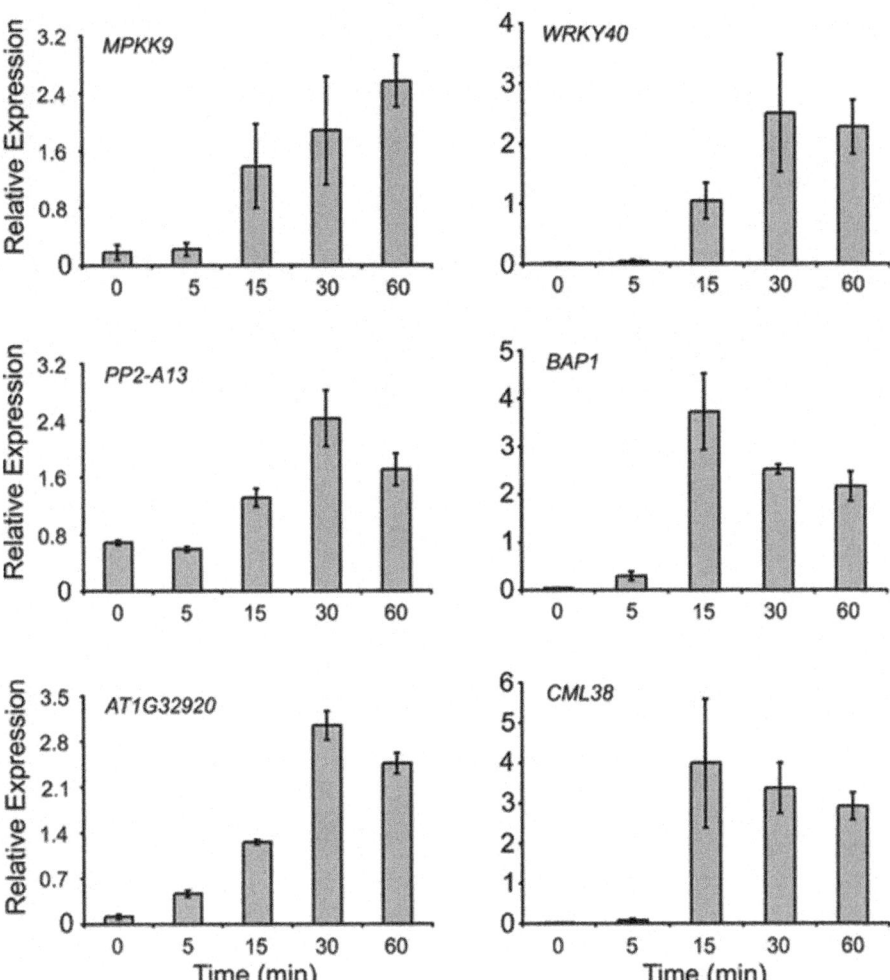

Figure: Selected RWR Genes Displaying a Rapid and Stable Expression Pattern Following Mechanical Wounding

Total RNA was extracted from 3-wk-old mechanically wounded tissue and subject to RT-qPCR analysis. *AtMPKK9*, *WRKY40*, *AtPP2-A13*, *BAP1*, *AT1G32920*, and *CML38* transcripts were normalised to the 60S ribosomal protein L14 measured in the same samples. Data are means of $n = 3 \pm$ SEM.

We also uncovered rapidly and transiently expressed RWR genes with a wide range of functions. One such example is the chromatin remodelling ATPase *SPLAYED* (*SYD*), which peaks 15 min post wounding. Because of the large changes in gene expression following stress it has been hypothesized that chromatin remodelling may be required to allow for stress-induced transcription to occur. Examples of stress-induced changes in histone acetylation state in plants have been described. Rapid and transient upregulation of *SYD* suggests that ATP-dependent chromatin remodelling may also take place in order to facilitate downstream stress-induced transcriptional changes.

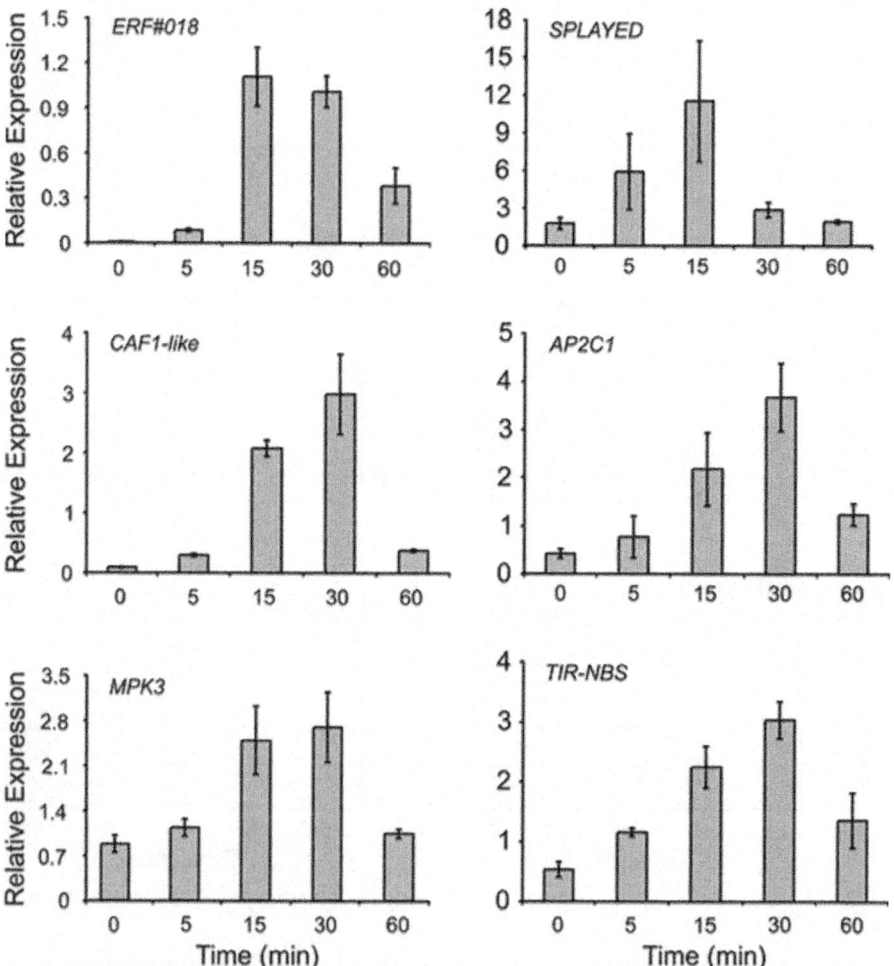

Figure: Selected RWR Genes Displaying a Rapid and Transient Expression Following Mechanical Wounding

Total RNA was extracted from 3-wk-old mechanically wounded tissue and subject to RT-qPCR analysis. *ERF#018*, *CAF1-like*, *AP2C1*, *MPK3*, and *TIR-NBS* transcripts were normalised to the 60S ribosomal protein L14 while *SPLAYED* transcripts were normalised to *TIP41-like* measured in the same samples. Data are means of $n = 3 \pm$ SEM.

Genes involved in signal transduction via reversible phosphorylation were also upregulated rapidly and transiently following wounding. One kinase identified was *MPK3*, a MAPK signal transduction component, which has been shown to function in innate immunity and stomatal development. *AP2C1*, a PP2C-type phosphatase with a MAPK interaction motif, was also shown to exhibit a rapid and transient expression pattern resulting from wounding. These results indicate that both phosphorylation and dephosphorylation of MAPK signaling components is involved in transduction of initial stress signaling events.

A third process implicated by genes identified in this study is that of mRNA turnover. Specifically, this process is demonstrated by the expression pattern of *CAF1-like*. In yeast, CAF1 has been shown to be a component of the major cytoplasmic deadenylase, which functions to remove the poly(A) tail, thereby initiating mRNA turnover. Additionally, the mouse CAF1 ortholog has been demonstrated to function as a 32 -52 -RNase with a preference for poly(A) substrates.

In *Arabidopsis*, RNA processing appears to play a role in response to cold stress. It is also of interest to note the difference in promoter and transcript expression patterns in response to wounding for *ERF#018* and *CAF1-like*.

When promoter activity was monitored using transcriptional luciferase fusions, activity peaked ~1 h 30 min after wounding. In contrast, transcript abundance measured via RT-qPCR peaked 15–30 min after wounding.

This discrepancy may be an artifact due to measuring promoter activity via luciferase protein activity, while RT-qPCR assayed transcript levels. An alternative explanation of these results is that mRNA levels of rapidly and transiently expressed genes such as *ERF#018* and *CAF1-like* are controlled post-transcriptionally, possibly via mRNA turnover. In support of this hypothesis, *CAF1-like* mRNA has a half-life of 38 min and has been classified as a gene with an unstable transcript. Furthermore, the mRNA of a second rapid and transient gene, *MPK3*, has a half-life of 43 min and is classified as a gene with an unstable transcript.

RWR Genes Are Regulated by Abiotic and Biotic Stresses

Examination of the RWR genes reveals a large number of known genes involved in abiotic and biotic stress responses (Table S1). Among the upregulated RWR genes were genes involved in ethylene signaling including *ACC synthase 6* (*ACS6*) as well as 15 of the 122 ethylene response factors (ERFs) in *Arabidopsis*, which have been shown to be involved in the response to both biotic and abiotic stresses. The transcriptional activators *CBF1* (*DREB1B*), *CBF2* (*DREB1C*), and *CBF3* (*DREB1A*), which confer tolerance to cold and drought, were also among the RWR genes. Additional RWR genes known to confer tolerance to a range of abiotic stresses include *STZ* (*ZAT10*) and *ZAT12*. RWR genes also include genes with a known function in response to biotic stress.

Examples include *BAP1*, a negative regulator of defence responses. *MPK3* and *FLS2* which function in response to pathogen-associated molecular patterns in the *Arabidopsis* innate immune response. The transcription factor *TGA3* which regulates *pathogenesis-related* (PR) genes and is required for basal pathogen resistance was also among the RWR genes. Finally, *ERD15* regulates not only cold and drought tolerance but also resistance to the bacterial necrotroph Erwinia carotovora subsp. *carotovora*.

The abundance of RWR genes with known abiotic and biotic stress tolerance functions led us to examine the role of wounding as a general stress perception mechanism on a global level. For this analysis, we compared the overlap in gene lists between the RWR genes and published transcript profiles for a number of stress conditions.

The statistical significance of the observed overlap in transcript profiles was then analysed using empirical permutation tests. We first compared RWR genes with published abiotic microarrays and found a strong overlap (unpublished data), which is in agreement with work recently published by Kilian et al..

For example, 49% of upregulated RWR genes have been previously shown to be upregulated upon cold treatment. Additionally, four of the nine genes (*At1g27730*, *At5g51190*, *At5g47230*, and *At5g04340*) found by Kilian et al. to be upregulated by 30 min of cold, drought, UV-B, salt, osmotic stress treatment, and wounding we discovered to be upregulated within 5 min of wounding. We next compared the RWR transcript profile with published transcript profiles of plants challenged with different biotic stresses.

For upregulated datasets there was a statistically significant overrepresentation of RWR genes in the transcript profile of all biotic stresses tested. Furthermore, the overrepresentation of RWR genes occurred only at early time points of P. rapae and OGA stressed plants. These data indicate that perception of mechanical stress may play a central role in the perception and initial response to a wide range of environmental stresses.

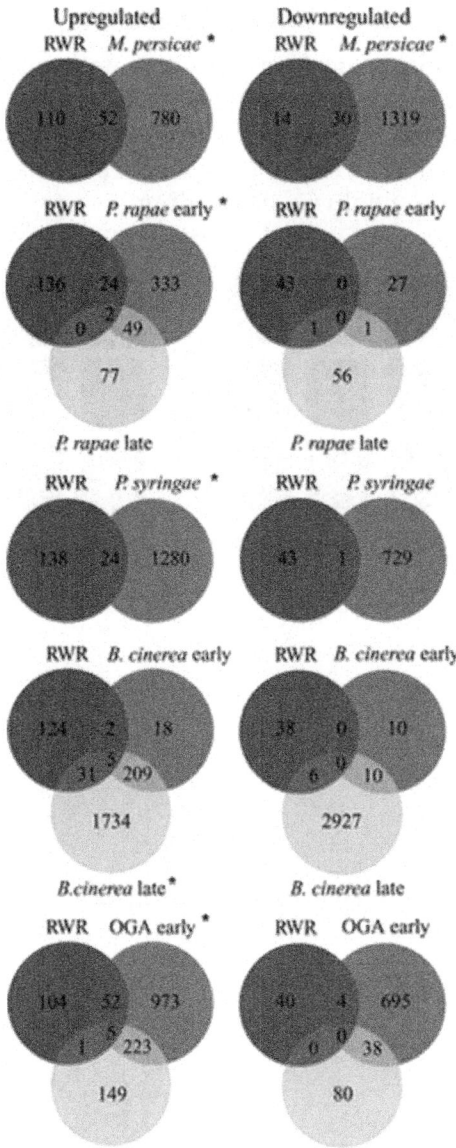

Figure: Comparison of RWR Genes with the Transcript Profile of Other Environmental Stimuli

An asterisk denotes a statistically significant overrepresentation of RWR genes in the transcript profile of the indicated stress ($p < 0.0001$). *Arabidopsis* plants were challenged with M. persicae for 48 h and 72 h; P. rapae early for 5 h; P. rapae late and P. syringae for 12 h and 24 h; *B cinerea* early and late for 18 h and 48 h, respectively; and OGA early and late for 1 h and 3 h, respectively.

RWR Genes Respond to Biological Elicitors

Various biological compounds are known to elicit stress-signaling networks. Due to the overlap between RWR genes and biotic stresses we tested whether the RWR genes *ERF#018* and *CAF1-like* respond to the biological elicitors OGA and insect regurgitant (IR) as well as cabbage looper (Trichoplusia ni) feeding.

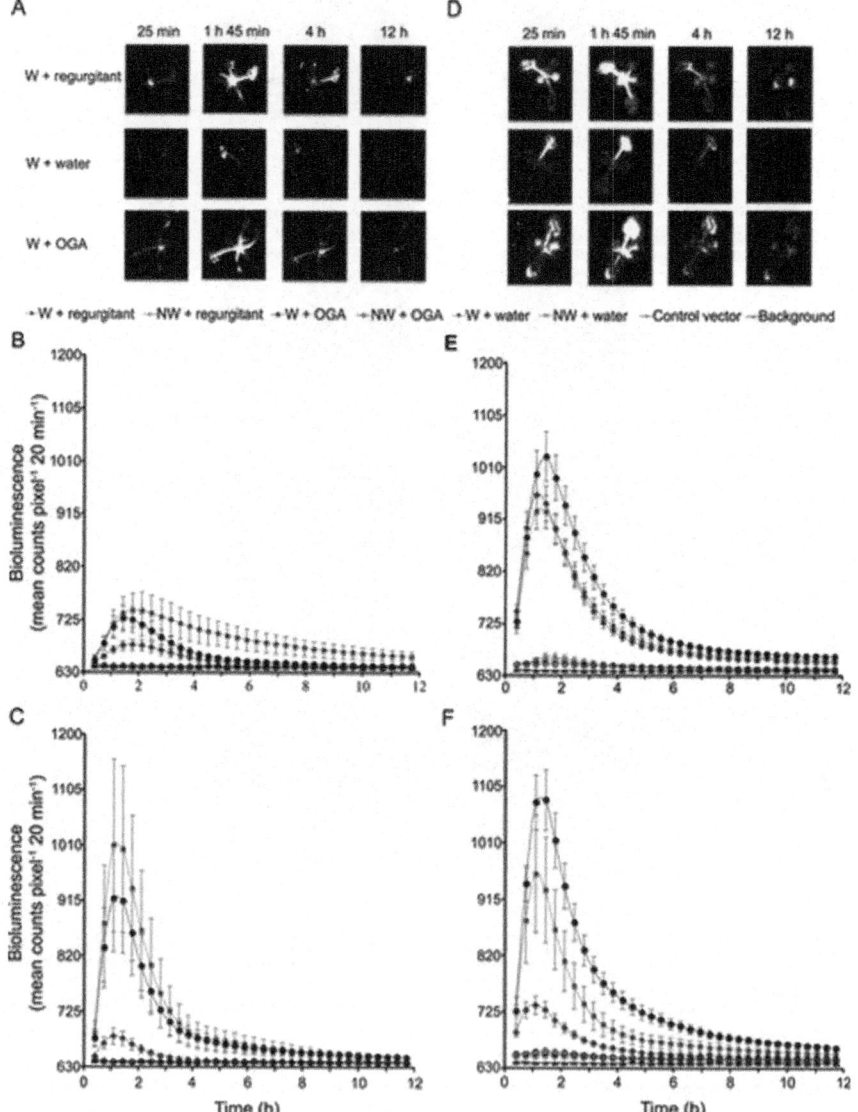

Figure: In Vivo Monitoring of RWR Gene Induction upon Addition of Biotic Elicitors

One leaf per plant was either NW or wounded (W) and then treated with oligouronides (OGA), IR, or double-distilled H_2O.

(A) Image of an individual $P_{ERF18}:LUC$ #3 transgenic plant over time.

(B) Local expression in the treated leaves of $P_{ERF18}:LUC$ #3 transgenic plants. Data are means of $n = 9 \pm$ SEM.

(C) Systemic expression of $P_{ERF18}:LUC$ #3 plants monitored in the shoot apex. Data are means of $n = 9 \pm$ SEM.

(D) Image of an individual $P_{CAF1L}:LUC$ #2 transgenic plant over time.

(E) Local expression in the treated leaves of $P_{CAF1L}:LUC$ #2 transgenic plants. Data are means of $n = 12 \pm$ SEM.

(F) Systemic expression of $P_{CAF1L}:LUC$ #2 plants monitored in the shoot apex. Data are means of $n = 12 \pm$ SEM.

We first tested whether $P_{ERF18}:LUC$ or $P_{CAF1-like}:LUC$ activity was induced by cabbage looper feeding. Indeed, cabbage looper feeding did result in enhanced luciferase activity, which verified that biological stress does induce *ERF#018* and *CAF1-like* (Videos S1 and S2). We therefore proceeded to test induction resulting from OGA and IR treatment. When OGA, IR, or H_2O were added to a nonwounded (NW) leaf, no induction of $P_{ERF18}:LUC$ or $P_{CAF1-like}:LUC$ activity was observed. The lack of induction is likely due to the application method we used (single droplet per plant) rather than an actual lack of response to the elicitors. When transcript profiling was performed on liquid cultured 10-d-old plants incubated with 50 µg ml^{-1} OGAs, *CAF1-like* was shown to be induced 1 h after addition of OGA to the media.

In $P_{ERF18}:LUC$-expressing plants, addition of OGA and IR to the wound site resulted in a significantly greater ($p < 0.05$) induction of luciferase activity than addition of H_2O in both local and systemic tissue. In contrast, addition of OGA or IR to the wound site did not result in a significant difference ($p > 0.05$) in $P_{CAF1-like}:LUC$ activity compared to addition of H_2O to the wound site in local tissue. However, a significantly greater induction, compared to H_2O, in $P_{CAF1-like}:LUC$ activity resulted from addition of OGA or IR to the wound site in systemic tissue.

There are a number of common second messengers downstream of mechanical wounding, cabbage looper feeding, and OGA treatment that may signal for the observed induction of RWR genes. One such secondary messenger is Ca^{2+}, which increases in intracellular concentration rapidly following wounding as well as OGA treatment. ROS are another secondary messenger that have been shown to increase in response to chewing insects, wounding, and OGA treatment.

Furthermore, while OGAs do not move systemically, ROS do accumulate systemically following wounding. This increase in ROS is likely through OGAs released by systemically induced polygalacturonase.

Finally, OGA, chewing insects, and wounding may all have a common mechanism of perception resulting in similarly induced secondary messengers. Both chewing insects and wounding have a physical effect on the plasma membrane. The perception of OGA has also been hypothesized to be a result of its physical effect on the plasma membrane, rather than through an actual receptor.

Circadian Regulation of the RWR Genes

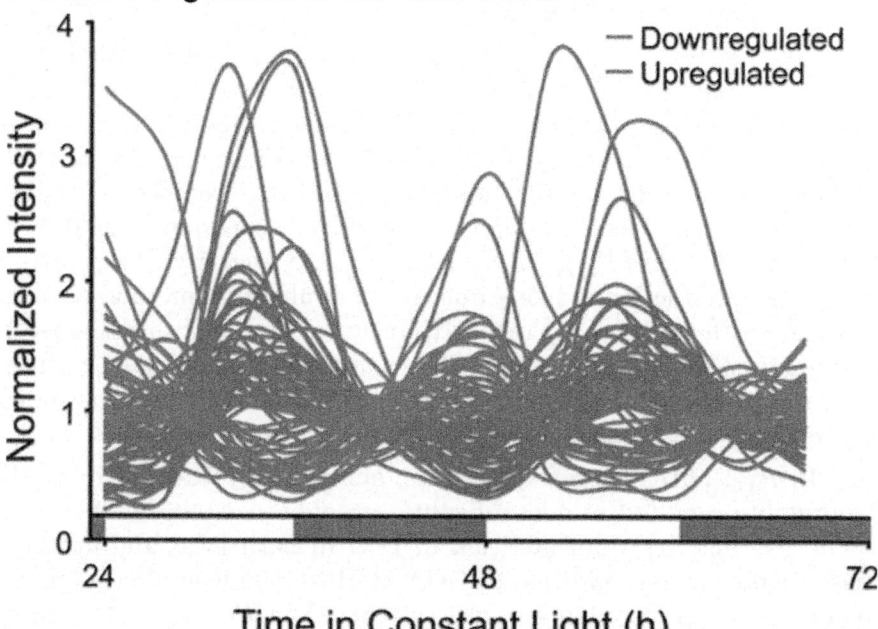

Figure: Circadian Regulation of RWR Genes

These data are comprised of circadian-regulated RWR genes. Upregulated RWR genes (blue) peak at dusk while downregulated RWR genes (pink) peak at dawn. Plants were entrained in light/dark cycles for 7 d and then released into constant light. Samples were collected every 4 h after plants were moved to constant light.

The circadian clock has been shown to regulate a number of environmentally regulated genes. Additionally, cold-induced expression of RWR genes *ZAT12*, *CBF1*, *CBF2*, and *CBF3* was recently reported to be gated by the circadian clock. These findings led us to examine globally whether RWR genes are under circadian regulation. Towards this aim, we compared the RWR genes with genes recently identified as circadian regulated.

Surprisingly, not only were RWR genes rhythmically expressed but the upregulated and downregulated genes also showed unexpected phase distributions. Forty-two percent of RWR upregulated genes are expressed at subjective dusk while 81% of downregulated RWR genes peak at subjective dawn (p d 0.0001). The circadian regulation of RWR genes may provide a mechanism to anticipate stresses caused by daily environmental changes.

Identification of a Novel Stress-Responsive *cis*-Regulatory Element

To begin dissecting the molecular mechanism underpinning the rapid stress response, we examined the promoters of the RWR genes for novel *cis*-regulatory elements.

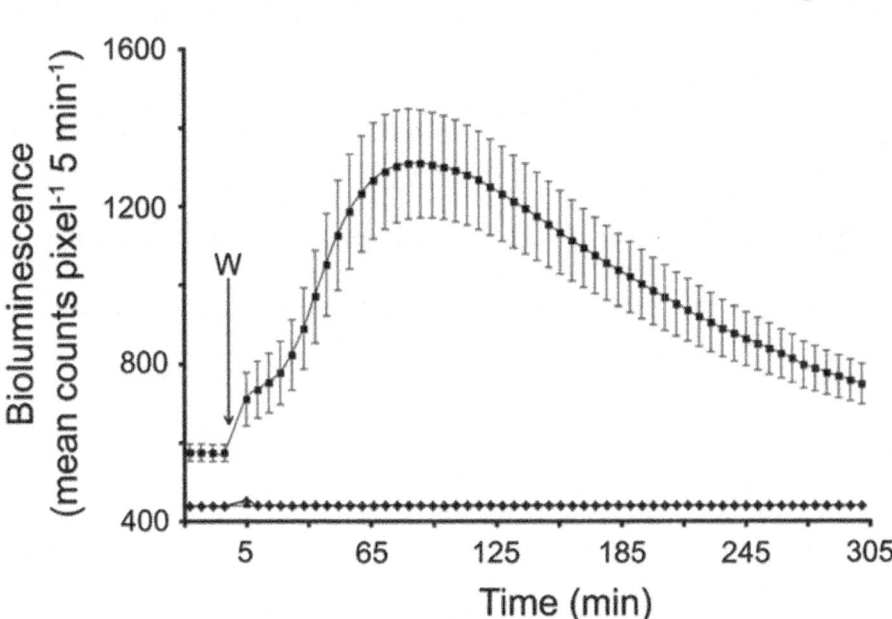

Figure: The RSRE Confers Wound-Induced Reporter Gene Expression

Independent T1 *4xRSRE:LUC* and *4xmtRSRE:LUC* lines were mechanically wounded. Luciferase activity was then monitored in the wounded leaf. Data are means of $n = 24 \pm$ SEM.

We identified the six-nucleotide repeat, CGCGTT, which we are terming the Rapid Stress Response Element (RSRE), as significantly over-represented (58 hits in 47 of the 162 upregulated promoters) in the promoters of upregulated RWR genes. To determine whether the RSRE is sufficient alone to confer stress-responsive transcription, we used luciferase reporter constructs.

Four tandem repeats of the RSRE and its consensus flanking sequence were separated by six nucleotides and cloned upstream of the minimal promoter region of the nopaline synthase (NOS) gene and modified luciferase coding region (*4xRSRE:LUC*). Additionally, to verify that the RSRE was the region conferring stress responsiveness, we mutated three of the six nucleotides in the RSRE (*4xmtRSRE:LUC*). The wound-induced expression of these constructs was then tested in 24 independent T1 plants to control for differences in expression resulting from the site of transgene insertion.

All 24 *4xRSRE:LUC* transgenic plants exhibited wound-induced luciferase expression. Furthermore, luciferase activity increased immediately following wounding and peaked ~80 min post wounding. Conversely, no *4xmtRSRE:LUC* transgenic plant exhibited luciferase activity before or after wounding. These data demonstrate that the RSRE is sufficient to confer a rapid response to stress.

Both abiotic and biotic stresses appear to share common signaling components with the RWR. We were therefore interested in whether the RSRE confers a rapid response to a range of stresses. To enable accurate quantification of the stress response we used a homozygous T3 *4xRSRE:LUC* line.

Because RWR genes respond to both OGAs and IR, we tested the expression of *4xRSRE:LUC* following treatment with these biotic elicitors. In the local leaf where the wound site was treated with OGA or IR, no induction over that of H_2O treatment was observed. Notably, in systemic tissues, a statistically significant ($p < 0.05$) synergistic enhancement of luciferase activity was detected in OGA- and IR-treated plants when compared to H_2O treatment. To further demonstrate that the RSRE responds to biotic stress, *4xRSRE:LUC* plants were challenged with cabbage loopers and B. cinerea. When *4xRSRE:LUC* plants were exposed to both forms of biotic stress, luciferase activity was induced, whereas no activity was observed in vector control lines (Videos S3 and S4). Additionally, in B. cinerea–infected plants, a low level of transient luciferase activity was first observed in the inoculated leaf. Luciferase activity was then observed at a greater level in systemic tissues (Video S4). These data clearly demonstrate that the RSRE responds to biotic stress.

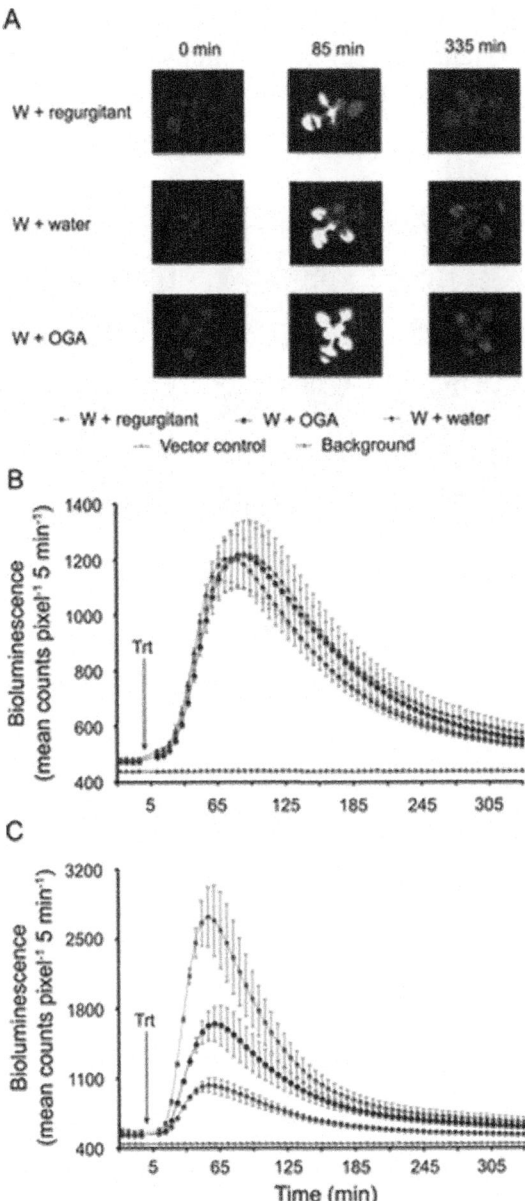

Figure: The RSRE Responds to Biotic Elicitors

(A) Image of an individual *4xRSRE:LUC* transgenic plant over time that was wounded and treated with OGA, IR, or H_2O.

(B) Local expression in the treated leaves of *4xRSRE:LUC* transgenic plants.

(C) Systemic expression of *4xRSRE:LUC* in the shoot apex. Data are means of 12 ± SEM.

While the RSRE responds to the abiotic stress of mechanical wounding, we wished to further demonstrate the role of the RSRE in response to abiotic stress.

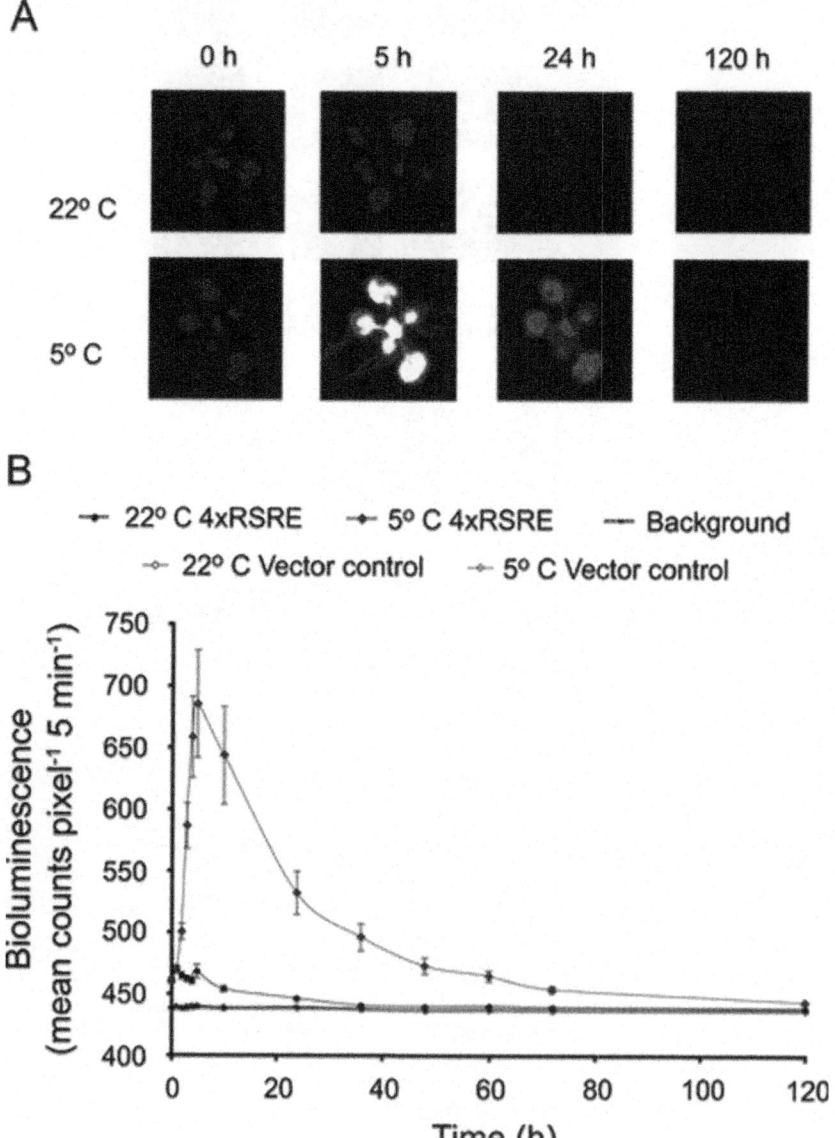

Figure: The RSRE Confers Cold-Induced Reporter Gene Expression

(A) Image of an individual *4xRSRE:LUC* transgenic plant incubated at 5 °C or 22 °C over time.

(B) Luciferase activity of *4xRSRE:LUC* plants. Data are means of 16 ± SEM.

Towards this aim, we exposed *4xRSRE:LUC* expressing plants to 5 °C. Plants were then removed from cold treatment at the indicated time for imaging. Additionally, control *4xRSRE:LUC* plants were also kept at 22 °C in equivalent light conditions and moved similarly to cold-treated plants to ensure that transfer to the imaging chamber did not result in induced luciferase activity. Induction of luciferase activity was observed after ~2 h of cold treatment. Furthermore, luciferase activity peaked after 5 h of cold treatment and then decreased towards basal expression levels. Notably, *4xRSRE:LUC* expression was also observed in the roots of cold-treated plants. To ensure that the *4xRSRE:LUC* plants were still competent to express luciferase, they were mechanically wounded after 120 h of cold treatment. Both cold- and 22 °C-treated *4xRSRE:LUC* plants exhibited luciferase induction following wounding (unpublished data). These data demonstrate that the RSRE is cold responsive and that initial signaling events leading to a cold response dampen even in continuous exposure to cold.

The rapid and transient response of the RSRE to multiple stress conditions is reminiscent of the yeast GSR promoter element STRE (stress response element; AGGGG). The STRE is responsible for rapid induction following various treatments such as heat, nitrogen starvation, low external pH, osmotic, and oxidative stress. Furthermore, even in the presence of continuous stress exposure, STRE-mediated gene induction dampens over time. An increase in unsaturated fatty acids upon stress appears to be responsible for the transient nature of STRE-mediated induction. When plants are exposed to abiotic and biotic stresses (cold and P. syringae, respectively), there is an increase in unsaturated fatty acids. Upon cold treatment, acyl-lipid desaturases are the enzymes that most efficiently introduce double bonds in membrane lipids, which results in the increased level of unsaturation. Similar to cold treatment, two of the RWR upregulated genes are acyl-lipid desaturases (*ADS1* and *ADS2*), which may increase the unsaturation of membranes upon wounding. It is therefore tempting to speculate that, as with the STRE in yeast, the increase in unsaturated membrane lipids resulting from both abiotic and biotic stresses may mediate the transient induction of RSRE-driven reporter gene expression.

We have shown that 5 min of mechanical stress is a sufficient amount of time for the plant to perceive the stress and mount a robust transcriptional response. The rapid transcriptional response to mechanical wounding shares a large overlap with both abiotic and biotic stresses and may therefore represent the initial GSR of *Arabidopsis*. In support of this view, the RWR upregulated genes comprised 25% of the genes identified as potential GSR genes via analysis of the AtGenExpress abiotic and biotic stress datasets. Additionally, in mammalian cells, physical stress to membranes during osmotic and UV radiation stress result in the nonspecific clustering of growth factor receptor tyrosine kinases and cytokine receptors. A similar nonspecific clustering of receptors during mechanical wounding and other environmental stresses may underlie the GSR of plants. We also show that the RWR genes are circadian regulated with consolidated phases of peak expression. Circadian regulation of RWR genes, which likely encompass initial components of the GSR, may enable plants to anticipate daily environmental changes and mount a general defence against these changes. Finally, we identified a *cis*-regulatory element (RSRE) over-represented in the promoters of RWR genes. The RSRE confers a rapid and transient response similar to the yeast GSR promoter element (STRE) and is a novel GSR *cis*-regulatory element in plants.

Since the RWR genes likely represent initial components of the GSR, they provide a valuable resource of candidate genes for engineering of multi-stress resistance. Similarly, the RSRE, which responds rapidly and transiently to abiotic and biotic stresses, may prove useful as a synthetic element for engineering of multi-stress tolerance. Finally, use of the RSRE in yeast one-hybrid and the *4xRSRE:LUC* line for mutant screens should help elucidate the upstream mechanisms of stress perception and initial signal transduction.

Biotic vs. Abiotic - Distinguishing Disease Problems from Environmental Stresses

Biotic plant problems are caused by living organisms, such as fungi, bacteria, viruses, nematodes, insects, mites, and animals. Abiotic disorders are caused by nonliving factors, such as drought stress, sunscald, freeze injury, wind injury, chemical drift, nutrient deficiency, or improper cultural practices, such as overwatering or planting too deep. Unfortunately, the damage caused by these various living and nonliving agents can appear very similar. Even with close observation, accurate diagnosis can be difficult. For example, browning of leaves on an oak tree caused by drought stress may appear similar to leaf browning caused by oak wilt, a serious vascular disease, or the browning cause by anthracnose, a fairly minor leaf disease.

When the cause of a plant health problem is not readily diagnosed, it's important to take a systematic approach and carefully consider site conditions, weather conditions, care of the plant, and the known biotic disease agents of that plant. The first important step is to determine the identity of the plant and its requirements for healthy growth.

There are a few clues to look for that will help you distinguish between abiotic and biotic disease problems.

- Abiotic damage often occurs on many plant species. Drought stress or chemical drift will likely cause damage on several types of plants in a yard or garden. In contrast, biotic disease problems are more limited to a certain species. The fungi that cause tomato leaf blight do not cause damage on sweet corn, for example.

- Abiotic damage does not spread from plant to plant over time. Biotic diseases can spread throughout one plant and also may spread to neighbouring plants of the same species. Wind-blown rain is a common way for disease agents to spread from plant to plant.

- Biotic diseases sometimes show physical evidence (signs) of the pathogen, such as fungal growth, bacterial ooze, or nematode cysts, or the presence of mites or insects. Abiotic diseases do not show the presence of disease signs.

An important take-home message is to remember that there may be several factors, abiotic and biotic, contributing to a plant health problem. For example, older trees that are stressed by drought conditions are often troubled by fungal canker diseases. Another example is the presence of decay fungus at the base of the tree. The primary problem may have been mower damage, which subsequently allowed entry of the fungus. The identification of the primary problem and other contributing factors is a necessary step in managing the problem or avoiding it in the future.

Abiotic and Biotic Stress Response Crosstalk in Plants

In the course of its evolution, plants have developed mechanisms to cope with and adapt to different types of abiotic and biotic stress imposed by the frequently adverse environment.

The biology of a cell or cells in tissues is so complicated that with any given stimulus from the environment, multiple pathways of cellular signaling that have complex interactions or crosstalk are activated;

these interactions probably evolved as mechanisms to enable the live systems to respond to stress with minimal and appropriate biological processes. The sensing of biotic and abiotic stress induces signaling cascades that activate ion channels, kinase cascades, production of reactive oxygen species (ROS), accumulation of hormones such as salicylic acid (SA), ethylene (ET), jasmonic acid (JA) and abscisic acid (ABA). These signals ultimately induce expression of specific subsets of defence genes that lead to the assembly of the overall defence reaction.

In plants, defence response genes are transcriptionally activated by different forms of environmental stress or by pathogens. The induction of expression of defence genes in the response against certain pathogens is further dependent on temperature and humidity, suggesting the existence of a complex signaling network that allows the plant to recognise and protect itself against pathogens and environmental stress. A body of research has shown that calcium and reactive oxygen species are second messengers in the early response to abiotic and biotic stress. For example, cytosolic calcium (Ca2+) levels increase in plant cells in response to various harsh environmental conditions, including pathogen attack, osmotic stress, water stress, cold and wounding. After the increase of Ca2+ concentration in the intracellular space, several simultaneous pathways are activated by calcium-interacting proteins such as Ca2+-dependent protein kinases (CDPKs), calmodulin and calcineurin Blike proteins (CBLs), all proteins with the structural 'EF-hand' calcium-binding motif.

It is also known that plants respond with an oxidative burst to avirulent microbial intruders or to the previously mentioned abiotic stress factors. In this response, NADPH oxidases generate O2– that is rapidly converted to H2O2. Recent evidence demonstrated that the NADPH oxidases are activated by Ca2+ signatures. ROS are generated by NADPH oxidases in the plasma membrane and increase in concentration in the cytoplasm; these species are also formed in mitochondrion and chloroplast.

The intricate and finely tuned molecular mechanisms activated in plants in response to abiotic and biotic environmental factors are not well understood, and less is known about the integrative signals and convergence points in different sets of partially overlapping reactions. It is now recognised that crosstalk between the second messengers Ca2+ and ROS modulates the activity of specific proteins that act at the nuclear level to control the expression of determinate defence genes. Recent studies exploring molecular players have identified and characterised several new genes, including kinases and transcription factors, that are involved in the crosstalk between signaling cascades involved in the responses against two or more types of stress. Phytohormones also play central roles in abiotic and biotic stress signaling. SA, JA and ET have central roles in biotic stress signaling. ABA is involved in the response to abiotic stress as low temperature drought and osmotic stress. ABA appears to function as a negative regulator in disease resistance, in opposite action to SA, ET and JA. Several transcription factors including AtMYC2, BOS1 and RD26 are mediators in multiple hormone signaling pathways.

In our recent studies of a *Phaseolus vulgaris/Colletotrichum lindemuthianum* pathosystem, genes such as *SUMO* (Small Ubiquitin-like MOdifier) and a calcium-binding like protein (*CaM*) were induced to different levels during the time course of the response to avirulent pathogen inoculation, ultraviolet (A-B) light or extreme temperatures. These findings indicate that these two molecules should be included in the category of integrative signals in abiotic and biotic stress response in plants.

Other well known players in plant response to abiotic and biotic stress are members of the WRKY transcription factor family. Expression patterns of *VvWRKY11*, *AtWRKY39* and *AtWRKY53* genes indicate that protein products of these genes are co-regulators of the plant response against pathogens, hydric stress and heat stress. In addition, some WRKY transcription factors (OsWRKY24 and OsWRKY45) antagonize ABA function by repression of ABA-inducible promoters, indicating that these molecules operate with versatile capabilities. Clearly, the signaling components in plant responses to different abiotic and biotic stress often overlap. Commonly the activated signaling cascades act via synergistic and antagonistic actions.

Powerful molecular tools, including transcriptome and proteome analysis, sequencing of entire genomes in plants, bioinformatic analysis and functional studies, are enabling the disection of networks and identification of key factors in abiotic and biotic signaling cascade crosstalk, and will reveal novel interplays between parallel signaling pathways in the plant responses to pathogens and abiotic stress.

Calcium (Ca2+) and Reactive Oxygen Species (ROS) as Second Messengers Common to Abiotic and Biotic Stress Responses

In plants, Ca2+ and ROS constitute important and common signaling molecules in the early response to abiotic and biotic stress.

Levels of Ca2+ and ROS rapidly increase in cells of local tissue soon after pathogen attack or stress exerted by environmental conditions. Calcium is perhaps the main signal transducer in the signaling cascades activated in plant response to any stimulus or stress, and the ubiquitous characteristic of this molecule in stress signaling justifies the role of the Ca2+ cation as an important node at which crosstalk between pathways can occur. Cytosolic Ca2+ levels increase in plant cells in response to various harsh environmental conditions, including pathogen challenge, osmotic stress, water stress, cold and wounding (Dey et al., 2010; Takahashi et al., 2011). For example, plant Ca2+ signals are involved in an array of intracellular signaling pathways after pest invasion. Upon herbivore feeding there is a dramatic Ca2+ influx, followed by the activation of Ca2+-dependent signal transduction pathways that include interacting downstream networks of kinases (Arimura and Maffei, 2010).

In the last three decades, it has become clear that Ca2+ is a universal message transducer that acts on sub-cellular and spatio-temporal patterns of accumulation and protein interaction. Ca2+ influx through membrane Ca2+ ion channels or carriers yields specific spatial and temporal sub-cellular calcium ion elevations (Errakhi et al., 2008). These signals are then transduced downstream through several simultaneous pathways by calcium-interacting proteins such as CDPKs and CBLs; these Ca2+-binding proteins all contain the 'EF-hand' calcium-binding motif (Kim et al., 2009). An example of Ca2+ concentration signatures related to specific signaling pathways is observed in tobacco stressed by wounding: Three calmodulin (CaM) isoforms (wound-inducible type I, hypersensitive response-inducible type III, and constitutive type II) are enabled at different cytosolic Ca2+ concentrations to activate the target enzymes NO synthase and NAD kinase (Karita et al., 2004).

There is ample evidence that ROS are also crucial second messengers involved in the response to diverse abiotic and biotic forms of stress. An oxidative burst takes place in response to avirulent microbial intruders (Lamb and Dixon, 1997) or to the previously mentioned abiotic stress factors including heat (Wahid et al., 2007), cold (Kwon et al., 2007), drought, salinity (Miller et al., 2010) and others. ROS production in plants by plasma membrane NADPH oxidases and apoplastic oxidases following pathogen recognition is well documented process (Allan and Fluhr, 1997; Lamb and Dixon, 1997; Bolwell et al., 2002; Torres et al., 2006; Galletti et al., 2008). Indeed, in plants a positive feedback mechanism involving NADPH oxidase, ROS and Ca2+ has been reported. Reduced levels of ROS stimulate Ca2+ influx into the cytoplasm and Ca2+ in turn activates NADPH oxidase to produce ROS (Takeda et al., 2008). Plant NADPH oxidases generate O2– that is converted to H2O2 by superoxide dismutase (SOD) and the peroxide diffuses through the cell wall to the extracellular medium and enters into the cell (Hammond-Kosack and Jones, 1996).

Reactive oxygen species are usually generated by NADPH oxidases in the plasma membrane, but in tobacco cells in response to abiotic stress as cadmium heavy metal, the anion superoxide is generated in mitochondria (Garnier et al., 2006). Mitochondria also serve as the site of ROS production upon abiotic stress exerted by copper in the marine alga *Ulva compressa* (Gonzalez et al., 2011). The NADPH oxidase is a multicomponent complex known as respiratory burst oxidase (RBO), initially described in mammals (Lambeth, 2004).

The RBO enzymatic subunit is the transmembrane gp91phox protein that transfers electrons to molecular oxygen to generate superoxide (Lherminier et al., 2009). In *Arabidopsis thaliana*, ten gp91phox homologs have been reported (Torres and Dangl, 2005). It has been shown that members of the Rboh family mediate the ROS production in defence responses to microorganisms, as well as in response to wounding or mechanical stress (Yoshioka et al., 2003; Torres and Dangl, 2005). In *Arabidopsis*, the NADPH oxidase *AtrbohD*, which contains two EF-hand calcium binding motifs, is synergistically activated by Ca2+ and phosphorylation. Phosphorylation levels are correlated with ROS production (Ogasawara et al., 2008).

In the early signaling pathways in the plant defence response to pathogens, the opening of Ca2+-associated of plasma membrane anion channels concomitant with the reactive oxygen species potential response have been described (Jurkowski et al., 2004; Dey et al., 2010). Crosstalk between these two signals in the plant response to abiotic stress has also been reported. In pea plants, the cellular response to long-term cadminum exposure consists of crosstalk between Ca2+- and ROS- activated pathways and signaling mediated by nitric oxide (NO) (Rodriguez-Serrano et al., 2009). In roots in *Arabidopsis thaliana*, mechanical stimulation triggers rapid and transient cytoplasmic Ca2+ concentration increases; this mechanical stimulation likewise elicites apoplastic ROS production with the same kinetics (Monshausen et al., 2009). Certainly, the ROS (specifically H2O2) production in a Ca2+-dependent manner and then the Ca2+ concentration regulation in cytoplasm by ROS through the activation of Ca2+ channels in the plasma membrane have been established (Takeda et al., 2008; Mazars et al., 2010). The co-occurrence and the levels of the induction of Ca2+ and ROS signatures vary greatly and is dependent on pathosystem and environmental situation. For example, in callose deposition in *Arabidopsis* in response to the flagelin epitope Flg22 and the polysaccharide chitosan, environmental variability that imposes differential growth conditions is correlated with levels of hydrogen peroxide production. This demonstrates that callose deposition is a multifaceted response controlled by multiple signaling pathways, depending of the environmental conditions and the challenging pathogen-associated molecular pattern (Luna et al., 2011). In another example, pharmacological studies indicate that acclimation to low temperatures requires Ca2+ influx across the plasma membrane and a transient increase of Ca2+ in the cytoplasm (White and Broadley, 2003), and in *Arabidopsis* mesophyll cells, cold transiently activates Ca2+-permeable channels (Carpaneto et al., 2007). The plant response to low temperature stress also includes production of reactive oxygen species (Heidarvand and Amiri, 2010).

Taking in account the aforementioned antecedents it is clear that responses to two or more forms of stress (biotic or abiotic) may overlap or converge in a common signaling element, for instance, Ca2+ or ROS or both, leading to similar downstream events. Calcium and ROS are ubiquitous second messengers in the abiotic and biotic stress signaling pathways and are in variable ways interconnected elements. There is strong evidence that Ca2+-dependent ROS production through respiratory burst oxidase homologue (RBOH) enzyme activation is the first link. Induction of Ca2+ plasma membrane channels through the increase of cytoplasmic ROS is a second connection. Although these signals co-occur, their magnitudes, spatial location and timing depend on the biological system. The fine signatures in Ca2+ and the recently introduced concept of signatures in ROS (sub-cellular and spatiotemporal patterns of ROS) (Mazars et al., 2010) explain the downstream signaling independence that results in unique molecular responses in plant systems to the environment constraints with specific and adaptive responses.

Calcium-Dependent Protein Kinases (CDPKs) and Mitogen-Activated Protein Kinases (MAPKs) Crosstalk in Response to Abiotic and Biotic Stress

The transient changes in cytosolic calcium content with their diverse spatio-temporal signatures observed under biotic or abiotic stress conditions require different calcium sensors. A larger and defined group of calcium sensors are the calcium-dependent protein kinases (CDPKs) which in turn have many different substrates. CDPKs possess a carboxyterminal calmodulin-like domain containing EF-hand calcium-binding sites plus a N-terminal protein kinase domain (Cheng et al., 2002). Thus, the signaling pathways activated in response to stress stand in part on CDPKs. The *Arabidopsis* genome encodes 34 CDPKs, but few substrates of these enzymes have been identified (Uno et al., 2009). Mitogen-activated protein kinases (MAPKs) are a family of Ser/Thr protein kinases widely conserved among eukaryotes. Them respond to extracellular stimuli and regulate various cellular activities, such as gene expression, mitosis, differentiation, proliferation, and cell survival/apoptosis. They work downstream of sensors/receptors and transmit extracellular stimuli into intracellular responses and at the same time amplifying the transducing signal (Ichimura et al., 2002). Amplification is accomplished by a MPK cascade of three hierarchically arranged, interacting types of kinases. MPK activity is induced upon phosphorylation by MPK kinases (MPKKs, MAPKKs, or MEKs), which are in turn phosphorylation activated by MPKK kinases (MPKKKs, MAPKKKs, or MEKKs). In *Arabidopsis*, there are 20 MPKs, 10 MPKKs, and 80 MPKKKs (Colcombet and Hirt, 2008). MAPKs act as last component in a protein kinase cascade, and one of their major tasks is to transducer an extracellular stimulus into a transcriptional response in the nucleus (Wurzinger et al., 2010).

In eukaryotes, CDPKs together with MAPKs are two signaling cascades widely activated in response to changing environmental abiotic and biotic stresses. In several pathosystems both cascades could be activated in response to the same stressing factor suggesting a crosstalk between those pathways (Wurzinger et al., 2010), or a specific CDPK or MAPK could be induced or activated in response to different biotic and abiotic stresses. Several studies in *Arabidopsis* demonstrate that: a) upon challenge exposure to biotic (bacterial pathogens) or abiotic (BTH, SA, and 4-chloro-SA) stress, MPK3 and MPK6 are activated and their respective mRNAs accumulate (Gerold et al., 2009); b) MKK2 is a key regulator of the cold- and salt-stress response (Teige et al., 2004) but, it was similarly involved in disease resistance to *Pseudomonas syringae* (Brader et al., 2007); c) the actived MKK9 protein in transgenic plants, induces the synthesis of ethylene and camalexin through the activation of the endogenous MPK3 and MPK6 kinases, moreover enhances the sensitivity to salt stress (Xu et al., 2008). In other hand, CDPKs CPK6 and CPK3 operate in ABA regulation of guard cell S-type anion- and Ca2+- permeable channels and stomatal closure (Mori et al., 2006), but besides its well-established role in abiotic stress adaptation, recent results in rice plants indicate that ABA is also involved in the regulation of pathogen defence responses, and mediates the repression of pathogen-induced ethylene signaling pathway in an MPK5- dependent manner (De Vleesschauwer et al., 2010). From the accumulated data, the biological significance of crosstalk among signaling pathways under stress conditions that operate by CDPKs alone or together with MAPKs and viceversa, demonstrate that these two groups of calcium-dependent enzymes and the mitogen-activated protein kinases are involved in signaling pathways that in plants, in some cases signify the establishment of cellular mechanisms that lead to the simultaneous reinforcement of the defence responses to pathogens as well to other forms of abiotic stress.

We are just to begin to uncover convergence points that command the crosstalk between these signaling pathways under various stress conditions.

Genetic Pathways Crosstalk in Response to Abiotic and Biotic Stress

A body of research demonstrates that plant defence response genes are transcriptionally activated by pathogens, as well by different forms of abiotic stress, or even more, the induction of specific defence genes in the response against certain pathogens, are dependent on specific environmental conditions, suggesting the existence of a complex signaling network that allows the plant to recognise and protect itself against pathogens and environmental stress. Similar induction patterns of members of the *14.3.3* gene family (*GF14b* and *GF14c*) by abiotic and biotic stresses such as salinity, drought, ABA and fungal inoculation have been documented in rice. The rice *GF14* genes contain *cis*-elements in their promoter regions that are responsive to abiotic stress and pathogen attack. The *14-3-3s* family genes are also subject to the regulation by certain transcript factors (Chen et al., 2006).

In rice, the *RO-292* gene is upregulated in roots by salt or drought stresses and by blast fungus infection (Hashimoto et al., 2004). Similarly, the *Mlo* gene in barley (*Hordeum vulgare*) act as modulators of defence and cell death in response to *Blumeria graminis* f. sp. *tritici* or *Magnaporte grisea* inoculation, and to wounding or the herbicide paraquat (Piffanelli et al., 2002). In *Arabidopsis*, at least five of the 29 *cytochrome P450* genes are induced by abiotic and biotic stress including *Alternaria brassicicola* or *Alternaria alternata*, paraquat, rose bengal, UV stress (UV-C), heavy metal stress (CuSO4), mechanical wounding, drought, high salinity, low temperature or hormones (salicylic acid, jasmonic acid, ethylene and abscisic acid).

These five *cytochrome P450* genes (*CYP81D11*, *CYP710A1*, *CYP81D8*, *Cyp71B6* and *CYP76C2*) are co-induced by metal stress (CuSO4), paraquat, salinity, ABA and pathogen inoculation.

A common characteristic shared by all of these induced genes, as in the 14.3.3 genes family, is the presence of *cis*-acting elements in regulatory regions of the gene; W-box (DNA binding sites for WRKY transcription factors), P-box (a positive *cis*-acting regulator of pathogen defence) and MYB recognition sites are common (Narusaka et al., 2004). A collection of genes, including transcription factors are co-activated by pathogen challenge and abiotic stress, examples of these genes mediating crosstalk between signaling pathways for biotic and abiotic stress responses are *DEAR1*, *BOS1* and *SlERF5*. *DEAR1* is a transcriptional repressor of DREB protein that mediates plant defence and freezing stress responses in *Arabidopsis*; the *DEAR1* mRNA accumulates in response to both pathogen infection (*Pseudomonas syringae*) and cold treatment (Tsutsui et al., 2009). *BOS1* codes for a R2R3MYB protein that acts as transcription factor that in *Arabidopsis* regulates responses to *Botrytis cinerea* infection and to water deficit, increased salinity and oxidative stress (Mengiste et al., 2003). *SlERF5* is highly expressed in response to the harpin protein coded in the *hrp* gene clusters in many Gram-negative phytopathogens; the over-expression of *SlERF5* is involved in the induction of the dehydration-responsive genes through the ABA-mediated abiotic stress response (Chuang et al., 2010).

Studies in our laboratory in common bean (*Phaseolus vulgaris*) leaves detached, inoculated with fungal pathogen, and maintained in humid chamber demonstrate that *chalcone synthase* (*CHS*), a gene implicated in the biosynthesis of phytoalexins in response to pathogen challenge (Ferrer et al., 1999), is also responsive to wounding at early times after stress. *CHS* mRNA is detected 6 hours post-wounding of leaves or at latter times post-inoculation with *Colletotrichum lindemuthianum*; the mRNA disappears by 12 hours post-wounding stress. In plants, following exposure to environmental stresses including pathogen attack and wounding, the phenylpropanoid pathway has important functions in the production of compounds including lignin, flavonoids and phytoalexins. Chalcone synthase (CHS) is a key enzyme in this pathway, catalysing the first step in flavonoid biosynthesis, whose expression can be induced in response to environmental stress (Richard et al., 2000). This evidence exhibits the importance of molecular events in induction of specific defence genes in the response against certain pathogens, are dependent on specific environmental conditions, suggesting the existence of a complex signaling network that allows the plant to recognise and protect itself against pathogens and environmental stress. Similar induction patterns of members of the *14.3.3* gene family (*GF14b* and *GF14c*) by abiotic and biotic stresses such as salinity, drought, ABA and fungal inoculation have been documented in rice. The rice *GF14* genes contain *cis*-elements in their promoter regions that are responsive to abiotic stress and pathogen attack. The *14-3-3s* family genes are also subject to the regulation by certain transcript factors (Chen et al., 2006).

In rice, the *RO-292* gene is upregulated in roots by salt or drought stresses and by blast fungus infection (Hashimoto et al., 2004). Similarly, the *Mlo* gene in barley (*Hordeum vulgare*) act as modulators of defence and cell death in response to *Blumeria graminis* f. sp. *tritici* or *Magnaporte grisea* inoculation, and to wounding or the herbicide paraquat (Piffanelli et al., 2002). In *Arabidopsis*, at least five of the 29 *cytochrome P450* genes are induced by abiotic and biotic stress including *Alternaria brassicicola* or *Alternaria alternata*, paraquat, rose bengal, UV stress (UV-C), heavy metal stress ($CuSO_4$), mechanical wounding, drought, high salinity, low temperature or hormones (salicylic acid, jasmonic acid, ethylene and abscisic acid).

These five *cytochrome P450* genes (*CYP81D11*, *CYP710A1*, *CYP81D8*, *Cyp71B6* and *CYP76C2*) are co-induced by metal stress ($CuSO_4$), paraquat, salinity, ABA and pathogen inoculation. A common characteristic shared by all of these induced genes, as in the 14.3.3 genes family, is the presence of *cis*-acting elements in regulatory regions of the gene; W-box (DNA binding sites for WRKY transcription factors), P-box (a positive *cis*-acting regulator of pathogen defence) and MYB recognition sites are common (Narusaka et al., 2004). A collection of genes, including transcription factors are co-activated by pathogen challenge and abiotic stress, examples of these genes mediating crosstalk between signaling pathways for biotic and abiotic stress responses are *DEAR1*, *BOS1* and *SlERF5*. *DEAR1* is a transcriptional repressor of DREB protein that mediates plant defence and freezing stress responses in *Arabidopsis*; the *DEAR1* mRNA accumulates in response to both pathogen infection (*Pseudomonas syringae*) and cold treatment (Tsutsui et al., 2009). *BOS1* codes for a R2R3MYB protein that acts as transcription factor that in *Arabidopsis* regulates responses to *Botrytis cinerea* infection and to water deficit, increased salinity and oxidative stress (Mengiste et al., 2003). *SlERF5* is highly expressed in response to the harpin protein coded in the *hrp* gene clusters in many Gram-negative phytopathogens; the over-expression of *SlERF5* is involved in the induction of the dehydration-responsive genes through the ABA-mediated abiotic stress response (Chuang et al., 2010).

Studies in our laboratory in common bean (*Phaseolus vulgaris*) leaves detached, inoculated with fungal pathogen, and maintained in humid chamber demonstrate that *chalcone synthase* (*CHS*), a gene implicated in the biosynthesis of phytoalexins in response to pathogen challenge (Ferrer et al., 1999), is also responsive to wounding at early times after stress. *CHS* mRNA is detected 6 hours post-wounding of leaves or at latter times post-inoculation with *Colletotrichum lindemuthianum*; the mRNA disappears by 12 hours post-wounding stress. In plants, following exposure to environmental stresses including pathogen attack and wounding, the phenylpropanoid pathway has important functions in the production of compounds including lignin, flavonoids and phytoalexins.

Chalcone synthase (CHS) is a key enzyme in this pathway, catalysing the first step in flavonoid biosynthesis, whose expression can be induced in response to environmental stress (Richard et al., 2000). This evidence exhibits the importance of molecular events in The signaling pathways in plants in response to microorganism intruders and to wound could be with a relevant level of crosstalk. In both cases, cytoplasmic Ca2+ increase and the reactive oxygen species production occur (Jurkowski et al., 2004; Karita et al., 2004; Dey et al., 2010), moreover the induction of *WRKY* and pathogenesis related (*PR*) gene expression (Leon et al., 2001; Takemoto et al., 2003; Huang et al., 2010). The level of crosstalk between different genetic pathways in the plant response to abiotic and biotic stress often vary, as expected, in accordance with the specificity of the stressors. On the biotic side, the response depends on the pathogen identity; on the abiotic side, it depends on the level of the stressing factor and the general environmental conditions. The commonality between different genetic pathways vary greater in relation with the species and the genotype in the plant species. In chickpea, the batteries of expressed genes identified in response to high salinity, drought, cold or pathogen inoculation show marked differential coincidences. It was found that the genes up-regulated in response to pathogens were more similar to these induced by high salinity than those up-regulated in response to cold or drought conditions. In 51 transcripts differentially expressed in plants inoculated with pathogen, 21 were common among *Ascochyta rabiei* inoculation and one or more of the other three abiotic conditions. It is noteworthy that no transcript was commonly differentially expressed across all the four stresses assessed. Conversely, other sets of genes were found to be specifically induced by only one treatment, indicating the existence of specific signaling routes in addition to shared pathways (Mantri et al., 2010). A similar convergence of signaling pathways was reported for systemin, oligosaccharide elicitors and UV-B radiation at the level of mitogen-activated protein kinases (MAPKs) in *Lycopersicon peruvianum* suspension-cultured cells. LeMPK1 and LeMPK2, were activated in response to systemin, four different oligosaccharide elicitors, and UV-B radiation, whereas LeMPK3, was only activated by UV-B radiation. The common activation of LeMPK1 and LeMPK2 by many stress signals is consistent with a substantial overlap among stress responses; while UV-B induces specific responses (Holley et al., 2003).

In our studies, in a *Phaseolus vulgaris/Colletotrichum lindemuthianum* pathosystem, the *SUMO* gene and the *EF-hand calcium-binding protein* gene were responsive to pathogen as well to the abiotic stresses UV light (UV-A and UV-B), and extreme temperatures (8° and 38°C). These two genes are induced to different levels by UV light and extreme temperatures conditions.

The highest expression for the *SUMO* mRNA upon UV treatment was lower than of the *EFhand calcium-binding protein* mRNA: After 4 hours of heat (38°C) treatment, the *EF-hand calcium-binding protein* mRNA levels surpass the *SUMO* mRNA levels (Alvarado-Gutiérrez et al., 2008). Thus, clearly the levels of individual defence genes are differentially regulated transcriptionally by abiotic and biotic forms of stress. In relation to *SUMO*, five WRKY transcription factors are SUMO1 targets (WRKY3, WRKY4, WRKY6, WRKY33, WRKY72); many WRKY transcription factors are commonly involved in plant defence reaction to pathogens, moreover several forms of abiotic stresses. Therefore, resistance protein signaling and SUMO conjugation also converge at transcription complexes. It is known that SUMO conjugation is essential to suppress defence signaling in non-infected plants, and recently was suggested a model in which SUMO conjugation can transform transcription activators into repressors, thereby preventing defence induction in the absence of a pathogen (Burg and Takken, 2010).

A complexity of the stress response in plants is evident when it is considered the natural fluctuating environmental conditions within a day or over longer periods of time. In the environment, changing states in light intensities, temperatures and pressures exerted by wind are normal. The dynamism inherent to factors that compose the environment impacts in changes in the profile of expression of some plant defence genes. As previously we reported, the *SUMO* and the *EF-hand calcium-binding protein* genes in the plant-pathogen interaction exhibit similar kinetics in the dark period, but not in the light period. For the *EFhand calcium-binding protein* gene, the transcript levels in light in the control treated (H2O sprinkled) leaves surpass those in the pathogen-treated leaves (Alvarado-Gutiérrez et al., 2008). Thus, these two genes, which are co-induced by two or more types of biotic and abiotic stresses, are also differentially regulated by the daily photoperiod advance and possibly by the circadian rhythm. These findings indicate that these two molecules should be included in the category of integrative signals in abiotic and biotic stress response in plants.

A number of *Arabidopsis thaliana* lesion-mimic mutants that show alterations in the responses to abiotic and biotic stresses have been reported. One class of these mutants exhibits constitutively increased *PR* gene expression, SA levels and heightened resistance to pathogen infection (Yoshioka et al., 2001; Jambunathan and McNellis, 2003; Jurkowski et al., 2004; Mosher et al., 2010); this class includes the *cpr22* mutant, which has mutations in two cyclic nucleotide-gated ion channels that impart the phenotype of spontaneous lesion formation, SA accumulation, constitutive *PR-1*, *PR-2* and *PR-5* gene expression and enhanced resistance to various pathogens (Mosher et al., 2010). Noteworthy, in the aforementioned mutants, the phenotypes exhibited are suppressed under high relative humidity and high temperature and are enhanced by low humidity and cold temperatures (Yoshioka et al., 2001; Mosher et al., 2010). Similarly, the effects on basal and resistance (*R*) gene-mediated resistance in *A. thaliana* and *Nicotiana benthamiana/Pseudomonas syringae* pathosystems are reduced at moderately elevated temperatures (Wang et al., 2009). In accordance with this data, a number of mutants in plants with de-regulated expression of R proteins have been shown temperature-dependent defence responses (Alcazar et al., 2009; Huang et al., 2010; Zhou et al., 2010). These data indicates that in these mutants, the resistance phenotypes are dependent on environmental conditions or that, at least, there are humidity and temperature sensitive steps (Mosher et al., 2010). Indeed, the resistance response mediated by R genes as well the basal resistance is attenuated when the temperature increases.

Collectively, these data suggest that specific batteries of defence genes are involved in different signaling cascades that converge with a degree of overlap in the response programs for pathogen defence and abiotic stress protection. There is a balanced interplay with fine-tuning between parallel signaling branches by different sets of partially overlapping reactions. Moreover, the genes that are the convergence points between different genetic pathways are differentially regulated, more evidently, when these genes are analysed in the time scale, and definitely, the genetic pathways activated by R genes are modulated in variable levels by environmental factors. There are common factors in the defence signaling pathways to abiotic (humidity and temperatures variable conditions) and biotic (pathogen infection) stresses. These convergence points expose the superimposed complexity levels in the response to environmental changes. A pending task is the deciphering of the specificity of the signal transduction processes that conduit to the establishment of the commonality among different stress responses.

Phytohormones have Central Roles in Abiotic and Biotic Stress Signaling

Plant hormones, also called phytohormones, were first defined as "a substance which, being produced in any one part of the organism, is transferred to another part and there influences a specific physiological process" in the classical book *Phytohormones* written by Frits Went and Kenneth in 1937. The five classical phytohormones: auxin, cytokinin, ET, gibberellins, ABA and the recently identified brassinosteroids, JA and SA, are chemical messengers present in trace quantities; their synthesis and accumulation are tightly regulated. Depending on the context, they are subject to positive or negative feedback control and often are affected by crosstalk due to environmental inputs. Phytohormones move throughout the plant body via the xylem or phloem transport stream, move short distances between cells or are maintained in their site of synthesis to exert their influence on target cells where they bind transmembrane receptors located at the plasma membrane or endoplasmic reticulum or interact with intracellular receptors. The downstream effects of hormonal signaling include alterations in gene expression patterns and in some cases nongenomic responses. Changes in plant hormones concentrations and tissue sensitivity to them regulate a whole range of physiological process that have profound effects on growth and development. The phytohormomes affect all phases of the plant life cycle and their responses to environmental stresses, both biotic and abiotic. Hormonal signalling is critical for plant defences against abiotic and biotic stresses (Crozier et al., 2000; Taiz and Zeiger, 2010; Williams, 2010).

Typically the phytohormones that regulate the responses against adverse cues are grouped into two types: those that play a major role in response to biotic stress (ET, JA and SA) and those that have pivotal roles regulating the abiotic stress responses (mainly ABA). Commonly the biotic defence signaling networks mediated by phytohormones are dependent on the nature of the pathogen and its mode of pathogenicity. SA plays a central role in the activation of defence responses against biotrophic and hemi-biotrophic pathogens as well as the establishment of systemic acquired resistance. By contrast, JA and ET are usually associated with defence against necrotrophic pathogens and herbivorous insects.

Concerning to abiotic stress, ABA is the most studied stress-responsive hormone; it is involved in the responses to drought, osmotic and cold stress (Peleg and Blumwald, 2011; Wasilewska et al., 2008; Bari and Jones, 2009; Vlot et al., 2009).

Salicylic Acid, Ethylene, Jasmonic Acid and Abscisic Acid: are they Working Alone?

In addition to roles in activation of defence responses against biotrophic and hemibiotrophic pathogens, SA is also important to the establishment of systemic acquired resistance (SAR) (Grant and Lamb, 2006; Vlot et al., 2009). When resistant tobacco and cucumber plants are inoculated with pathogens, the levels of SA increase (Malamy et al., 1990; Rasmussen et al., 1991). Exogenous applications of this chemical messenger result in the induction of *PR* genes increasing resistance to a broad range of pathogens (Vlot et al., 2009). In addition, transgenic plants and mutants of tobacco and *Arabidopsis* in which endogenous SA levels are reduced, fail to develop SAR or express *PR* genes; instead, they displayed heightened susceptibility to both virulent and avirulent pathogens. When these plants are treated with the SA synthetic analogue, 2,6-dichloro-isonicotinic acid, resistance and *PR* genes expression are restored (Gaffney et al., 1993; Delaney et al., 1994; Vernooij et al., 1995.; Nawrath and Métraux, 1999; Nawrath et al., 2002; Genger et al., 2008; Vlot et al., 2009). By contrast, over-expression of bacterial SA biosynthetic genes in transgenic tobacco confers highly elevated SA levels, *PR* gene expression, and enhanced resistance (Verberne et al., 2000). The SAR is induced systemically by a signal generated in the inoculated leaf; this signal is transmitted via the phloem to the uninfected portions of the plant (Grant and Lamb, 2006; Parker, 2009). SA levels rise coincidently with or just prior to SAR and systemic *PR* gene expression or peroxidase activation in pathogen-infected tobacco or cucumber, also was detected in the phloem of pathogen-infected cucumber and tobacco, and radio-tracer studies suggest that a significant amount of SA in the systemic leaves of pathogen-infected tobacco and cucumber is transported from the inoculated leaf.

This was initially proposed to serve as signal in systemic acquired resistance; however, leaf detachment assays show that the mobile signal moves out of the infected leaf before increased SA levels are detected in petiole exudates from that leaf (Malamy et al., 1990; Rasmussen et al., 1991; Vlot et al., 2009). SA can be methylated to form methyl salicylate, in tobacco by the esterase SABP2 (an SAbinding protein). Recently, it has been shown that, methyl salicylate, which is induced upon pathogen infection, acts as an internal plant signal and also as an airborne defence signal (Forouhar et al., 2005; Park et al., 2007).

In plant defence responses against insects and microbial pathogens, JA is a crucial component. In *Arabidopsis* leaves, jasmonates control the expression of an estimated 67-85% of wound- and insect-regulated genes. Treatment of plants with JA results in enhanced resistance to herbivore challenge. Mutants defective in the biosynthesis or perception of JA show compromised resistance to herbivore attackers (Bari and Jones, 2009). Attack of herbivores such as *Manduca sexta* in tobacco induces the JA signaling activity (Paschold et al., 2007). Similarly, JA signaling is induced in tomato and *Arabidopsis* by *Tetranychus urticae* and *Pieris rapae*, respectably (Li et al., 2002; Reymond et al., 2004; De Vos et al., 2005). However, not all herbivores activate JA signaling in plants (Bari and Jones, 2009). The production of proteinase inhibitors (PIs) and other anti-nutritive compounds such as polyphenol oxidase (PPO), threonine deaminase (TD), leucine amino peptidase and acid phosphatase (VSP2) are mediated by JA in order to deter, sicken or kill the attacking insect (Howe and Jander, 2008). Also terpenoids and other volatile compounds produced by an herbivore-attacked plant are recognised by other carnivorous and parasitoid insects. The blends of compounds are specific to the particular plant/herbivore interaction, and the discerning carnivore uses this information to find its favourite meal (Howe and Jander, 2008; Williams, 2011).

Phytohormone Signaling Networks Act Together

Necrotrophic pathogens include most fungi and oomycetes as well as some bacteria. Defences to these types of pathogens are often mediated by JA and ET. JA and ethylene operate synergistically to activate the expression of a subset of defence genes following pathogen inoculation in *Arabidopsis* (Thomma et al., 2001; Glazebrook, 2005). Experimental data confirm that JA and ethylene signaling pathways act together. Analysis of the mutants *coi1* (jasmonate insensitive) and *ein2* (ethylene insensitive) revealed that the induction of JA response marker gene *PDF1.2* by *Alternaria brassicicola* requires both JA and ethylene signaling pathways (Penninckx et al., 1998; Thomma et al., 2001). Genes acting as point controls between these two pathways have been described: *CEV1* acts as a negative regulator and *ERF1* (ethylene response factor 1) is a positive regulator (Ellis et al., 2002; Lorenzo et al., 2003).

SA and JA are mutually antagonistic. Mutations that disrupt JA signalling (*coi1*) lead to the enhanced basal and inducible expression of the SA marker gene *PR1*, whereas mutations that disrupt SA signaling (*npr1*) lead to concomitant increases in the basal or induced levels of the JA marker gene *PDF1.2* (Kazan and Manners, 2008). Plants inoculated with virulent strains of *Pseudomonas syringae* pv. tomato treated with SA show compromised resistance to necrotrophic pathogen *Alternaria brassicicola*, which is sensitive to JA-dependent defences (Spoel et al., 2007). The non-expresser of PR genes 1 (NPR1) is a master regulator of SA signaling. *Arabidopsis npr1* mutants fail in SA-mediated suppression of JA responsive genes suggesting that NPR1 plays an important role in the SA-JA interaction (Spoel et al., 2007). Acting downstream from NPR1, WRKY70, a transcription factor (TF) acts as a positive regulator of SA-dependent defences and a negative regulator of JA-dependent defences and plays central role in determining the balance between these two pathways. Suppression of *WRKY70* expression allows increased expression from JA-responsive genes and increased resistance to a pathogen sensitive to JA-dependent defences. In contrast, over-expression of WRKY70 results in the constitutive expression of SA-responsive *PR* genes and enhanced resistance to SA-sensitive pathogens but reduces resistance to JA-sensitive pathogens (Li et al., 2004). Recently, WRKY6, WRKY53, mitogen activated protein kinase 4 (MPK4) and GRX480 (glutaredoxin) were reported to affect antagonism between SA- and JA-mediated signaling (Petersen et al., 2000; Brodersen et al., 2006; Mao et al., 2007; Miao and Zentgraf, 2007; Ndamukong et al., 2007). As we explained, plant hormone signaling pathways extensively interact during plant defence again pathogens and herbivores. Lifestyles of different pathogens are not often readily classifiable as purely biotrophic or necrotrophic.

Therefore, those interacting points or crosstalk between SA and JA/ET pathways may be regulated in a pathogen-specific manner (Adie et al., 2007; Bari and Jones, 2009).

Abscisic Acid in Abiotic and Biotic Responses Cross Talk in Plants

As sessile organisms, plants often have to cope with multiple environmental stresses; therefore most plants employ complex regulatory mechanisms to trigger effective responses against various biotic and abiotic stresses.

In this scenario, phytohormones are the main players regulating these responses. To coordinate the complex interactions, an intense crosstalk among the regulatory networks is necessary. ABA is involved in the regulation of many aspects of plant growth and development and also is the major hormone that controls plant responses to abiotic stresses (Wasilewska et al., 2008).

In the last decade, our understanding of ABA involvement to pathogen susceptibility and its relationship to other phytohormones involved in biotic stress response have increased.

Exogeneous ABA treatment increases the susceptibility of various plant species to bacterial and fungal pathogens (Heinfling et al., 1980; McDonald and Cahill, 1999; Thaler and Bostock, 2004; Mohr and Cahill, 2007)(Henfling et al., 1980; McDonald & Cahill., 1999; Mohr & Cahill, 2003; Thaler & Bostock, 2004; Ward et al., 1989). ABA-deficient tomato mutants show a reduction in susceptibility to the necrotroph *Botrytis cinerea* (Audenaert et al., 2002) and virulent isolates of *Pseudomonas syringae* pv tomato DC3000 (Thaler and Bostock, 2004; de Torres-Zabala et al., 2007), and ABA-deficient *Arabidopsis* has reduced susceptibility to the oomycete *Hyaloperonospora parasitica* (Mohr and Cahill, 2003). In general, ABA is involved in the negative regulation of plant defences against various biotrophic and necrotrophic pathogens. However, the role of ABA appears to be complex and may vary depending on the pathosystem. The role of ABA as a positive regulator of defence has also been reported (Mauch-Mani and Mauch, 2005). ABA activates stomatal closure that acts as a barrier against bacterial infection (Melotto et al., 2006). As a result, ABA-deficient mutants show more susceptibility to *Pseudomonas syringae* pv. tomato. In addition, treatment with ABA protects plants against *Alternaria brassicicola* and *Plectosphaerella cucumerina* indicating that ABA acts as a positive signal for defence against some necrotrophs (Ton and Mauch- Mani, 2004). Pathogen challenge results in the alteration of ABA levels in plants. For example, tobacco plants infected with tobacco mosaic virus (TMV) have increased ABA levels, and treatment with ABA enhances TMV resistance in tobacco (Whenham et al., 1986). Similarly, *Arabidopsis* plants challenged with virulent isolates of *Pseudomonas syringae* pv tomato DC3000, accumulate higher levels of ABA and JA than unchallenged plants (de Torres-Zabala et al., 2007).

Additionally, mutants deficient in ABA are more sensitive to infection by the fungal pathogens *Pythium irregulare* (Adie et al., 2007) and *Leptosphaeria maculans* (Kaliff et al., 2007). The situation becomes even more complicated when pathogens are tested on ABA signaling mutants, such as *abi4,* which displays opposite resistance responses towards these two fungi. Along the same line, the mutations *abi1-1* and *abi2-1* actually foster differential resistance responses against *Leptosphaeria maucans* (Kaliff et al., 2007; Wasilewska et al., 2008). Transcriptome and meta analyses of expression profiles altered by infection with the necrotroph *Pythium irregulare* identified many JA-induced genes but also highlighted the importance of ABA as a regulator, as the ABA responsive element (ABRE) appears in the promoters of many of the defence genes (Adie et al., 2007; Wasilewska et al., 2008). This indicates that ABA plays an important role in the activation of plant defence through transcriptional reprogramming of plant cell metabolism. Moreover, ABA is required for JA biosynthesis and the expression of JA responsive genes after *Phytium irregular* infection (Adie et al., 2007). Recently, it has been identified the first molecular component in crosstalk between biotic and abiotic stress, the rice *MAP* gene *OsMPK5*. ABA antagonize pathogen-activated ET signaling via OsMPK5 (De Vleesschauwer et al., 2010).

The exact molecular mechanism of ABA action on plant defence responses against diverse pathogens started to be elucidated. Identification of more factors involved in ABA-mediated crosstalk between biotic and abiotic stress signaling merits extensive future study.

WRKY and Other Transcription Factors as players in Plant Response to Abiotic and Biotic Stress

Plant responses to environmental stimuli involve a network of molecular mechanisms that vary depending on the nature of environmental signal. In the signal transduction network that leads from the perception of stress signals to the expression of stress-responsive genes, transcription factors play an essential role. TFs are a group of master proteins that interact with *cis*-elements present in promoter regions upstream of genes and regulate their expression. Most TFs impact multiple physiologic processes such as metabolism, cell cycle progression, growth, development and reproduction. Several transcription factors are mediators of multiple phytohormone signaling networks.

Transcription Factors in Crosstalk Stress Responses

The TFs are involved in responses against biotic and abiotic stress, and they play an esential role in regulation of plant adaptation to environmental changes. A few TFs have been reported to take part in the crosstalk between abiotic and biotic stress signaling networks.

The basic helix-loop-helix (bHLH) domain-containing transcription factor AtMYC2 is a positive regulator of ABA signaling. The genetic lession of *AtMYC2* results in elevated levels of basal and activated transcription from JA-ethylene responsive defence genes (Abe et al., 2003; Anderson et al., 2004). MYC2 differentially regulates two branches of JA mediated responses; it positively regulates wound-responsive genes, including *VSP2, LOX3,* and *TAT,* but represses the expression of pathogen-responsive genes such as *PR4, PR1,* and *PDF1.2.* These complex interactions are co-mediated by the ethylene-responsive transcription factor ERF1 (Lorenzo et al., 2003; Lorenzo et al., 2004). The botrytis susceptible 1 *(BOS1)* gene of *Arabidopsis* encodes an R2R3MYB transcription factor that mediates responses to certain signals, possibly through ROS intermediates from both biotic and abiotic stress agents (Mengiste et al., 2003). There are also four members of the *NAC* family of genes that encode plant-specific transcription factors involved in diverse biological processes. *OsNAC6, Arabidopsis transcription activation factor 1 (ATAF1), ATAF2* and *dehydration 26 (RD26)* are potentially involved in regulation of responses to abiotic and biotic stresses (Wu et al., 2009).

WRKY Transcription Factors

WRKY proteins are a recently identified class of DNA-binding proteins that recognise the TTGAC(C/T) W-box elements found in the promoters of a large number of plant defence related genes (Dong et al., 2003). These TFs contain WRKY domains that appear to be unique to plants (Eulgem and Somssich, 2007). The name of the WRKY family is derived from its highly conserved 60 amino acid long WRKY domain, comprising highly conserved WRKYGQK at N-terminus and a novel metal chelating zinc finger signature at C-terminus. *WRKY* genes thought to be plant-specific TFs that have been subject to a large plant-specific diversification. Phylogenetic analysis shows that the *WRKY* genes are clustered into several different groups on the basis of their amino acid sequences (Yamasaki et al., 2005; Eulgem and Somssich, 2007). *WRKY* genes probably originated concurrently with the major plant phyla.

Current information suggests that WRKY factors play a key role in regulating the pathogen induced defence program. From the beginning of research into WRKY transcription factors, it was evident that they play roles in regulating several different plant processes. It is common for a single WRKY transcription factor to regulate transcriptional reprogramming associated with multiple plant processes. The dynamic web of signaling in which WRKY factors operate has multiple inputs and outputs (Rushton et al., 2010). It is expected that a single WRKY transcription factor has activity on both abiotic and biotic stress pathways and cross talks with different signal transduction pathways. The rice *WRKY45 (OsWRKY45)* gene expression is markedly induced in response to ABA and various abiotic stress factors such as NaCl, dehydration; in addition expression is induced by pathogens such as *Pyricularia oryzae Cav.* and *Xanthomonas oryzae* pv. oryzae. Moreover, *OsWRKY45* over-expressing plants exhibited several changes: a) the constitutive expression of ABA-induced responses and abiotic-related stress factors, b) markedly enhanced drought resistance and c) increased expression of *PR* genes and resistance to the bacterial pathogen *Pseudomonas syringae*. Thus, OsWRKY45 shows a dual role, acting as a regulator and as a protective molecule upon water deficit and pathogen attack (Qiu and Yu, 2009). *VvWRKY11* from *Vitis vinnifera* is a nuclear protein that is expressed rapidly and transiently in response to treatment with SA or pathogen *Plasmopara viticola*. Transgenic *Arabidopsis* seedlings over-expressing *VvWRKY11* have higher tolerance to water stress induced by mannitol than wild-type plants. These results demonstrate that the *VvWRKY11* gene is involved in the response to dehydration and biotic stress (Liu et al., 2011). Other well known players in plant responses to abiotic and biotic stresses are members of the WRKY transcription factor family. Expression patterns of *VuWRKY11, AtWRKY39* and *AtWRKY53* indicate that these genes are coregulator of the plant response against pathogens and hydric and heat stress. In addition, some WRKY transcription factors (OsWRKY24 and OsWRKY45) antagonize ABA function, repressing an ABA-inducible promoter, indicating that these molecules operate with versatile capabilities.

Crop growth and crop yield are affected by environmental cues. There is a need of greater understanding of plant physiological responses to the abiotic and biotic stresses. We can understand stress as a stimulus or influence that is outside the normal range of homeostatic control in a given organism: If a stress tolerance is exceeded, mechanisms are activated at molecular, biochemical, physiological and morphological levels; once stress is controlled, a new physiological state is established, and homeostasis is re-established. When the stress is retired the plant may return to the original state or a new physiological state.

Plants continually encounter stress even under environmental conditions that we think of as normal. The environment changes during the day, day to day and throughout the year, thus plants must respond to stress over the course of each day and often must respond to several stresses at the same time. Study of stress responses show that there is much crosstalk among signaling networks during specific stress responses. Thus, plants may respond to stress perception by an initial global response and follow with specific stress responses.

As we discussed in this chapter, convergence points between biotic and abiotic stress signaling pathways have begun to be analysed. Specific factors including transcription factors such as *WRKYs, ATAF1 and 2, MYC2, RD2, BOS1, OsNAC6* and OsMPK5 kinase are molecular player, common to multiple networks or involved in crosstalk between stress signaling pathways regulated by abscisic acid, salicylic acid, jasmonic acid and ethylene as well as ROS signaling. Powerful molecular tools, including transcriptome and proteome analyses, sequencing of entire genomes in plants, bioinformatic analyses and functional studies, will enable the dissection of networks and identification of key factors in abiotic and biotic signaling cascade crosstalk, which will reveal novel interplays between parallel signaling pathways in the plant responses to biotic and abiotic stress.

Chapter 4

Crop Plant Resistance to Biotic and Abiotic Factors

Changes in crop production and the impact of new food and environmental legislation are having an influence on the significance of pests and diseases which attack plants and reduce yields. Climate change will also have an impact on pests and pathogens, and may increase exposure to abiotic stresses such as drought and heat. With the expected increase in world population outstripping the land available for cropping, maximising utilisable yields by breeding for resistance to biotic and abiotic stresses is becoming an imperative. It is therefore important that plant breeders can identify the most important constraints on production in a particular crop and region.

In recent years, food concerns in Europe have been largely centred on safety and quality, on how food is grown and on the impact of agriculture on the environment. The food shortages of two years ago raised the spectre of basic food security after many years when this was not regarded as top priority. With predictions that climate change will reduce the land available for cropping, maximising yields has increased as a priority but not at the expense of quality and safety. This paper explores this tension and examines the role for plant breeding, especially breeding for biotic and abiotic stresses, within modern agriculture.

Food Demand

For the last half a century, global grain demand and production have more than tripled for wheat, and for maize (corn) the increase is even greater, with a recent rapid acceleration due to the demand in the USA for biofuels. However, throughout the world, approximately 800 million people are malnourished and the demand for food will increase as the population increases. It is predicted that by 2050 the world population will increase from approximately 6.7 billion to over

9 billion and that the current trend for more resource-intensive diets, which include more dairy and meat products, will continue. It is estimated that current global production of wheat must increase annually by about 2% (Singh & Trethowan 2007). At the same time, the demand for land from uses other than agriculture is increasing. Biofuels have already been mentioned; other demands include housing and industrial buildings, timber and forest conservation. (Evans 2009). The recognition that biodiversity and the quality of the environment need to be preserved for future generations also leads to competition for land use and can also negatively affect crop yields through a reduced use, or total exclusion of, inorganic fertilisers and pesticides as in, for example, organic systems.

A comprehensive study of pest, disease and weed losses in eight crops which occupy half the world's cropped land to date was published by a team of German crop scientists (Orke *et al.* 1994). The study found that, overall, pests accounted for preharvest losses of 42% of the potential value of output, with 15% attributable to insects and 13% each to weeds and pathogens. An additional 10% of the potential value was lost postharvest. Losses in Europe alone were lower (for example a loss of 9.7% of production caused by plant diseases) but were still very significant, in spite of a substantial use of pesticides. Abiotic stresses such as drought, heat and salinity add considerably to these losses, and are likely to increase with climate change. In a world demanding more food from a limited amount of land, improved resistance to biotic and abiotic stresses is a priority.

Climate Change

At the same time as the demand for food is intensifying, the climate is changing, with inevitable consequences for agriculture and the world's food supply. The potential consequences have been discussed by Rosenzweig & Hillel (1995). They state that "vulnerability to climate change is systematically greater in developing countries, which in most cases are located in lower, warmer latitudes. In those regions, cereal grain yields are projected to decline under climate change scenarios, across the full range of expected warming.

Agricultural exporters in middle and high latitudes ...stand to gain, as their national production is predicted to expand, and particularly if grain supplies are restricted and prices rise. Thus, countries with the lowest income may be the hardest hit." In Europe, predicted changes in climatic conditions depend on location, with the greatest levels of warming predicted for Mediterranean and northeastern areas, increased precipitation in northern areas (particularly in winter) and decreased precipitation in southern areas (Brooker & Young 2006). The challenge in many areas of the world will be to produce more food with limited supplies of water, and breeding for drought tolerance and water use efficiency are key to this (Ober 2008). Globally, drought already results in greater yield loss than any other single biotic or abiotic factor (Boyer 1982) and even in the UK drought losses are estimated to be 1-2 t/ha (Foulkes *et al.* 2007). However, Semenov (2008) considers that, in England and Wales, heat stress around flowering might represent a greater risk to wheat production in England than drought, because although the summer is predicted to be drier in the 2050s, winter is predicted to be wetter and water might still be available to the growing crop in late spring and summer. Collier *et al.* (2008) evaluated the potential impact of future extreme weather events on horticultural crops in the UK using a stochastic weather generator linked with UKCIP02 (Hulme *et al.* 2002) projections of future climate. This study indicated that episodes of summer drought severe enough to interrupt the continuity of supply of salads and other vegetables will increase and there will be a requirement for winter cauliflowers with different temperature sensitivities from those used currently. Important pests, such as cutworm (*Agrotis segetun*) and diamond-back moth (*Plutella xylostella*) could become a greater threat: in the case of the former, the number surviving to third instar increased with time in the model; in the latter, there was an increase in the number of generations. The impact of climate change on other pests due to changes in life cycles might be expected throughout Europe and may pose a serious threat to production systems.

Climate change is also expected to impact on pathogens and pathosystems. Turner (2008) reported modelling work to predict potential levels of disease under climate change for the years 2081-2090. She concluded that wheat brown rust (*Puccinia recondita*) will become the primary target for disease control strategies, because it is favoured by warmer, drier summers.

Conversely, Septoria leaf blotch (*Mycosphaerella graminicola*), currently the most important disease of wheat in the UK, was predicted to decline. However, Turner acknowledged that the model only accounted for effects on the pathogen and climate will also affect the host, making risk prediction more challenging. Roche *et al.* (2008) modelled the potential impact of climate change on wheat brown rust for four contrasting French sites, taking into account a range of climatic factors and their effect on the pathogen and plant-pathogen interactions. Surprisingly they found no clear trend in infection rates, which they concluded was due to opposing effects, for example an increase in temperature accelerated the disease cycle but was counteracted by a reduction in leaf surface wetness duration. Also, when plant development changes were taken into account, although temperature accelerated the disease cycle it also had the same effect on the development of the crop, maintaining the status quo. The spread of diseases may well also be affected by climate change. Insect vectors of pathogens such as the fungi causing Dutch elm disease (*Ophiostoma ulmi* and *O. novo-ulmi*) are likely to respond to warmer summers by extending their geographic ranges and hence the ranges of disease incidence. Another important pathogen of trees, *Phytophthora cinnamomi*, an aggressive introduced fungus which causes root and stem-base diseases of oaks, chestnuts and many other tree species, is predicted to become more active across coastal areas of the UK and Europe (Lonsdale & Gibbs 2002).

Agriculture is potentially very sensitive to climate change but there are clearly many uncertainties, which create difficulties for plant breeders who are making a long-term investment. However, breeding for disease resistance may not only be beneficial in adaptation to climate change, it may have a role in limiting greenhouse gas emissions. Berry *et al.* (2008) calculated the reductions in emissions that could be achieved in the UK from disease control: with current cultivars and fungicide use, there is the potential to save up to 1.14Mt CO_2 eq. per annum. This saving could be improved through the use of more effective disease resistance, providing it is not associated with a yield penalty. A general increase in resistance to Septoria leaf blotch of one point on the 1-9 scale used by the HGCA Recommended Lists (Anon. 2009) would decrease greenhouse gas emissions by approximately 13kg CO_2 eq.

Land Use

In an area as diverse as Europe, it is not possible to generalise on changes in land use. However, a desk-based study has been carried out on cropping on the chalkland of the East Anglian region of the UK, which is nowadays primarily arable (Parry *et al.* 2006) Over recent years there has been a decline in mixed farming and a switch from spring to autumn cropping.

The study suggested that if market forces determine land use in this area in the future, the landscape would become more aggregated, with oilseed rape and wheat dominating. Wheat would be farmed on the larger fields on average, with oilseed rape on smaller fields. Economies of scale and greater use of contractors would lead to block-cropping, with large areas (>200ha) of a single crop.

The report was principally commissioned to study the impact of land use changes on the environment but clearly disease and pest pressure will increase if fewer crop types are grown in larger blocks. Since 2006, when the report was published, set-aside has been abolished within the EU, reducing the amount of land left fallow.

EU Legislation

EU legislation introduced in recent years as a response to concerns about food safety and agriculture's impact on the environment may have a impact on our ability to control diseases and pests, especially in those countries which have a considerable reliance on pesticides, such as the UK where the mild and wet climate encourages the development of many diseases.

Pesticide Legislation

Currently the EU is in the final negotiation phase of a new legislative package on pesticides (a revision of Directive 91/414/EEC). Among other things, this introduces cut-off criteria based on hazards rather than risks. It is still not clear how many pesticides will eventually be withdrawn as a result of this legislation. It has been suggested that the list may include fungicides such as some of the triazoles (which control powdery mildews, rusts, and many leaf spotting fungi on a wide range of crops), mancozeb (which controls downy mildews and potato blight (Phytophthora infestans) and is widely seen as vital to prevent resistance developing in other fungicides) and quinoxyfen (which controls powdery mildews on, for example, cereals and grape vines). They will add to the 60% which have already been withdrawn from the European market over the last ten years (Anon. 2008a).

Water Framework Directive

The implementation of the Water Framework Directive is likely to have an impact on a number of pesticides. Although herbicides are particularly vulnerable, many insecticides are also at risk, and metaldehyde, used for controlling slugs, is already under scrutiny as it has been found in water at concentrations above the EU limit (Twining et al. 2009). Metaldehyde is particularly important for growers of vegetables, potatoes and oilseed rape. Some crops, such as potatoes (Johnston & Pearce 2008), differ significantly in cultivar resistance to attack by slugs, whereas others, such as oilseed rape, do not.

In addition to these products, important pesticides have been lost in recent years due to pesticide resistance and there is concern that, with a reduction in the number of active chemicals, there will be an increase in selection pressure for resistance. Many organisations have been advocating a greater use of integrated pest management systems for a number of years and enhanced disease and pest resistance clearly has a significant part to play in this (Anon 2008a).

Mycotoxin Legislation

In 2006 the European Commission introduced regulation No. 881/2006 setting maximum levels for certain contaminants in foodstuffs. This included a number of important mycotoxins, including aflotoxins in dried fruit and cereals, Ochratoxin A in cereals, dried vine fruit and wine, patulin in apples, cider and fruit juices, deoxynivalenol (DON) and zearalenone in processed and unprocessed cereals including maize, and fumonisins in maize (Anon. 2006). Maximum levels for the *Fusarium* mycotoxins (DON, zearalenone and fumonisins) in maize were lowered the following year, in order to avoid a disruption of the market whilst maintaining a high level of public health protection.

Mycotoxin regulations are having a significant effect on food production. For example, in the UK, control of Fusarium head blight has become a much more important issue, because of the production of DON. In the field, infection of ears by some *Fusarium* species can result in the production of mycotoxins when weather is warm and wet at flowering but there is little correlation between Fusarium-damaged grain and mycotoxin occurrence. In the past two seasons, until January 2008, millers relied on a risk assessment developed by the HGCA and completed by farmers (Anon. 2008b). However, concentrations of DON above the EU accepted limit have been found in batches of grain for milling classified as at low risk which has led to compulsory testing for mycotoxin and a review of the risk assessment system.

Fungicide control of head blight is not easy, as fungicides have to be applied at precisely the right time. Improved resistance would be of great benefit as it has been demonstrated that DON levels are lower in wheat varieties with higher disease resistance (Edwards 2007). Until recently, this disease was not regarded as of high priority for UK plant breeders but this has changed because of the legislation.

The Challenge for Plant Breeding

Even with modern techniques which can speed up the breeding process, plant breeders have to determine their objectives many years in advance of a cultivar being released. This has never been easy but with the pressures on production systems described above and the largely unpredictable effect of a number of them, determining the importance of specific traits 10-15 years ahead of the release of a new cultivar is a real challenge. To add to the difficulty, a whole range of traits alongside resistance to biotic and abiotic stresses have to be considered. For example, the HGCA Recommended List 2009/10 for winter wheat (Anon. 2009) measures treated and untreated yield plus eight grain quality characteristics and five agronomic features alongside resistance to seven diseases and one pest. There is also a need to try and ensure that the sources of resistance to pathogens employed are durable, to escape from the 'boom and bust' cycle common with some pathosystems. Many sources of durable resistance are controlled by a number of genes which makes breeding much more challenging.

Molecular biology has developed rapidly in the past two decades and this is benefiting resistance breeding programmes, although the impact is not as great as it could be if genetic modification was more acceptable in Europe. The production of markers for resistance genes and other traits is developing rapidly and will enable much more effective selection to take place. In addition to this, there is a greater understanding of the mechanisms of resistance, through the cloning and characterisation of resistance genes and an understanding of biotic and abiotic stress signalling pathways. These scientific developments have the potential to provide new strategies for effective breeding for host resistance to stresses in the future.

FAO Methodologies on Crop Water Use and Crop Water Productivity

The FAO Land and Water Development Division has played an active role during the past three decades in developing and promoting guidelines and methodologies on crop water management at the field level that have become widely-used standards. This particularly applies to the methodologies for the calculation of crop water requirements and crop water productivity in irrigated and rainfed agriculture. These methodologies include guidelines on Crop Water Requirements by Doorenbos and Pruitt (1973) published as Irrigation and Drainage (I&D) Paper No. 24 and on Yield Response to Water by Doorenbos and Kassam (1979) as I&D No. 33.

A revision of the procedures for estimating crop water requirements was undertaken following an expert meeting in Rome in 1990. New procedures were elaborated including a revised method for estimating reference evapotranspiration (ETo) and crop evapotranspiration (ETc), published in 1998 as I&D No. 56 by Allen, Pereira, Raes and Smith (1998).

Calculation procedures for crop water management and applications for planning and management in irrigated and rainfed agriculture were further facilitated by the development of computerized procedures in CROPWAT, published as I&D No. 46 by Smith (1992).

This note provides an overview of the FAO methodologies that have been developed and promoted for the computation of crop water requirements and crop water productivity under adequate and deficit water supply, and for irrigation requirements and scheduling.

Guidelines on Crop Water Requirements: I&D No. 24

I&D No. 24 on Crop Water Requirements was first published in 1973 and subsequently revised in 1977. It became an international standard for calculating crop water requirements, extensively used by irrigation engineers, agronomists, hydrologists, economists and environmentalists.

Owing to the difficulty of obtaining readily available and accurate field measurements, the estimation of crop water requirements were derived from estimating crop evapotranspiration according to standardised crop and climatic conditions. A range of empirical methods was developed to estimate potential crop evapotranspiration (ETc) from readily available climatic parameters. The water requirements of a given crop was derived through a crop coefficient that integrated the combined effects of crop transpiration and soil evaporation into a single crop coefficient, according the following relationship:

$$ETcrop = Kc \times ETo$$

where: ETo is reference crop evapotranspiration Kc is crop coefficient

ETcrop is the crop evapotranspiration, computed for optimal conditions (ETc)

ETc was defined as the evapotranspiration from a disease-free, well fertilized crop, grown in large fields, under optimum soil water conditions, and achieving full production under the given ecological environment.

I&D No. 24 provided four methods for computing ETo, according the availability of climatic data, as well as procedures to determine Kc for different crops and growth stages, as briefly reviewed below.

Reference Crop Evapotranspiration

Evapotranspiration (ET) comprises the simultaneous movement of water from the soil and vegetation surfaces into atmosphere through evaporation (E) and transpiration (T).

Reference crop evapotranspiration ETo is the evapotranspiration from a reference crop with the specific characteristics of grass, fully covering the soil and not short of water and represents the evaporative demand of the atmosphere at a specific location and the time of the year independently of crop type, crop development and management practices, and soil factors. The only factors affecting ETo are climatic parameters. Consequently, ETo is a climatic parameter and can be computed from weather data.

Relating ET to a specific surface provides a reference to which ET from other crop surfaces can be related. It obviates the need to define a separate ET level for each crop and stage of growth. ETo values measured or calculated at different locations or in different seasons are comparable as they refer to the ET from the same reference surface.

A large number of more or less empirical methods were developed by numerous scientists and specialists worldwide to estimate evapotranspiration from different climatic variables. Relationships were often subject to rigorous local calibrations and proved to have limited global validity.

The calculation procedure included in I&D No. 24 for ETc consists of:

1. identifying the crop growth stages, determining their lengths, and selecting the corresponding Kc coefficients;
2. adjusting the selected Kc coefficients for frequency of wetting or climatic conditions during the stage;
3. constructing the crop coefficient curve (allowing one to determine Kc values for any period during the growing period); and
4. calculating ETc as the product of ETo and Kc.

I&D No. 24 provided a standardized range of crop coefficients for a large number of crops.

Revised Guidelines on Crop Evapotranspiration: I&D No 56

Advances in crop water research and more accurate procedures in determining crop water use revealed the need for a revision of I&D

No. 24, particularly in relation to the estimation of ETo. An expert consultation was held in May 1990 in Rome to advise on options to revise the procedures introduced in I&D No. 24.

The study by Jensen, Burman and Allen (1990) comparing a range of 20 different ETo estimation methods demonstrated clearly the superior performance of the procedures introduced by Monteith (1965) in the Penman equation.

By introducing the aerodynamic and canopy resistance in the Penman-Monteith combination method, a better simulation of wind and turbulence effects and of the stomatal behaviour of the crop canopy was achieved (Monteith 1965). The earlier difficulties in the use of the method related to the estimation of the canopy resistance values was largely overcome by progress in research and reliable estimates of the two parameters for a range of crops including the reference crops grass and alfalfa.

FAO Penman-Monteith Method

The FAO expert consultation in 1990 reached unanimous agreement in recommending the Penman-Monteith approach as the best performing method to estimate evapotranspiration of a reference crop ETo and adopted the estimates for bulk surface and aerodynamic resistance as elaborated by Allen et al. (1989) as standard values for the reference crop

The adoption of fixed values for crop surface resistance and crop height, required an adjustment of the concept of reference evapotranspiration which was redefined as follows:

Reference evapotranspiration is the rate of evapotranspiration from a hypothetical reference crop with an assumed crop height (12 cm), a fixed crop surface resistance (70 s m$^{-1)}$ and albedo (0.23), closely resembling the evapotranspiration from an extensive surface of green grass cover of uniform height, actively growing, completely shading the ground and with adequate water.

To further standardize the use of the FAO Penman-Monteith method, studies were undertaken to provide recommendations when limited meteorological data are available. Procedures were developed to estimate values for vapour pressure, solar radiation and wind speed. This allowed the use of the Penman-Monteith even if only the temperature data are available. This excluded the need to maintain any other empirical ETo estimation method as a standard, and only one method for estimating ETc is presently recommended, which has largely contributed to the transparency and consistency in reference evapotranspiration and crop water requirement studies.

I&D No. 56 provides detailed procedures for the calculation of ETo with different time steps, ranging from hours to months, and includes computations by hand with the help of a calculation sheet, or by means of a computer.

Crop Evapotranspiration: Dual step – Single Crop Coefficient

The adoption of the Penman-Monteith method would in theory allow the possibility of calculating ETc directly from climatic data and by integrating the crop resistance, albedo and air resistance factors in the Penman-Monteith approach. However, as there is still considerable lack of information for different crops, the dual step procedure is maintained using the crop coefficient Kc to derive crop evapotranspiration from the evapotranspiration of a reference crop. The Penman-Monteith method is used for the estimation of the evapotranspiration of a hypothetical reference crop with fixed crop parameters, i.e. ETo.

Experimentally determined ratios of ETc/ETo, called crop coefficients (Kc) are used to relate ETc to ETo or ETc=KcETo. This is referred to as the ETc using single crop coefficient.

Difference in leaf anatomy, stomata characteristics, aerodynamic properties and even albedo causes the crop transpiration to differ from the reference crop evapotranspiration under the same climatic conditions. Due to variations in the crop characteristics throughout its growing season, Kc for a given crop changes from sowing till harvest.

A review of crop coefficients given in (FAO 1975) resulted in an update of Kc values to be applied to the FAO Penman-Monteith method and procedures to arrive at better estimates under various climatic conditions and crop height and expanding the range of crops and crop types (FAO 1998).

Crop Evaporation and Transpiration : Dual Crop Coefficient

For more detailed calculations in crop simulation studies, ETc values are needed on a daily basis, requiring a more accurate estimation of crop transpiration and soil evaporation. The effect of specific wetting events on the value Kc and ETc needs to be taken into account for more accurate estimations of ETc. This is done by splitting Kc into two separate coefficients, one for crop transpiration, i.e., the basal crop coefficient (Kcb) representing the transpiration of the crop, and one for soil evaporation, the soil water evaporation coefficient (Ke). Thus, a dual crop coefficient approach was introduced in I&D No. 56 for ETc as ETc=(Kcb+Ke)ETo.

The dual crop coefficient approach is more complicated and more computationally intensive than the single crop coefficient approach. The procedure is conducted on a daily basis and intended for applications using computers. It is recommended that the approach be followed when, improved estimates for Kc are needed, for example to schedule irrigation for individual fields on a daily basis.

The calculation procedure for ETc using dual crop coefficient approach consists of:

1. identifying the lengths of crop growth stages, and selecting the corresponding Kcb coefficient;
2. adjusting the selected Kcb coefficients for climatic conditions during the growth stage;
3. constructing the basal crop coefficient curve (allowing one to determine Kcb values for any period during the growing period);
4. determining daily Ke values for surface evaporation; and
5. calculating ETc as the product of ETo and (Kcb+Ke).

Standard values of revised Kc and Kcb for crop stages initial, mid-season and end-season are given in FAO (1998).

Crop Evapotranspiration Under Non-standard Conditions (ETc-adj)

When cultivating crops in the field, the real crop evapotranspiration may deviate from ETc due to non-optimal conditions such as water shortage, the presence of pests and diseases, soil salinity, low soil fertility or waterlogging. This may result in low plant density and poor growth and may reduce the evapotranspiration rate below ETc.

When the soil water potential drops below a threshold value, the crop is said to be water stressed. The effects of soil water stress are described by multiplying the basal crop coefficient by the water stress coefficient, Ks. Thus, ETc-adj=(KsKcb+Ke)ETo. For soil water limiting conditions, Ks<1. Where there is no soil water stress, Ks=1. Ks describes the effect of water stress on crop transpiration. Where the single crop coefficient is used, the effect of water stress is incorporated into Kc as ETc-adj=KsKcETo. Ks is given as:

$$Ks = (TAW\text{-}Dr)/(TAW\text{-}RAW) \text{ or } (TAW\text{-}Dr)/[(1\text{-}p)TAW]$$

where TAW is total available water RAW is readily available water Dr is root zone depletion, and p is fraction of TAW that a crop can extract from the root zone without suffering water stress

The empirically derived values for p are given in FAO (1998) and they differ from one crop to another; p is a function of the evaporation power of the atmosphere because the rate of root water uptake is influenced by soil matrix potential and the associated hydraulic conductivity. The estimation of Ks requires a daily soil water balance computation for the root zone as described in FAO (1998).

Yield Response to Water: I&D No. 33

The uptake of soil water by the crop to meet evaporative demand results in reduced water content in the soil and crop. When full crop water requirement is not met, water deficit in the plant can develop to a point when stomatal closure would occur to reduce further water loss and water stress, if soil and plant water content is not restored by either rain water or irrigation. The closure of the stomata results in a parallel reduction in the uptake of CO_2, photosynthesis and biomass production. Plant water deficits can therefore develop into reduced crop growth, and crop development and yields may be effected depending on the extent of the water deficit and the impact on vital phenological and growth processes. The manner in which crop water deficit affects growth, development and yield varies with crop and crop type, and crop growth stage.

To evaluate the effect of crop water deficit on yield decrease through the quantification of relative evapotranspiration (ETc-adj/ETc), FAO undertook an analysis of research results from a large amount of crop water studies. The findings were published in the FAO I&D No. 33 (FAO 1979) in which a linear crop-water production functions was introduced to predict the reduction of crop yield when crop stress was caused by a shortage of soil water according to the following relationship:

$$(1-Ya/Ym)=Ky(1-ETc\text{-}adj/ETc)$$

where: Ky is yield response factor

ETc-adj is adjusted (actual) evapotranspiration (actual evapotranspiration ETa in I&D No. 33)

ETc is crop evapotranspiration for standard conditions (no water stress) (maximum evapotranspiration ETm in I&D No. 33)

Ya is actual crop yield.

Ym is maximum expected or agronomically attainable crop yield under no biotic or abiotic stress.

Ky is a factor that describes the reduction of relative yield according to the reduction in ETc caused by soil water shortage. Ky values are crop specific and vary over the growing season according to growth stage as shown for maize in the following figures for total growing period and for different growth stages.

To compute Ym, two methods were suggested: a method based on Kassam (1977) and applicable to many crops, and a method from Slabbers et al. (1978) applicable to four crops. The former method was extensively applied in computing yields of many crops in the agro-ecological zones project (FAO 1978-81), and the approach has been widely used by others to develop similar methods. For a given crop type, Ym is derived from net crop biomass, taking into account harvest index. Net crop biomass is derived from gross biomass, taking into account the effect of radiation and temperature on crop photosynthesis, and of temperature on crop respiration. A knowledge of length of growth cycle and leaf area index at maximum cover is required.

I&D 33 provides Ky values for a range of crops based on an analysis in the seventies of an extensive amount of available research studies on crop water yield relationships. The linear crop-water production functions developed in I&D No. 33 have shown over the years that the empirical approach of determining yield response to water is robust and fairly insensitive to variations in the specific characteristics of crops. In situations where the availability of site-specific data is limited, the estimated yield results compare well with those derived from more sophisticated approaches at modelling water-limited crop production.

However, since the development of the I&D No. 33 method, a large number of studies and research results have enhanced the information and knowledge base on yield response to water (e.g., Kirda, Moutonnet, Hera and Nielsen 1999; Moutonnet 2000). A comparison of Ky factors from I&D No. 33 and IEAE-CRP field studies. Also, there have been significant changes in the genetic characteristics of many crops and crop types. Further, more advanced techniques and simulation studies have allowed a more precise and in-depth analysis of crop-soil-water relationship and yield water responses.

There is a need therefore to review the methodology and to confirm – with possible adjustments – that the I&D No. 33 water production function method should be maintained as one of the principal practical methods to estimate crop water productivity and to update the crop information and parameters. The update should include a review of the various simulation models of crop production that have been developed over the last 15 years and which, given specific soil and weather information, produce predictions of water-limited yields of economically important crops.

CROPWAT: I&D Nos. 46 and 49

CROPWAT is a water balance-based computer programme to calculate crop water requirements and irrigation water requirements from climatic and crop data. The programme also allows the development of irrigation schedules for different management conditions and the calculation of scheme water supply for varying cropping patterns.

The procedures for the calculation of crop water requirements and irrigation requirements are based on methodologies presented in I&D Nos. 56 and 33. The programme is meant as a practical tool to help carry out standard calculations for design and management of irrigation schemes, and for improving irrigation practices and the planning of irrigation schedules under varying water supply conditions. Water balance procedures also allow an assessment of effective rainfall and an evaluation of rainfed production through calculated yield decreases through water balance procedures.

To calculate water requirements and irrigation scheduling to avoid water stress, the knowledge of the p factor is essential.

CROPWAT can compute water requirements and irrigation schedules for deficit irrigation. For this, it uses the yield response factors derived from the crop-water production functions synthesized in FAO (1979) for 23 crops. This is described in the next section. For planning and managing deficit irrigation production, it is necessary to know the relationship between yield and water deficit of crops. This allows yield reduction predictions to be made and taken into account in making irrigation management decisions to optimise crop water productivity and return to investment. An example of the impact of water stress applied during flowering for cotton is given in the following figure (Smith, in prep., 2001).

In order to allow the ready calculation from a wide range of countries a climatic database CLIMWAT for CROPWAT has been developed compiling agrometerological data from over 3200 station in 144 countries and published in I&D No. 49. (FAO 1993).

Some Methodological Issues for Water Management in the Context of Water Use Efficiency and Crop Water Productivity

Precise knowledge on crop response to water is essential in a range of applications for policies and investment strategies at national and regional level, as well as in practical management tools at basin, scheme and farm level, as follows:

- To assess the impact of drought, rainfall variability and climatic change on yield, production and environment;

- to evaluate water use efficiency and crop water productivity under prevailing rain patterns and traditional farm practices and define with farmers options for improvement and appropriate strategies to optimise yields and to reduce risks of crop failure related to crop choice, planting time, soil cultivation and crop cultural practices (weeding, density, fertility,) and to define options for water conservation and supplemental irrigation;

- to define under irrigate crop conditions water supply strategies for optimal crop production and economic returns under conditions of reduced water supply and to advise farmers to optimise timing and application rate of crop irrigation for optimal yields and income also under limited water supply;

- to define national and regional policies, plans and strategies to meet food requirements under conditions of drought and limited water supply in rainfed and irrigated agriculture;

- to identify research programmes in crop improvement and natural resources management for improved water productivity in both rainfed and irrigated crop production, including identifying opportunities for biotechnology.

Thus, it would be appropriate to review the FAO methodologies in I&D Nos. 24, 33, 46 and 56 in the context of the above framework of water management applications and practices.

The overall aim of agricultural water management is to enable farm managers to achieve high levels of irrigation efficiencies water use efficiencies and crop productivities that will maximise return on investments in rainfed and irrigated conditions under adequate or deficit water supply. This requires accurate predictions of crop water requirements and crop response to water for determining irrigation requirements and planning and implementation of irrigation schedules to achieve desired objectives. In the case of rainfed condition, such predictions are required to assess impact of drought and rainfall variability on yield and production, to define options for improvement and appropriate strategies to optimise yields and to reduce risks of crop failure and to define options for water conservation and supplemental irrigation.

For irrigation scheduling that aim at meeting full crop water requirements, the empirical factor p is an important parameter in computing irrigation water requirements and in the planning and implementation of irrigation schedules.

Productivity Improving Technologies

Productivity improving technologies date back to antiquity, with rather slow progress until the late Middle Ages. Technological progress was aided by literacy and the diffusion of knowledge that accelerated after the spinning wheel spread to Western Europe in the 13th century. The spinning wheel increased the supply of rags used for pulp in paper making, whose technology reached Sicily sometime in the 12th century. Cheap paper was a factor in the development of the moveable type printing press, ca. 1440, which lead to a large increase in the number of books and titles published. Books on science and technology eventually began to appear, such as the mining technical manual De Re Metallica.

Later, near the beginning of the Industrial Revolution, came publication of the Encyclopédie, written by numerous contributors and edited by Denis Diderot and Jean le Rond d'Alembert (1751-72). It contained many articles on science and was the first general encyclopedia provide in depth coverage on the mechanical arts, but far more celebrated for it's presentation of thoughts of the Enlightenment.

An important mechanism for the transfer of technical knowledge were scientific societies, such as The Royal Society of London for Improving Natural Knowledge, better known as the Royal Society and technical colleges, such as the École Polytechnique.

Technological and economic progress did not proceed at a significant rate until the English Industrial Revolution in late 18th century and even then productivity grew about 0.5% annually. High productivity growth began during late 19th century in what is sometimes call the Second Industrial Revolution. Most major innovations of the Second Industrial Revolution were based on the modern scientific understanding of chemistry, electromagnetic theory and thermodynamics. Productivity gains were not just the result of inventions, but also of continuous improvements to those inventions which greatly increased output in relation both capital and labour compared to the original inventions.

Since the beginning of the Industrial Revolution, some of the major contributors to productivity have been as follows:

1. Replacing human and animal power with water and wind power, steam, electricity and internal combustion and greatly increasing the use of energy

2. Energy efficiency in the conversion of energy to: useful work, process heat or chemical energy in the manufacture of materials

3. Infrastructures: canals, railroads, highways and pipelines

4. Mechanization, both production machinery and agricultural machines

5. Work practices and processes: The American system of manufacturing, Taylorism or scientific management, mass production, assembly line, modern business enterprise

6. Materials handling: bulk materials, palletization and containerization

7. Scientific agriculture: fertilizers and the green revolution, livestock and poultry management

8. New materials, new process for their production and dematerialisation.

9. Communications: Telegraph, telephone, radio, satellites, fiber optic network and the Internet

10. Home economics: Public water supply, household gas, appliances

11. Automation and process control

12. Computers and software, data processing.

In recent decades there here have been a number of excellent books and papers published on the history of technology, the role of energy in economics and related issues such as resource depletion, some of which are referenced herein.

Details of Productivity Improving Technologies

A description of economic events and technologies that created the great productivity growth that began in the period from 1870-90 is given by David Ames Wells (1891). "The economic changes that have occurred during the last quarter of a century -or during the present generation of living men- have unquestionably been more important and more varied than during any period of the world's history". David Ames Wells, 1889

Replacing Human and Animal Power and Greatly Increasing Overall Power

Before the industrial revolution the only sources of power were water, wind and muscle. Most good water power sites (those not requiring massive modern dams) in Europe were developed during medieval period. In the 1750s John Smeaton, the "father of civil engineering," significantly improved the efficiency of the water wheel by applying scientific principles, thereby adding badly needed power for the Industrial Revolution.

In 1711 a Newcomen steam engine was installed for pumping water from a mine, a job that typically was done by large teams of horses, of which some mines used as many as 500. Fossil fuel energy first exceeded all animal and water power in 1870. The role energy and machines replacing physical work is discussed in Ayres-Warr (2004).

By about 1870 steam power first exceeded all water, wind and muscle power. While steamboats were used in some areas, as recently as the late 19th Century thousands of workers pulled barges. Until the late 19th century most coal and other minerals were mined with picks and shovels and crops were harvested and grain threshed using animal power or by hand. Heavy loads like 382 pound bales of cotton were handled on hand trucks until the early 20th century.

Excavation was done with shovels until the late 19th century when steam shovels came into use. It was reported that a labourer on the western division of the Erie Canal was expected to dig 5 cubic yards per day in 1860; however, by 1890 only 3-1/2 yards per day were expected. Today's large electric shovels have buckets that can hold 168 cubic metres and consume the power of a city of 100,000.

Dynamite, a safe to handle blend of nitroglycerin and diatomaceous earth was patented in 1867 by Alfred Nobel. Dynamite increased productivity of mining, tunneling, road building, construction and demolition and made projects such as the Panama Canal possible.

Steam power was applied to threshing machines in the late 19th century. There were steam engines that moved around on wheels under their own power that were used for supplying temporary power to stationary farm equipment such as threshing machines. These were called *road engines,* and Henry Ford seeing one as a boy was inspired to build an automobile. Steam tractors were used but never became popular.

With internal combustion came the first mass produced tractors (Fordson ca. 1917). Tractors replaced horses and mules for pulling reapers and combine harvesters, but in the 1930s self powered combines were developed. Output per man hour in growing wheat rose by a factor of about 10 from the end of World War II until about 1985, largely because of powered machinery, but also because of increased crop yields. Corn manpower showed a similar but higher productivity increase.

One of the greatest periods of productivity growth coincided with the electrification of factories which took place between 1900 and 1930 in the U.S.

Energy Efficiency and Productivity: The Useful Work Growth Theory

Energy efficiency has played a significant role in increasing productivity in the past; however, most industrial processes have exhausted the easy efficiency gains. The early Newcomen steam engine was about than 0.5% efficient and was improved to slightly over 1% by John Smeaton before Watt's improvements. Watt's improvements increased thermal efficiency to 2%, and today's steam turbines have efficiencies in the 40% range.

More efficient steam and internal combustion engines have higher power-to-weight ratios. The Newcomen and Watt engines operated near atmospheric pressure and used atmospheric pressure, or actually a vacuum caused by condensing steam, to do work. Higher pressure engines were light enough, and efficient enough to be used for powering ships and locomotives. Multiple expansion (multi-stage) engines were developed in the 1870s and were efficient enough for the first time to allow ships to carry more freight than coal, leading to great increases in international trade. The most efficient prime mover is the two stroke marine diesel engine developed in the 1920s, now ranging in size to over 100,000 horsepower with a thermal efficiency of 50%. Steam locomotives that used up to 20% of the U.S. coal production were replaced by diesel locomotives after World War II, saving a great deal of energy and reducing manpower for handling coal, boiler water and mechanical maintenance.

Improvements in steam engine efficiency caused a large increase in the number of steam engines and the amount of coal used, as noted by William Stanley Jevons in *The Coal Question*. This is called the Jevons paradox.

Electric lights were far more efficient than oil or gas lighting and did not generate heat, smoke and fumes. Electric light extended the work day, making factories, businesses and homes more productive. Electric light was not a great fire hazard like oil and gas light.

When anti-friction bearings were introduced in locomotives three female office workers demonstrated their efficiency by manually pulling the Timken 1111 locomotive. Industrial process have been continuously improved to reduce the energy consumption per unit of production.

The Ayres-Warr Model(2004) analysed the production function and explained part of the Solow residual by electrical generation efficiency.

Infrastructures

Sailing ships could transport goods for over a thousand miles for the cost of 30 miles by wagon. A horse that could pull a one ton wagon could pull a 30 ton barge. During the English or First Industrial Revolution, supplying coal to the furnaces at Manchester was difficult because there were few roads and because of the high cost of using wagons. However, canal barges were known to be workable, and this was demonstrated by building the Bridgewater Canal, which opened in 1761, bringing coal from Worsley to Manchester. The Bridgewater Canal's success started a frenzy of canal building that lasted until the appearance of railroads in the 1830s.

Railroads greatly reduced the cost of overland transportation. It is estimated that by 1890 the cost of wagon freight was U.S. 24.5 cents/ton-mile versus 0.875 cents/ton-mile by railroad.

Electric street railways (trams, trolleys or streetcars) were the final phase of railroad building from the late 1890s and first two decades of the 20th century. Street railways were soon displaced by motor buses and automobiles after WW I.

Highways with internal combustion powered vehicles completed the mechanization of overland transportation. When trucks appeared ca. 1920 the price transporting farm goods to market or to rail stations was greatly reduced. Motorized highway transport also reduced inventories.

Before iron and steel were in widespread use, wooden pipelines were used, such as those once supplying water to London from springs located away from the city. Iron and steel pipelines came into use during latter part of the 19th century, but only became a major infrastructure during the 20th century. Centrifugal pumps and centrifugal compressors are efficient means of pumping liquids and natural gas.

The relative energy required for transport of a tonne-km for various modes of transport are: pipelines=1(basis), water 2, rail 3, road 10, air 100.

Mechanization (General) and in Agriculture

The most important mechanical devices before the Industrial Revolution were water and wind mills. Water wheels date to Roman times and windmills somewhat later. Water and wind power were first used for grinding grain into flour, but were later adapted to power trip hammers for pounding rags into pulp for making paper and for crushing ore. Just before the Industrial revolution water power was applied to bellows for iron smelting. Wind and water power were also used in sawmills. The technology of building mills and mechanical clocks was important to the development of the machines of the Industrial Revolution.

The spinning wheel was a medieval invention that increased thread making productivity by a factor greater than ten. One of the early developments that preceded the Industrial Revolution was the stocking frame (loom) of ca. 1589. Later in the Industrial Revolution came the flying shuttle, a simple device that doubled the productivity of weaving. Spinning thread had been a limiting factor in cloth making requiring 10 spinners using the spinning wheel to supply one weaver. With the spinning jenny a spinner could spin eight threads at once. The water frame (Ptd. 1768) adapted water power to spinning, but it could only spin one thread at a time. The water frame was easy to operate and many could be located in a single building. The spinning mule (1779) allowed a large number of threads to be spun by a single machine using water power. A change in consumer preference for cotton at the time of increased cloth production resulted in the invention of the cotton gin (Ptd. 1794). Steam power eventually was used as a supplement to water during the Industrial Revolution, and both were used until electrification. A graph of productivity of spinning technologies can be found in Ayres (1989), along with much other data related this article.

With a cotton gin (1792) in one day a man could remove seed from as much upland cotton as would have previously taken a woman working two months to process at one pound per day. The sewing machine, invented and improved during the early 19th century and produced in large numbers by the 1870s, increased productivity by more than 500%.

Machine tools, which cut, grind and shape parts, were another important mechanical innovation of the Industrial Revolution. Perhaps the best early example of a productivity increase by machine tools and special purpose machines is the ca. 1803 Portsmouth Block Mills. With these machines 10 men could produce as many blocks as 110 skilled craftsmen. However, around 1900, it was the combination of small electric motors, speciality steels and new cutting and grinding materials that allowed machine tools to mass produce steel parts. Production of the Ford Model T required 32,000 machine tools.

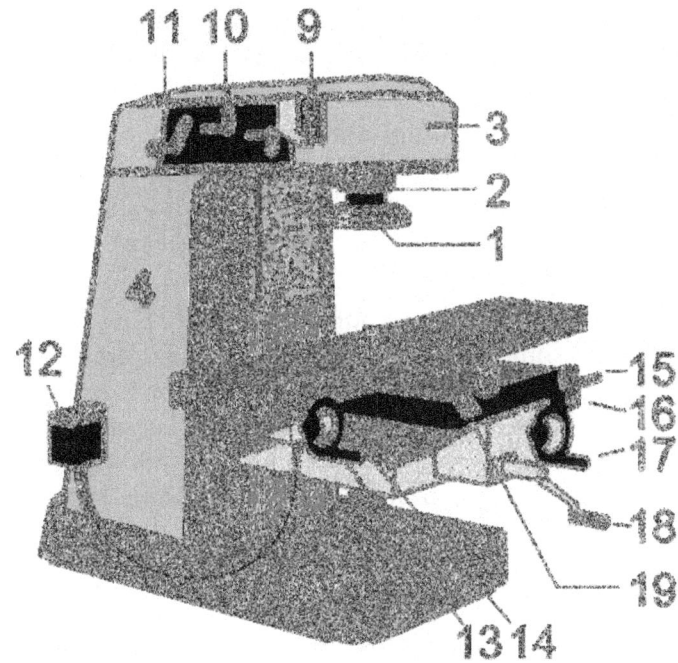

Figure: Vertical milling machine, an important machine tool. 1: milling cutter 2: spindle 3: top slide or overarm 4: column 5: table 6: Y-axis slide 7: knee 8: base

Modern manufacturing began around 1900 when machines, aided by electric, hydraulic and pneumatic power, began to replace hand methods in industry. An early example is the Owens'es automatic glass bottle blowing machine, which reduced labour in making bottles by over 80%.

Mechanical stokers for feeding coal to locomotives were in use in the 1920s. A completely mechanized and automated coal handling and stoking system was first used to feed pulverized coal to an electric utility boiler in 1921.

Coal seam undercutting machines appeared around 1890 and were used for 75% of coal production by 1934. Coal loading was still being done manually with shovels around 1930, but mechanical pick up and loading machines were coming into use. The use of the coal boring machine improved productivity of sub-surface coal mining by a factor of three between 1949 and 1969.

Jethro Tull's seed drill (ca. 1701) was a mechanical seed spacing and depth placing device that increased crop yields and saved seed, which was important when yields were measured in terms of seeds harvested per seed planted, which was typically between 3 and 6. The seed drill was an important factor in the British Agricultural Revolution.

Since the beginning of agriculture threshing was done by hand with a flail, requiring a great deal of labour. The threshing machine (ca. 1794) simplified the operation and allowed it to use animal power. Threshing machines displaced thousands of workers in Europe, many of who were driven to the brink of starvation.

Before ca. 1790 a worker could harvest 1/4 acre per day with a scythe. It was estimated that for each of Cyrus McCormick's horse pulled reapers (Ptd. 1834) freed up five men for military service in the U.S. Civil War. By 1890 two men and two horses could cut, rake and bind 20 acres of wheat per day. In the 1880s the reaper and threshing machine were combined into the combine harvester. These machines required large teams of horses or mules to pull. Over the entire 19th century the output per man hour for producing wheat rose by about 500% and for corn about 250%.

Farm machinery and higher crop yields reduced the labour to produce 100 bushels of corn from 35 to 40 hours in 1900 to 2 hours 45 minutes in 1999. The conversion of agricultural mechanization to internal combustion power began after 1915. The horse population began to decline in the 1920s after the conversion of agriculture and transportation to internal combustion. In addition to saving labour, this freed up much land previously used for supporting draft animals.

The peak years for tractor sales in the U.S. were the 1950s. There was a large surge in horsepower of farm machinery in the 1950s.

Work Practices and Processes

Changes to traditional work processes that were done after analysing the work and making it more systematic greatly increased the productivity of labour and capital.

This was the changeover from the European system of craftsmanship, where a crafstman made a whole item, to the American system of manufacturing which used special purpose machines and machine tools that made parts with precision so as to be interchangeable.

The process took decades to perfect at great expense because interchangeable parts were more costly at first. Interchangeable parts were achieved by using fixtures to hold and precisely align parts being machined, jigs to guide the machine tools and gauges to measure critical dimensions of finished parts.

Other work processes involved minimising the amount of steps in doing individual tasks, such as bricklaying, by performing time and motion studies to determine the one best method, the system becoming known as Taylorism after Fredrick Winslow Taylor who is the best known developer of this method, which is also known as *scientific management* after his work *The Principles of Scientific Management*. Electrification allowed the placement of machinery such as machine tools in a systematic arrangement along the flow of the work. The assembly line, which used motorized conveyors to transfer parts and assemblies to workers, was a key step leading to mass production.

Business administration, which includes management practices and accounting systems is another important form of work practices. Business administration as we know it arose from the mass production era.

Work processes are well described at the following links:

The American system of manufacturing, Taylorism or scientific management, mass production, assembly line, containerized freight.

Modern business enterprize (MBE) is the organisation and management of businesses, particularly large ones. MBE's employ professionals who use knowledge based techniques such areas as engineering, research and development, information technology, business administration, finance and accounting. MBE's typically benefit from economies of scale.

"Before railroad accounting we were moles burrowing in the dark." Andrew Carnegie

Materials Handling

Dry bulk materials handling systems use a variety of stationary equipment such as conveyors, stackers, reclaimers and mobile equipment such as loaders to handle high volumes of ores, coal, grains, sand, gravel, crushed stone, etc. Bulk materials are systems are used at mines, for loading and unloading ships and at factories that process bulk materials into finished goods, such as steel and paper mills.

Around 1900 various types of conveyors (belt, slat, bucket, screw or auger), overhead cranes and industrial trucks began being used for handling materials and goods in various stages of production in factories. A well known application of conveyors is Ford. Motor Co.'s assembly line (ca. 1913), although Ford used various industrial trucks, overhead cranes, slides and whatever devices necessary to minimise labour in handling parts in various stages of production.

Liquids and gases are handled with centrifugal pumps and compressors, respectively.

Conversion to powered material handling increased during WW 1 as shortages of unskilled labour developed and unskilled wages rose relative to skilled labour.

Handling goods on pallets was a significant improvement over using hand trucks or carrying sacks or boxes by hand and greatly speeded up loading and unloading of trucks, rail cars and ships. Pallets can be handled with pallet jacks or forklift trucks. Loading docks built to architectural standards allow trucks or rail cars to load and unload at the same elevation as the warehouse floor.

Containerization was used in both world wars, particularly WW II, but became commercial in the 1960s. Containerization left large numbers of warehouses at wharves in port cities vacant, freeing up land for other development.

Scientific Agriculture

Losses of agricultural products to spoilage, insects and rats contributed greatly to productivity. Much hay stored outdoors was lost to spoilage before indoor storage or some means of coverage became common. Pasteurisation of milk allowed it to be shipped by railroad. (It was noted that calves fed pasteurised milk were less likely to develop tuberculosis, and soon it was found that pasteurisation reduced the incidences of several other diseases in humans.) Keeping livestock indoors in winter reduces the amount of feed needed. Also, feeding chopped hay and ground grains, particularly corn (maize), was found to improve digestibility. The amount of feed required to produce a kg of live weight chicken fell from 5 in 1930 to 2 by the late 1990s and the time required fell from three months to six weeks.

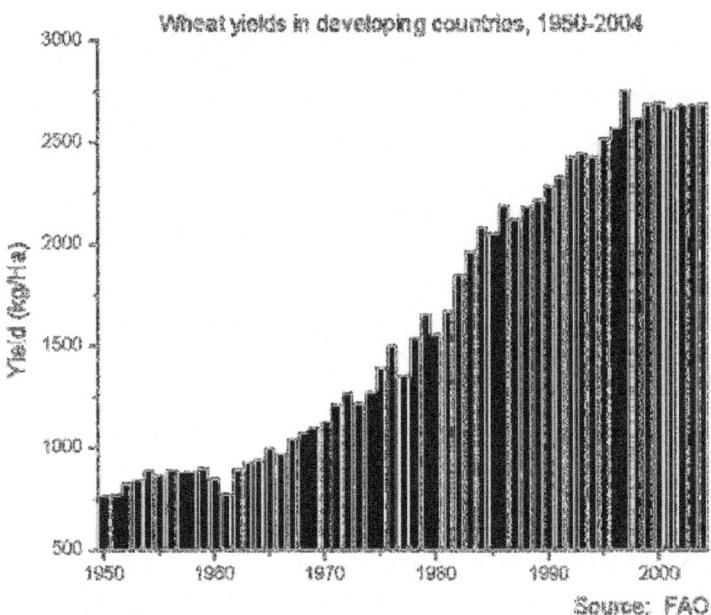

Figure: Wheat yields in developing countries, 1950 to 2004, kg/HA baseline 500

The Green Revolution increased crop yields by a factor of 3 for soybeans and between 4 and 5 for corn (maize), wheat, rice and some other crops. Using data for corn (maize) in the U.S., yields increased about 1.7 bushels per acre from the early 1940s until the first decade of the 21st century when concern was being expressed about reaching limits of photosynthesis. Because of the constant nature of the yield increase, the annual percentage increase has declined from over 5% in the 1940s to 1% today, so while yields for a while outpaced population growth, yield growth now lags population growth.

High yields would not be possible without significant applications of fertilizer particularly nitrogen fertilizer which was made affordable by the Haber-Bosch ammonia process. Nitrogen fertilizer is applied in many parts of Asia in amounts subject to diminishing returns, which however does still give a slight increase in yield. Crops in Africa are in general starved for NPK and much of the world's soils are deficient in zinc, which leads to deficiencies in humans.

The greatest period of agricultural productivity growth in the U.S. occurred from World War 2 until the 1970s.

Land is considered a form of capital, but otherwise has received little attention relative to its importance as a factor of productivity by modern economist, although it was important in classical economics. However, higher crop yields effectively multiplied the amount of land.

New Materials, Processes and De-materialisation

Production of steel and other metals was hampered by the difficulty in producing sufficiently high temperatures for melting. An understanding thermodynamic principles such as recapturing heat from flue gas by preheating combustion air resulted in higher energy efficiency and higher temperatures. Preheated combustion air was used in iron production and in the Siemens-Martin furnace. Today many industrial processes use preheated combustion air for fuel economy.

The Bessemer (Ptd.1855) and the Siemens-Martin (ca. 1865) processes greatly reduced the cost of steel. Steel has much higher strength than wrought iron and allowed long span bridges, high rise buildings, automobiles and other items. Steel also made superior threaded fasteners (screws, nuts, bolts), nails, wire and other hardware items. Steel rails lasted 17 times longer than wrought iron rails. The cheapness and superiority of steel to wrought iron led to cessation of practically all iron production by WW II.

Today a variety of alloy steels are available that have superior properties for special applications like automobiles, pipelines and drill bits. High speed or tool steels, whose development began in the late 19th century, allowed machine tools to cut steel at much higher speeds. High speed steel and even harder materials were an essential component of mass production of automobiles.

Some of the most important speciality materials are steam turbine and gas turbine blades, which have to withstand extreme mechanical stress and high temperatures.

The size of blast furnaces grew greatly over the 20th century and innovations like additional heat recovery and pulverized coal, which displaced coke and increased energy efficiency.

By the end of the 19th century the Bessemer process was displaced by the open hearth furnace (OHF). After World War II the OHF was displaced by the basic oxygen furnace (BOF), which used oxygen instead of air and required about 35–40 minutes to produce a batch of steel compared to 8 to 9 hours for the OHF. The BOF also was more energy efficient.

By 1913, 80% of steel was being made from molten pig iron directly from the blast furnace, eliminating the step of casting the "pigs" (ingots) and remelting.

After 1950 continuous casting contributed to productivity of converting steel to structural shapes by eliminating the intermittent step of making slabs, billets (square cross-section)) or blooms (rectangular) which then usually have to be reheated before rolling into shapes.

As a result of these innovations, between 1920 and 2000 labour requirements in the steel industry decreased by a factor of 1,000, from more than 3 worker-hours per tonne to just 0.003.

Paper was made one sheet at a time by hand until development of the Fourdrinier paper machine (ca. 1801) which made a continuous sheet. Paper making was severely limited by the supply of cotton and linen rags from the time of the invention of the printing press until the development of wood pulp (ca. 1840s). The sulfite process for making wood pulp was developed in the 1860s and 1870s. Paper made from sulfite pulp had superior strength properties than the previously used ground wood pulp (ca. 1840). The kraft (Swedish for *strong*) pulping process was commercialised in the 1930s. Pulping chemicals are recovered and internally recycled in the kraft process, also saving energy and reducing pollution. Kraft paperboard is the material that the outer layers of corrugated boxes are made of. Until Kraft corrugated boxes were available, and even for some decades after, packaging consisted largely of wooden crates and boxes. Corrugated boxes required much less labour to manufacture and offered good protection to their contents.

Plastics can be inexpensively made into everyday items and have significantly lowered the cost of a variety of goods including packaging, containers, parts and household piping.

Seismic exploration, beginning in the 1920s, uses reflected sound waves to map subsurface geology to help locate potential oil reservoirs. This was a great improvement over previous methods, which involved mostly luck and good knowledge of geology, although luck continued to be important in several major discoveries. Rotary drilling was a faster and more efficient way of drilling oil and water wells. It became popular after being used for the initial discovery of the East Texas field in 1930.

Dematerialisation is the reduction of use of materials in manufacturing, construction, packaging or other uses. It is made possible by substitution with better materials and by engineering to reduce weight while maintaining function. Modern examples are plastic beverage containers replacing glass and paperboard, plastic shrink wrap used in shipping and light weight plastic packing materials. Dematerialisation has been occurring in the U. S. steel industry where the peak in consumption occurred in 1973 on both an absolute and per capita basis. Optical fiber began to replace copper wire in the telephone network during the 1980s.

Communications

The telegraph appeared around the beginning of the railroad era and railroads typically installed telegraph lines along their routes for communicating with the trains.

Teleprinters appeared in 1910 and had replaced between 80 and 90% of Morse code operators by 1929. It is estimated that one teletypist replaced 15 Morse code operators.

The early use of telephones was primarily for business. Monthly service cost about one third of the average worker's earnings. The telephone along with trucks and the new road networks allowed businesses to reduce inventory sharply during the 1920s.

Telephone calls were handled by operators using switchboards until the 1920s when the automatic (dial) telephone and automatic switchboard came into use, and by 1929, 31.9% of the Bell system was automatic.

After WWII microwave transmission began being used for long distance telephony and television.

The diffusion of telephony to households was mature by the arrival of fiber optic communications in the late 1970s. Fiber optics greatly increased the transmission capacity of information over previous copper wires and further lowered the cost of long distance communication.

Communications satellitess came into use in the 1960s and today carry a variety of information including credit card transaction data, radio, television and telephone calls. The Global Positioning System (GPS) operates on signals from satellites.

Fax (short for facsimile) machines of various types had been in existence since the early 1900s but became widespread beginning in the mid 1970s.

Home Economics: Public Water Supply Household Gas Supply and Appliances

Before public water was supplied to households it was necessary for someone to haul up to 10,000 gallons of water to the average household.

Gas utilities first supplied synthetic gas, mainly for lighting. In the late 19th century natural gas began being supplied to households. This saved many hours of feeding wood fires for heating and cooking.

Household appliances followed household electrification in the 1920s', with consumers buying electric ranges, toasters, refrigerators and washing machines. As a result of appliances and convenience foods, time spent on meal preparation and clean up, laundry and cleaning decreased from 58 hours/week in 1900 to 18 hours/week by 1975. Less time spent on housework allowed more women to enter the labour force.

Automation

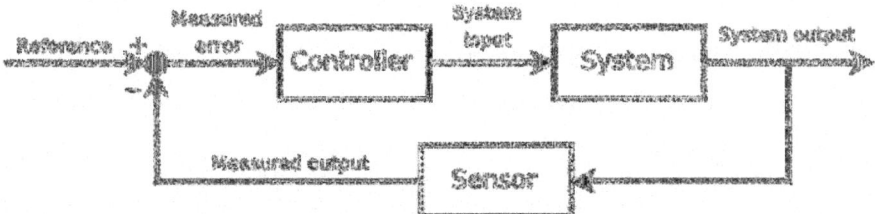

Figure: The concept of the feedback loop to control the dynamic behaviour of the system: this is negative feedback, because the sensed value is subtracted from the desired value to create the error signal, which is processed by the controller, which provides proper corrective action. A typical example would be to control the opening of a valve to hold a liquid level in a tank. Process control is a widely used form of automation.

Automation in the original sense means automatic control, meaning a process is run with minimum operator intervention. An simple analogy is cruise control on a car, which applies continuous correction when a sensor on the controlled variable (Speed in this example) deviates from a set-point and can respond in a corrective manner to hold the setting. Process control is the usual form of automation that allows industrial operations like oil refineries, steam plants generating electricity or paper mills to be run with a minimum of manpower, usually from a number of control rooms.

The earliest applications of process control were mechanisms that adjusted the gap between mill stones for grinding grain and for keeping windmills facing into the wind. The centrifugal governor used for adjusting the mill stones was copied by James Watt for controlling speed of steam engines; however, it took much development work to achieve the degree of steadiness necessary to operate textile machinery. Mathematical analysis of control theory was first developed by James Clerk Maxwell.

Automation of the telephone system allowed dialing local numbers instead of having calls placed through an operator. Further automation allowed callers to place long distance calls by direct dial. Eventually almost all operators were replaced with automation.

Machine tools were automated with Numerical control (NC) in the 1950s. This soon evolved into computerized numerical control (CNC).

Industrial robots were used on a limited scale from the 1960s but began their rapid growth phase in the mid 1980s after the widespread availability of microprocessors used for their control. The diffusion curve of robots went through the build out phase over the next decade with the saturation approach inflection point in the early 1990s. By 2000 there were over 700,000 robots world-wide.

The ultimate objective of automation is autonomous machines, that is, machines that run themselves, without operator attention. While this has been achieved to some extent in some industries, in many industries it is necessary to have operators because of the large amount of defective product than can be produced in a short time when things go wrong. Also, operators are necessary for safety and protection of valuable equipment.

Computers, Semiconductors, Data Processing and Information Technology

Early electric data processing was done by running punched cards through tabulating machines, the holes in the cards allowing electrical contact to increment electronic counters. Tabulating machines were in a category called unit record equipment, through which the flow of punched cards was arranged in a program-like sequence to allow sophisticated data processing. They were widely used before the introduction of computers.

The usefulness of tabulating machines was demonstrated by compiling the 1890 U.S. census, allowing the census to be processed in less than a year and with great labour savings compared to the estimated 13 years by the previous manual method.

The first digital computers were more productive than tabulating machines, but not by a great amount. Early computers used thousands of vacuum tubes (thermionic valves) which used a lot of electricity and constantly needed replacing. By the 1950s the vacuum tubes were replaced by transistors which were much more reliable and used relatively little electricity. By the 1960s thousands of transistors and other electronic components could were being manufactured on silicon semiconductor wafers as integrated circuits, which are universally used in today's computers.

Computers used paper tape and punched cards for data and programming input until the 1980s when it was still common to receive monthly utility bills printed on a punched card that was returned with the customer's payment.

In 1973 IBM introduced point of sale (POS) terminals in which electronic cash registers were networked to the store mainframe computer. By the 1980s bar code readers were added. These technologies automated inventory management. Wal-Mart was an early adopter of POS.

Data storage became better organised after the development of relational database software that allowed data to be stored in different tables. For example, a theoretical airline may have numerous tables such as: airplanes, employees, maintenance contractors, caterers, flights, airports, payments, tickets, etc. each containing a narrower set of more specific information than would a flat file, such as a spreadsheet. These tables are related by common data fields called *keys*. Data can be retrieved in various specific configurations by posing a *query* without having to pull up a whole table. This, for example, makes it easy to find a passenger's seat assignment by a variety of means such as ticket number or name.

Since the mid 1990s, interactive web pages have allowed users to access various servers over Internet to engage in e-commerce such as online shopping, paying bills, trading stocks, managing bank accounts and renewing auto registrations. This is the ultimate form of back office automation because the transaction information is transferred directly to the database.

Computers also greatly increased productivity of the communications sector, especially in areas like the elimination of telephone operators. In engineering, computers replaced manual drafting with CAD, with a 500% average increase in a draftsman's output. Software was developed for calculations used in designing electronic circuits, stress analysis, heat and material balances. Process simulation software has been developed for both steady state and dynamic simulation, the latter able to give the user a very similar experience to operating a real process like a refinery or paper mill, allowing the user to optimise the process or experiment with process modifications.

Automated teller machines (ATM's) became popular in recent decades and self checkout at retailers appeared in the 1990s.

The Airline Reservations System and banking are areas where computers are practically essential. Modern military systems also rely on computers.

Computers did not revolutionize manufacturing because automation, in the form of control systems, had already been in existence for decades, although they did allow more sophisticated control, which led to improved product quality and process optimisation.

Secular Decline in Productivity Growth

"The years 1929-1941 were, in the aggregate, the most technologically progressive of any comparable period in U.S. economic history." Alexander J. Field

U.S. productivity growth has been in long term decline since the early 1970s. Part of the early decline was attributed to increased governmental regulation since the 1960s, including stricter environmental regulations. However, most of the decline in productivity growth is due to exhaustion of opportunities. Robert J. Gordon considered productivity to be "One big wave" that crested and is now receding to a lower level, while M. King Hubbert called the phenomenon of the great productivity gains preceding the Great Depression a "one time event."

Because of reduced population growth in the U.S. and a peaking of productivity growth, U.S. GDP growth has never returned to the 4% plus rates of the pre-World War 1 decades.

The computer and computer like semiconductor devices used in automation are the most significant productivity improving technologies developed in the final decades of the twentieth century; however, their contribution was disappointing. Economist Robert J. Gordon is among those who questioned whether computers lived up to the great innovations of the past, such as electrification. This issue is known as the Productivity paradox. Gordon's analysis of productivity in the U.S. gives two possible high points, one between World War

1 and World War 2 and the other between the 1920s and the early post World War 2 decades, depending on how government capital is treated.

Whereas lack of knowledge of scientific principles and efficient work methods was the norm before the mid-19th century, today we have trained professionals in civil, structural, mechanical, chemical, electrical, industrial and other fields of engineering, computer science, information technology, medicine and medical technology and management and business. Opportunities to improve productivity are no longer overlooked and incremental improvements are made wherever possible, but rarely do they create dramatic savings that can be widely applied throughout the economy.

Typically productivity gains are highest in the early years of a new technology or product. The development of the steam engine is rather unique because there was no knowledge of thermodynamics until after Watt's improvements, so it took over 50 years from the time of the Newcomen engine (1712) until Watt's condenser and other improvements increased efficiency by 400% ca. 1765. The study of the steam engine and the simultaneous development of thermodynamics led to continued improvements, at a decelerating rate, until efficiency approached theoretical limits in the 1960s with high pressure steam turbines.

Another example of productivity increases with a new process is a new, mechanized factory producing light bulbs that started operating in 1925. After six years of operation output per worker hour increased fivefold.

The early automobile industry struggled with producing enough automobiles to achieve economies of scale that were thought to be necessary to bring costs down so as to be affordable. Ford Motor Co. solved the problem with a totally new manufacturing concept which became known as mass production. The amount of labour, and consequently the price of the Ford Model T did fall dramatically after the development of the assembly line in 1914, and further with the factory designed for mass production, but after those new processes productivity gains were much slower.

The recent example of high productivity in a new industry occurred in the computer and related industries in the late 1990s, during which time computer related industries were responsible for most of the overall productivity growth.

Diminishing Marginal Returns on Technology

The 99.9% reduction in labour required to produce steel (Item 8 above) between 1920 and 2000 is an example of the exhaustion of savings opportunities. Exhaustion or saturation limits can be illustrated by the logistic function, discussed in the International Institute for Applied Systems Analysis (IIASA) work of Cesare Marchetti, and by Carlota Prez and others. Most basic materials, agricultural commodities, automobiles and appliances are like the steel example in that by far the greatest amount of labour has already been saved so that removing the remaining labour would result infinitely high output per hour in terms of physical product but would only slightly lower cost. The largest productivity gains in absolute and relative terms typically occurred soon after the introduction of a new technology or product. Examples include the assembly line, which came a decade and a half after the manufacturing of automobiles and in the manufacture of electric light bulbs. A modern example is the performance of semiconductors, and in fact, most of the productivity gains of the last decades were in in semiconductor, computer and Internet related industries.

Robert Ayres, Benjamin Warr and Vaclav Smil have all written that the processes for making many basic materials such as steel, aluminum and various chemicals and electricity generation have reduced energy consumption to where it is approaching theoretical minimums. Resource depletion decreases productivity as more effort in the form of labour, materials and energy are required for extraction and processing. For example, early U.S. onshore oil production yield has shown a consistent decline in the number of barrels of oil produced per foot drilled. Ore grades of copper and other important minerals have significantly declined in concentration, requiring much higher volumes of low grade ore to be handled and processed.

Based on the exhaustion of opportunities and resource depletion, Ayres-Warr (2009) are forecasting that economic growth in developed countries will end sometime after 2030. Economic theory that deals with historical long term business cycles and their relationship to technology refers to these cycles as Kondratiev waves. Resource depletion is in the field of ecological economics.

Improvement in Living Standards

Chronic hunger and malnutrition were the norm for the majority of the population of the world including England and France, until the latter part of the 19th century. Until about 1750, in large part due to malnutrition, life expectancy in France was about 35 years, and only slightly higher in England. The U.S. population of the time was adequately fed, were much taller and had life expectancies of 45–50 years. A vivid description of living standards of the mill workers in England in 1844 was given by Fredrick Engels.

The gains in standards of living have been accomplished largely through increases in productivity. In the U.S. the amount of personal consumption that could be bought with one hour of work was about $3.00 in 1900 and increased to about $22 by 1990, measured in 2010 dollars. For comparison, a U. S. worker today earns more (in terms of buying power) working for ten minutes than subsistence workers, such as the English mill workers that Fredrick Engels wrote about in 1844, earned in a 12 hour day.

As a result of productivity the work week declined considerably over the 19th century. By the 1920s the average work week was 49 hours, but the work week was reduced to 40 hours (after which overtime premium was applied) as part of the National Industrial Recovery Act of 1933. At the time of the Great Depression of the 1930s it was understood that with the enormous productivity gains due to electrification, mass production and agricultural mechanization, there was no need for a large number of previously employed workers. M. King Hubbert, the namesake of peak the oil curve, advocated a four work day in his prescient paper *Man Hours and Distribution*.

Early Productivity Data

Data on productivity is not reliable before the 20th century. Most data from before the 20th century comes from more recent attempts at reconstruction, which is the speciality of new economic history.

One of the earlier sources of 20th century productivity data is the 1940 study by the Brookings Institution which gives productivity by major U.S. industries from 1919 to 1939.

John W. Kendrick of the National Bureau of Economic Research published data series on output, labour, inputs and capital for major industry divisions over the period between 1870 to 1953.

Chapter 5

Crop Evapotranspiration

Evaporation is the process whereby liquid water is converted to water vapour (vapourisation) and removed from the evaporating surface (vapour removal). Water evaporates from a variety of surfaces, such as lakes, rivers, pavements, soils and wet vegetation.

Energy is required to change the state of the molecules of water from liquid to vapour. Direct solar radiation and, to a lesser extent, the ambient temperature of the air provide this energy. The driving force to remove water vapour from the evaporating surface is the difference between the water vapour pressure at the evaporating surface and that of the surrounding atmosphere. As evaporation proceeds, the surrounding air becomes gradually saturated and the process will slow down and might stop if the wet air is not transferred to the atmosphere. The replacement of the saturated air with drier air depends greatly on wind speed. Hence, solar radiation, air temperature, air humidity and wind speed are climatological parameters to consider when assessing the evaporation process.

Where the evaporating surface is the soil surface, the degree of shading of the crop canopy and the amount of water available at the evaporating surface are other factors that affect the evaporation process. Frequent rains, irrigation and water transported upwards in a soil from a shallow water table wet the soil surface. Where the soil is able to supply water fast enough to satisfy the evaporation demand, the evaporation from the soil is determined only by the meteorological conditions. However, where the interval between rains and irrigation becomes large and the ability of the soil to conduct moisture to pear the surface is small, the water content in the topsoil drops and the soil surface dries out. Under these circumstances the limited availability of water exerts a controlling influence on soil evaporation. In the absence of any supply of water to the soil surface, evaporation decreases rapidly and may cease almost completely within a few days.

Atmosphere

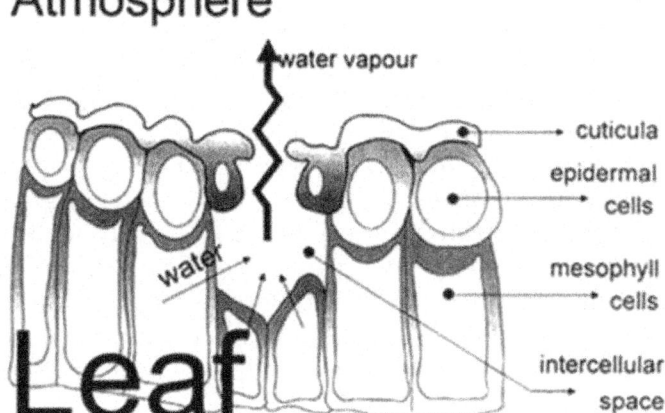

Figure: Schematic representation of a stoma

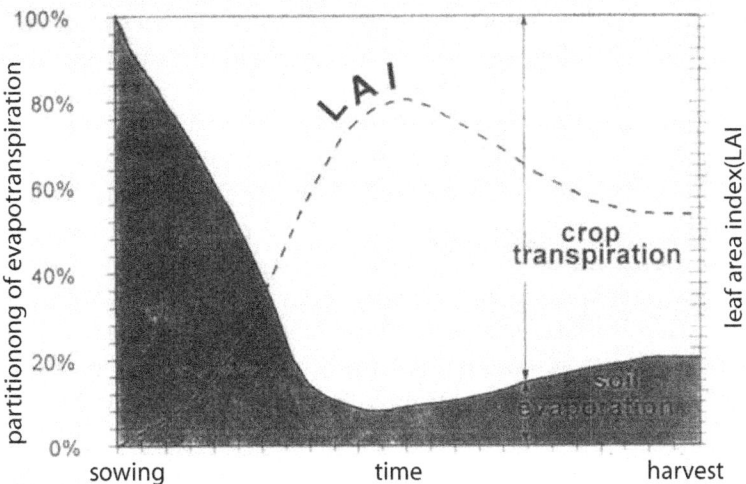

Figure: The partitioning of evapotranspiration into evaporation and transpiration over the growing period for an annual field crop

Transpiration

Transpiration consists of the vapourisation of liquid water contained in plant tissues and the vapour removal to the atmosphere. Crops predominately lose their water through stomata. These are small openings on the plant leaf through which gases and water vapour pass. The water, together with some nutrients, is taken up by the roots and transported through the plant. The vapourisation occurs within the leaf, namely in the intercellular spaces, and the vapour exchange with the atmosphere is controlled by the stomatal aperture. Nearly all water taken up is lost by transpiration and only a tiny fraction is used within the plant. Transpiration, like direct evaporation, depends on the energy supply, vapour pressure gradient and wind. Hence, radiation, air temperature, air humidity and wind terms should be considered when assessing transpiration. The soil water content and the ability of the soil to conduct water to the roots also determine the transpiration rate, as do waterlogging and soil water salinity. The transpiration rate is also influenced by crop characteristics, environmental aspects and cultivation practices. Different kinds of plants may have different transpiration rates. Not only the type of crop, but also the crop development, environment and management should be considered when assessing transpiration.

Evapotranspiration (ET)

Evaporation and transpiration occur simultaneously and there is no easy way of distinguishing between the two processes. Apart from the water availability in the topsoil, the evaporation from a cropped soil is mainly determined by the fraction of the solar radiation reaching the soil surface. This fraction decreases over the growing period as the crop develops and the crop canopy shades more and more of the ground area. When the crop is small, water is predominately lost by soil evaporation, but once the crop is well developed and completely covers the soil, transpiration becomes the main process. In the given figure the partitioning of evapotranspiration into evaporation and transpiration is plotted in correspondence to leaf area per unit surface of soil below it. At sowing nearly 100% of ET comes from evaporation, while at full crop cover more than 90% of ET comes from transpiration.

Units

The evapotranspiration rate is normally expressed in millimetres (mm) per unit time. The rate expresses the amount of water lost from a cropped surface in units of water depth. The time unit can be an hour, day, decade, month or even an entire growing period or year. As one hectare has a surface of 10000 m^2 and 1 mm is equal to 0.001 m, a loss of 1 mm of water corresponds to a loss of 10 m^3 of water per hectare. In other words, 1 mm day^{-1} is equivalent to 10 m^3 ha^{-1} day^{-1}. Water depths can also be expressed in terms of energy received per unit area. The energy refers to the energy or heat required to vapourise free water. This energy, known as the latent heat of vapourisation (l), is a function of the water temperature. For example, at 20°C, l is about 2.45 MJ kg^{-1}. In other words, 2.45 MJ are needed to vapourise 1 kg or 0.001 m^3 of water. Hence, an energy input of 2.45

MJ per m^2 is able to vapourise 0.001 m or 1 mm of water, and therefore 1 mm of water is equivalent to 2.45 MJ m^{-2}. The evapotranspiration rate expressed in units of MJ m^{-2} day^{-1} is represented by 1 ET, the latent heat flux.

Table below summarises the units used to express the evapotranspiration rate and the conversion factors.

Table 1. Conversion factors for evapotranspiration

	Depth	*volume per unit area*	*energy per unit area* *	
	mm day^{-1}	*m^3 ha^{-1} day^{-1}*	*l s^{-1} ha^{-1}*	*MJ m^{-2} day^{-1}*
1 mm day^{-1}	1	10	0.116	2.45
1 m^3 ha^{-1} day^{-1}	0.1	1	0.012	0.245
1 l s^{-1} ha^{-1}	8.640	86.40	1	21.17
1 MJ m^{-2} day^{-1}	0.408	4.082	0.047	1

* For water with a density of 1000 kg m^{-3} and at 20°C.

Example 1. Converting evaporation from one unit to another

On a summer day, net solar energy received at a lake reaches 15 MJ per square metre per day. If 80% of the energy is used to vapourise water, how large could the depth of evaporation be?

From Table 1:	1 MJ m^{-2} day^{-1} =	0.408 mm day^{-1}
Therefore:	0.8 x 15 MJ m^{-2} day^{-1} = 0.8 x 15 x	
	0.408 mm d^{-1} =	4.9 mm day^{-1}

The evaporation rate could be 4.9 mm/day

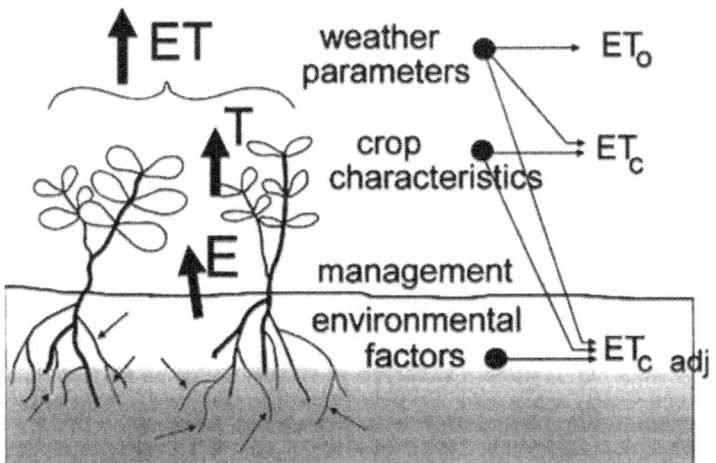

Figure: Factors affecting evapotranspiration with reference to related ET concepts

Factors Affecting Evapotranspiration

Weather parameters, crop characteristics, management and environmental aspects are factors affecting evaporation and transpiration.

Weather Parameters

The principal weather parameters affecting evapotranspiration are radiation, air temperature, humidity and wind speed. Several procedures have been developed to assess the evaporation rate from these parameters. The evaporation power of the atmosphere is expressed by the reference crop evapotranspiration (ET_o). The reference crop evapotranspiration represents the evapotranspiration from a standardized vegetated surface.

Crop Factors

The crop type, variety and development stage should be considered when assessing the evapotranspiration from crops grown in large, well-managed fields. Differences in resistance to transpiration, crop height, crop roughness, reflection, ground cover and crop rooting characteristics result in different ET levels in different types of crops under identical environmental conditions. Crop evapotranspiration under standard conditions (ET_c) refers to the evaporating demand from crops that are grown in large fields under optimum soil water, excellent management and environmental conditions, and achieve full production under the given climatic conditions.

Management and Environmental Conditions

Factors such as soil salinity, poor land fertility, limited application of fertilizers, the presence of hard or impenetrable soil horizons, the absence of control of diseases and pests and poor soil management may limit the crop development and reduce the evapotranspiration. Other factors to be considered when assessing ET are ground cover, plant density and the soil water content. The effect of soil water content on ET is conditioned primarily by the magnitude of the water deficit and the type of soil. On the other hand, too much water will result in waterlogging which might damage the root and limit root water uptake by inhibiting respiration.

When assessing the ET rate, additional consideration should be given to the range of management practices that act on the climatic and crop factors affecting the ET process. Cultivation practices and the type of irrigation method can alter the microclimate, affect the crop characteristics or affect the wetting of the soil and crop surface. A windbreak reduces wind velocities and decreases the ET rate of the field directly beyond the barrier.

The effect can be significant especially in windy, warm and dry conditions although evapotranspiration from the trees themselves may offset any reduction in the field. Soil evaporation in a young orchard, where trees are widely spaced, can be reduced by using a well-designed drip or trickle irrigation system.

The drippers apply water directly to the soil near trees, thereby leaving the major part of the soil surface dry, and limiting the evaporation losses. The use of mulches, especially when the crop is small, is another way of substantially reducing soil evaporation. Anti-transpirants, such as stomata-closing, film-forming or reflecting material, reduce the water losses from the crop and hence the transpiration rate.

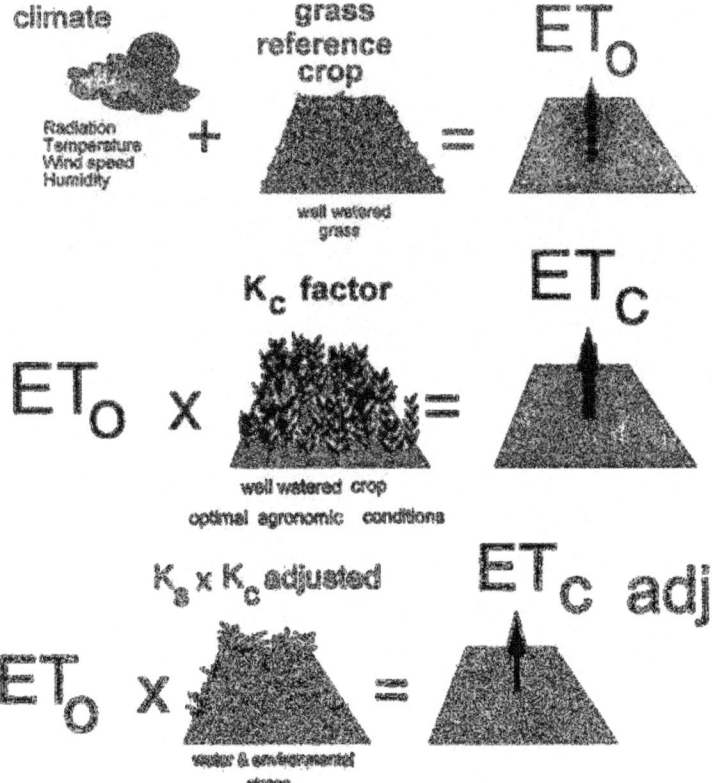

Figure: Reference (ET_o), crop evapotranspiration under standard (ET_c) and non-standard conditions ($ET_{c\ adj}$)

Where field conditions differ from the standard conditions, correction factors are required to adjust ET_c. The adjustment reflects the effect on crop evapotranspiration of the environmental and management conditions in the field.

Evapotranspiration Concepts

Distinctions are made between reference crop evapotranspiration (ET_o), crop evapotranspiration under standard conditions (ET_c) and crop evapotranspiration under non-standard conditions ($ET_{c\,adj}$). ET_o is a climatic parameter expressing the evaporation power of the atmosphere. ET_c refers to the evapotranspiration from excellently managed, large, well-watered fields that achieve full production under the given climatic conditions. Due to sub-optimal crop management and environmental constraints that affect crop growth and limit evapotranspiration, ET_c under non-standard conditions generally requires a correction.

Reference Crop Evapotranspiration (ET$_o$)

The evapotranspiration rate from a reference surface, not short of water, is called the reference crop evapotranspiration or reference evapotranspiration and is denoted as ET_o. The reference surface is a hypothetical grass reference crop with specific characteristics. The use of other denominations such as potential ET is strongly discouraged due to ambiguities in their definitions.

The concept of the reference evapotranspiration was introduced to study the evaporative demand of the atmosphere independently of crop type, crop development and management practices. As water is abundantly available at the reference evapotranspiring surface, soil factors do not affect ET. Relating ET to a specific surface provides a reference to which ET from other surfaces can be related. It obviates the need to define a separate ET level for each crop and stage of growth. ET_o values measured or calculated at different locations or in different seasons are comparable as they refer to the ET from the same reference surface.

The only factors affecting ET_o are climatic parameters. Consequently, ET_o is a climatic parameter and can be computed from weather data. ET_o expresses the evaporating power of the atmosphere at a specific location and time of the year and does not consider the crop characteristics and soil factors. The FAO Penman-Monteith method is recommended as the sole method for determining ET_o. The method has been selected because it closely approximates grass ET_o

at the location evaluated, is physically based, and explicitly incorporates both physiological and aerodynamic parameters. Moreover, procedures have been developed for estimating missing climatic parameters.

Typical ranges for ET_0 values for different agroclimatic regions are given in Table. These values are intended to familiarize inexperienced users with typical ranges, and are not intended for direct application. The calculation of the reference crop evapotranspiration is discussed in Part A of this handbook.

Crop Evapotranspiration under Standard Conditions (ET_c)

The crop evapotranspiration under standard conditions, denoted as ET_c, is the evapotranspiration from disease-free, well-fertilized crops, grown in large fields, under optimum soil water conditions, and achieving full production under the given climatic conditions.

Table. Average ET_0 for different agroclimatic regions in mm/day

Regions	Mean Daily Temperature (°C)		
	Cool ~10°C	Moderate 20°C	Warm > 30°C
Tropics and subtropics			
- humid and sub-humid	2 - 3	3 - 5	5 - 7
-arid and semi-arid	2 - 4	4 - 6	6 - 8
Temperate region			
- humid and sub-humid	1 - 2	2 - 4	4 - 7
-arid and semi-arid	1 - 3	4 - 7	6 - 9

The amount of water required to compensate the evapotranspiration loss from the cropped field is defined as crop water requirement. Although the values for crop evapotranspiration and crop water requirement are identical, crop water requirement refers to the amount of water that needs to be supplied, while crop evapotranspiration refers to the amount of water that is lost through evapotranspiration. The irrigation water requirement basically represents the difference between the crop water requirement and effective precipitation. The irrigation water requirement also includes additional water for leaching of salts and to compensate for non-uniformity of water application. Calculation of the irrigation water requirement is not covered in this publication, but will be the topic of a future Irrigation and Drainage Paper. Crop evapotranspiration can be calculated from climatic data and by integrating directly the crop resistance, albedo and air resistance factors in the Penman-Monteith approach. As there is still a considerable lack of information for different crops, the Penman-Monteith method is used for the estimation of the standard reference crop to determine its evapotranspiration rate, i.e., ET_o. Experimentally determined ratios of ET_c/ET_o, called crop coefficients (K_c), are used to relate ET_c to ET_o or $ET_c = K_c\ ET_o$.

Differences in leaf anatomy, stomatal characteristics, aerodynamic properties and even albedo cause the crop evapotranspiration to differ from the reference crop evapotranspiration under the same climatic conditions. Due to variations in the crop characteristics throughout its growing season, K_c for a given crop changes from sowing till harvest. The calculation of crop evapotranspiration under standard conditions (ET_c) is discussed in Part B of this handbook.

Crop Evapotranspiration under Non-standard Conditions ($ET_{c\ adj}$)

The crop evapotranspiration under non-standard conditions ($ET_{c\ adj}$) is the evapotranspiration from crops grown under management and environmental conditions that differ from the standard conditions. When cultivating crops in fields, the real crop evapotranspiration may deviate from ET_c due to non-optimal conditions such as the presence of pests and diseases, soil salinity, low soil fertility, water shortage or waterlogging. This may result in scanty plant growth, low plant density and may reduce the evapotranspiration rate below ET_c.

The crop evapotranspiration under non-standard conditions is calculated by using a water stress coefficient K_s and/or by adjusting K_c for all kinds of other stresses and environmental constraints on crop evapotranspiration. The adjustment to ET_c for water stress, management and environmental constraints is discussed in Part C of this handbook.

Determining Evapotranspiration

ET Measurement

Evapotranspiration is not easy to measure. Specific devices and accurate measurements of various physical parameters or the soil water balance in lysimeters are required to determine evapotranspiration. The methods are often expensive, demanding in terms of accuracy of measurement and can only be fully exploited by well-trained research personnel. Although the methods are inappropriate for routine measurements, they remain important for the evaluation of ET estimates obtained by more indirect methods.

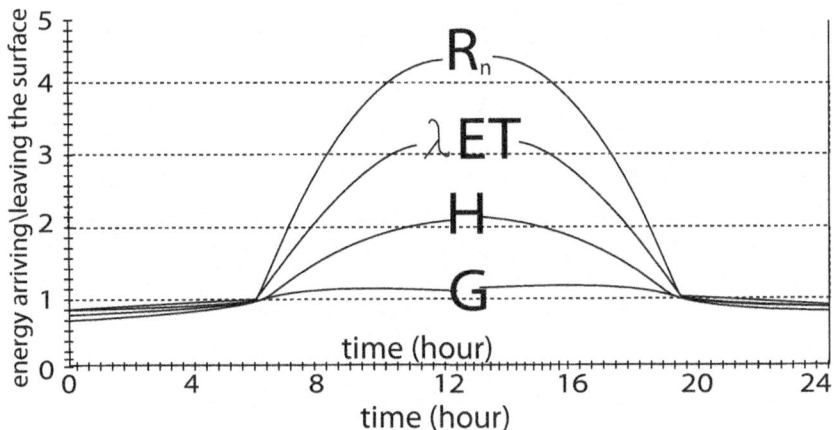

Figure: Schematic presentation of the diurnal variation of the components of the energy balance above a well-watered transpiring surface on a cloudless day

Energy Balance and Microclimatological Methods

Evaporation of water requires relatively large amounts of energy, either in the form of sensible heat or radiant energy. Therefore the evapotranspiration process is governed by energy exchange at the vegetation surface and is limited by the amount of energy available. Because of this limitation, it is possible to predict the evapotranspiration rate by applying the principle of energy conservation. The energy arriving at the surface must equal the energy leaving the surface for the same time period.

All fluxes of energy should be considered when deriving an energy balance equation. The equation for an evaporating surface can be written as:

$$R_n - G - l\,ET - H = 0 \qquad\qquad (1)$$

where R_n is the net radiation, H the sensible heat, G the soil heat flux and l ET the latent heat flux. The various terms can be either positive or negative. Positive R_n supplies energy to the surface and positive G, l ET and H remove energy from the surface.

In Equation 1 only vertical fluxes are considered and the net rate at which energy is being transferred horizontally, by advection, is ignored. Therefore the equation is to be applied to large, extensive surfaces of homogeneous vegetation only. The equation is restricted to the four components: R_n, l ET, H and G. Other energy terms, such as heat stored or released in the plant, or the energy used in metabolic activities, are not considered These terms account for only a small fraction of the daily net radiation and can be considered negligible when compared with the other four components.

The latent heat flux (l ET) representing the evapotranspiration fraction can be derived from the energy balance equation if all other components are known. Net radiation (R_n) and soil heat fluxes (G) can be measured or estimated from climatic parameters. Measurements of the sensible heat (H) are however complex and cannot be easily obtained. H requires accurate measurement of temperature gradients above the surface.

Another method of estimating evapotranspiration is the mass transfer method. This approach considers the vertical movement of small parcels of air (eddies) above a large homogeneous surface. The eddies transport material (water vapour) and energy (heat, momentum) from and towards the evaporating surface.

By assuming steady state conditions and that the eddy transfer coefficients for water vapour are proportional to those for heat and momentum, the evapotranspiration rate can be computed from the vertical gradients of air temperature and water vapour via the Bowen ratio. Other direct measurement methods use gradients of wind speed and water vapour. These methods and other methods such as eddy covariance, require accurate measurement of vapour pressure, and air temperature or wind speed at different levels above the surface. Therefore, their application is restricted to primarily research situations.

Soil Water Balance

Evapotranspiration can also be determined by measuring the various components of the soil water balance. The method consists of assessing the incoming and outgoing water flux into the crop root zone over some time period. Irrigation (I) and rainfall (P) add water to the root zone. Part of I and P might be lost by surface runoff (RO) and by deep percolation (DP) that will eventually recharge the water table. Water might also be transported upward by capillary rise (CR) from a shallow water table towards the root zone or even transferred horizontally by subsurface flow in (SF_{in}) or out of (SF_{out}) the root zone. In many situations, however, except under conditions with large slopes, SF_{in} and SF_{out} are minor and can be ignored.

Soil evaporation and crop transpiration deplete water from the root zone. If all fluxes other than evapotranspiration (ET) can be assessed, the evapotranspiration can be deduced from the change in soil water content (D SW) over the time period:

$$ET = I + P - RO - DP + CR \pm D\ SF \pm D\ SW \qquad (2)$$

Some fluxes such as subsurface flow, deep percolation and capillary rise from a water table are difficult to assess and short time periods cannot be considered. The soil water balance method can usually only give ET estimates over long time periods of the order of week-long or ten-day periods.

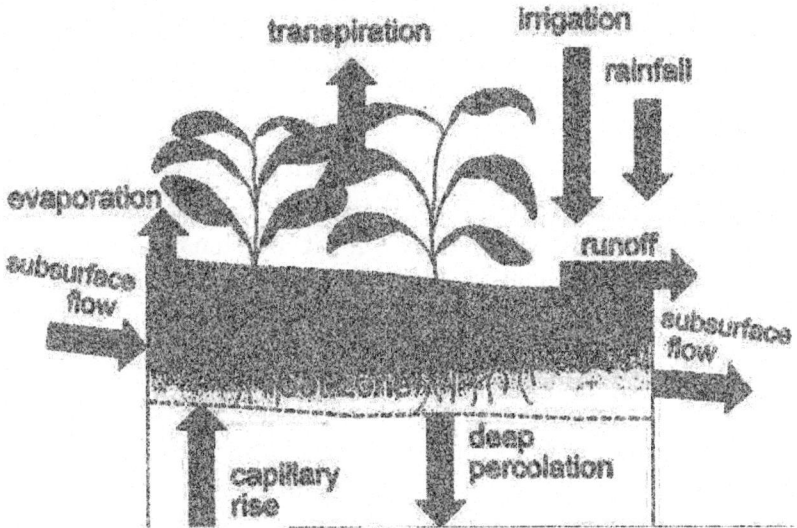

Figure: Soil water balance of the root zone

Lysimetres

By isolating the crop root zone from its environment and controlling the processes that are difficult to measure, the different terms in the soil water balance equation can be determined with greater accuracy. This is done in lysimetres where the crop grows in isolated tanks filled with either disturbed or undisturbed soil. In precision weighing lysimetres, where the water loss is directly measured by the change of mass, evapotranspiration can be obtained with an accuracy of a few hundredths of a millimetre, and small time periods such as an hour can be considered. In non-weighing lysimetres the evapotranspiration for a given time period is determined by deducting the drainage water, collected at the bottom of the lysimetres, from the total water input.

A requirement of lysimetres is that the vegetation both inside and immediately outside of the lysimetre be perfectly matched (same height and leaf area index). This requirement has historically not been closely adhered to in a majority of lysimetre studies and has resulted in severely erroneous and unrepresentative ET_c and K_c data.

As lysimetres are difficult and expensive to construct and as their operation and maintenance require special care, their use is limited to specific research purposes.

ET Computed from Meteorological Data

Owing to the difficulty of obtaining accurate field measurements, ET is commonly computed from weather data. A large number of empirical or semi-empirical equations have been developed for assessing crop or reference crop evapotranspiration from meteorological data. Some of the methods are only valid under specific climatic and agronomic conditions and cannot be applied under conditions different from those under which they were originally developed.

Numerous researchers have analysed the performance of the various calculation methods for different locations. As a result of an Expert Consultation held in May 1990, the FAO Penman-Monteith method is now recommended as the standard method for the definition and computation of the reference evapotranspiration, ET_0. The ET from crop surfaces under standard conditions is determined by crop coefficients (K_c) that relate ET_c to ET_0. The ET from crop surfaces under non-standard conditions is adjusted by a water stress coefficient (K_s) and/or by modifying the crop coefficient.

ET Estimated from Pan Evaporation

Evaporation from an open water surface provides an index of the integrated effect of radiation, air temperature, air humidity and wind on evapotranspiration. However, differences in the water and cropped surface produce significant differences in the water loss from an open water surface and the crop. The pan has proved its practical value and has been used successfully to estimate reference evapotranspiration by observing the evaporation loss from a water surface and applying empirical coefficients to relate pan evaporation to ET_0.

Need for a Standard ET_o Method

A large number of more or less empirical methods have been developed over the last 50 years by numerous scientists and specialists worldwide to estimate evapotranspiration from different climatic variables. Relationships were often subject to rigorous local calibrations and proved to have limited global validity. Testing the accuracy of the methods under a new set of conditions is labourious, time-consuming and costly, and yet evapotranspiration data are frequently needed at short notice for project planning or irrigation scheduling design. To meet this need, guidelines were developed and published in the FAO Irrigation and Drainage Paper No. 24 'Crop water requirements'. To accommodate users with different data availability, four methods were presented to calculate the reference crop evapotranspiration (ET_0): the Blaney-Criddle, radiation, modified Penman and pan evaporation methods. The modified Penman method was considered to offer the best results with minimum possible error in relation to a living grass reference crop. It was expected that the pan method would give acceptable estimates, depending on the location of the pan. The radiation method was suggested for areas where available climatic data include measured air temperature and sunshine, cloudiness or radiation, but not measured wind speed and air humidity. Finally, the publication proposed the use of the Blaney-Criddle method for areas where available climatic data cover air temperature data only.

These climatic methods to calculate ET_0 were all calibrated for ten-day or monthly calculations, not for daily or hourly calculations. The Blaney-Criddle method was recommended for periods of one month or longer. For the pan method it was suggested that calculations should be done for periods of ten days or longer. Users have not always respected these conditions and calculations have often been done on daily time steps.

Advances in research and the more accurate assessment of crop water use have revealed weaknesses in the methodologies. Numerous researchers analysed the performance of the four methods for different locations. Although the results of such analyses could have been influenced by site or measurement conditions or by bias in weather data collection, it became evident that the proposed methods do not behave the same way in different locations around the world. Deviations from computed to observed values were often found to exceed ranges indicated by FAO. The modified Penman was frequently found to overestimate ET_0, even by up to 20% for low evaporative conditions. The other FAO recommended equations showed variable adherence to the reference crop evapotranspiration standard of grass.

To evaluate the performance of these and other estimation procedures under different climatological conditions, a major study was undertaken under the auspices of the Committee on Irrigation Water Requirements of the American Society of Civil Engineers (ASCE). The ASCE study analysed the performance of 20 different methods, using detailed procedures to assess the validity of the methods compared to a set of carefully screened lysimetre data from 11 locations with variable climatic conditions. The study proved very revealing and showed the widely varying performance of the methods under different climatic conditions. In a parallel study commissioned by the European Community, a consortium of European research institutes evaluated the performance of various evapotranspiration methods using data from different lysimetre studies in Europe.

The studies confirm the overestimation of the modified Penman introduced in FAO Irrigation and Drainage Paper No. 24, and the variable performance of the different methods depending on their adaptation to local conditions. The comparative studies may be summarised as follows:

- The Penman methods may require local calibration of the wind function to achieve satisfactory results.

- The radiation methods show good results in humid climates where the aerodynamic term is relatively small, but performance in arid conditions is erratic and tends to underestimate evapotranspiration.

- Temperature methods remain empirical and require local calibration in order to achieve satisfactory results. A possible exception is the 1985 Hargreaves' method which has shown reasonable ET_o results with a global validity.

- Pan evapotranspiration methods clearly reflect the shortcomings of predicting crop evapotranspiration from open water evaporation. The methods are susceptible to the microclimatic conditions under which the pans are operating and the rigour of station maintenance. Their performance proves erratic.

- The relatively accurate and consistent performance of the Penman-Monteith approach in both arid and humid climates has been indicated in both the ASCE and European studies.

The analysis of the performance of the various calculation methods reveals the need for formulating a standard method for the computation of ET_o. The FAO Penman-Monteith method is recommended as the sole standard method. It is a method with strong likelihood of correctly predicting ET_o in a wide range of locations and climates and has provision for application in data-short situations. The use of older FAO or other reference ET methods is no longer encouraged.

Formulation of the Penman-Monteith Equation

Penman-Monteith Equation

In 1948, Penman combined the energy balance with the mass transfer method and derived an equation to compute the evaporation from an open water surface from standard climatological records of sunshine, temperature, humidity and wind speed.

This so-called combination method was further developed by many researchers and extended to cropped surfaces by introducing resistance factors.

The resistance nomenclature distinguishes between aerodynamic resistance and surface resistance factors. The surface resistance parameters are often combined into one parameter, the 'bulk' surface resistance parameter which operates in series with the aerodynamic resistance.

The surface resistance, r_s, describes the resistance of vapour flow through stomata openings, total leaf area and soil surface. The aerodynamic resistance, r_a, describes the resistance from the vegetation upward and involves friction from air flowing over vegetative surfaces. Although the exchange process in a vegetation layer is too complex to be fully described by the two resistance factors, good correlations can be obtained between measured and calculated evapotranspiration rates, especially for a uniform grass reference surface.

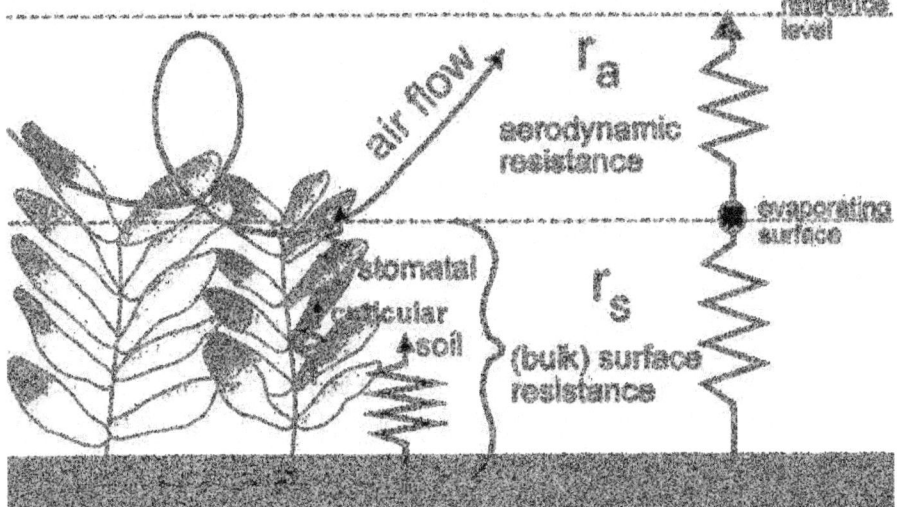

Figure: Simplified representation of the (bulk) surface and aerodynamic resistances for water vapour flow

The Penman-Monteith form of the combination equation is:

$$\lambda ET = \frac{\Delta(R_n - G) + \rho_a C_p \frac{(e_s e_a)}{r_a}}{\Delta + \gamma\left(1 + \frac{r_s}{r_a}\right)} \quad (3)$$

where R_n is the net radiation, G is the soil heat flux, $(e_s - e_a)$ represents the vapour pressure deficit of the air, r_a is the mean air density at constant pressure, c_p is the specific heat of the air, D represents the slope of the saturation vapour pressure temperature relationship, g is the psychrometric constant, and r_s and r_a are the (bulk) surface and aerodynamic resistances.

The Penman-Monteith approach as formulated above includes all parameters that govern energy exchange and corresponding latent heat flux (evapotranspiration) from uniform expanses of vegetation. Most of the parameters are measured or can be readily calculated from weather data. The equation can be utilised for the direct calculation of any crop evapotranspiration as the surface and aerodynamic resistances are crop specific.

Aerodynamic Resistance (r_a)

The transfer of heat and water vapour from the evaporating surface into the air above the canopy is determined by the aerodynamic resistance:

$$r_a = \frac{\ln\left[\frac{Z_m d}{Z_{om}}\right] \ln\left[\frac{Z_{h-d}}{Z_{oh}}\right]}{k^2 u_z} \quad (4)$$

where

r_a aerodynamic resistance [s m^{-1}],

z_m height of wind measurements [m],

z_h height of humidity measurements [m],

d zero plane displacement height [m],

z_{om} roughness length governing momentum transfer [m],

z_{oh} roughness length governing transfer of heat and vapour [m],

k von Karman's constant, 0.41 [-],

u_z wind speed at height z [m s^{-1}].

The equation is restricted for neutral stability conditions, i.e., where temperature, atmospheric pressure, and wind velocity distributions follow nearly adiabatic conditions (no heat exchange). The application of the equation for short time periods (hourly or less) may require the inclusion of corrections for stability. However, when predicting ET_0 in the well-watered reference surface, heat exchanged is small, and therefore stability correction is normally not required.

Many studies have explored the nature of the wind regime in plant canopies. Zero displacement heights and roughness lengths have to be considered when the surface is covered by vegetation. The factors depend upon the crop height and architecture. Several empirical equations for the estimate of d, z_{om} and z_{oh} have been developed. The derivation of the aerodynamic resistance for the grass reference surface.

(Bulk) Surface Resistance (r$_s$)

The 'bulk' surface resistance describes the resistance of vapour flow through the transpiring crop and evaporating soil surface. Where the vegetation does not completely cover the soil, the resistance factor should indeed include the effects of the evaporation from the soil surface. If the crop is not transpiring at a potential rate, the resistance depends also on the water status of the vegetation. An acceptable approximation to a much more complex relation of the surface resistance of dense full cover vegetation is:

$$r_s = \frac{r_l}{LAI_{active}} \qquad (5)$$

where

r_s (bulk) surface resistance [s m^{-1}],

r_l bulk stomatal resistance of the well-illuminated leaf [s m^{-1}],

LAI_{active} active (sunlit) leaf area index [m^2 (leaf area) m^{-2} (soil surface)].

The Leaf Area Index (LAI), a dimensionless quantity, is the leaf area (upper side only) per unit area of soil below it. It is expressed as m^2 leaf area per m^2 ground area. The active LAI is the index of the leaf area that actively contributes to the surface heat and vapour transfer. It is generally the upper, sunlit portion of a dense canopy. The LAI values for various crops differ widely but values of 3-5 are common for many mature crops. For a given crop, green LAI changes throughout the season and normally reaches its maximum before or at flowering. LAI further depends on the plant density and the crop variety. The bulk stomatal resistance, r_l, is the average resistance of an individual leaf.

This resistance is crop specific and differs among crop varieties and crop management. It usually increases as the crop ages and begins to ripen. There is, however, a lack of consolidated information on changes in r_l over time for the different crops. The information available in the literature on stomatal conductance or resistance is often oriented toward physiological or ecophysiological studies.

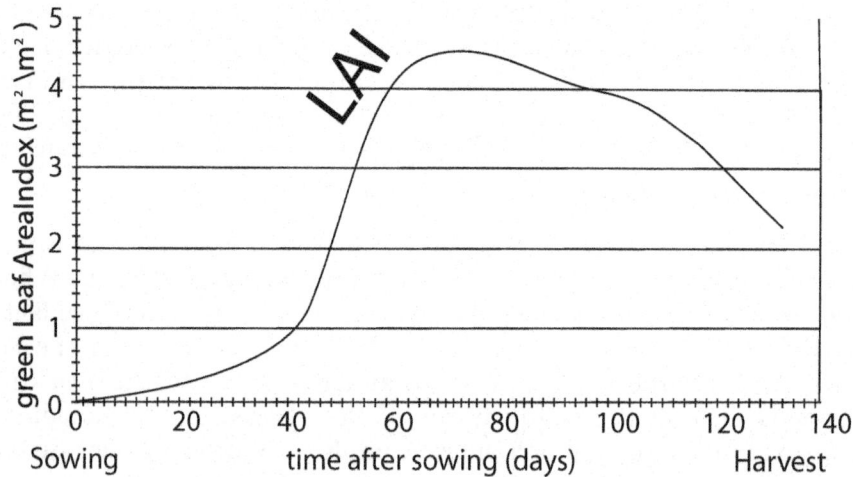

Figure: Typical presentation of the variation in the active (green) Leaf Area Index over the growing season for a maize crop

The stomatal resistance, r_l, is influenced by climate and by water availability. However, influences vary from one crop to another and different varieties can be affected differently. The resistance increases when the crop is water stressed and the soil water availability limits crop evapotranspiration. Some studies indicate that stomatal resistance is influenced to some extent by radiation intensity, temperature, and vapour pressure deficit.

Reference Surface

To obviate the need to define unique evaporation parameters for each crop and stage of growth, the concept of a reference surface was introduced. Evapotranspiration rates of the various crops are related to the evapotranspiration rate from the reference surface (ET_o) by means of crop coefficients.

In the past, an open water surface has been proposed as a reference surface. However, the differences in aerodynamic, vegetation control and radiation characteristics present a strong challenge in relating ET to measurements of free water evaporation. Relating ET_0 to a specific crop has the advantage of incorporating the biological and physical processes involved in ET from cropped surfaces.

Grass, together with alfalfa, is a well-studied crop regarding its aerodynamic and surface characteristics and is accepted worldwide as a reference surface.

Because the resistance to diffusion of vapour strongly depends on crop height, ground cover, LAI and soil moisture conditions, the characteristics of the reference crop should be well defined and fixed. Changes in crop height result in variations in the roughness and LAI. Consequently, the associated canopy and aerodynamic resistances will vary appreciably with time. Moreover, water stress and the degree of ground cover have an effect on the resistances and also on the albedo.

To avoid problems of local calibration which would require demanding and expensive studies, a hypothetical grass reference has been selected.

Difficulties with a living grass reference result from the fact that the grass variety and morphology can significantly affect the evapotranspiration rate, especially during peak water use. Large differences may exist between warm-season and cool-season grass types. Cool-season grasses have a lower degree of stomatal control and hence higher rates of evapotranspiration. It may be difficult to grow cool-season grasses in some arid, tropical climates.

The FAO Expert Consultation on Revision of FAO Methodologies for Crop Water Requirements accepted the following unambiguous definition for the reference surface:

"A hypothetical reference crop with an assumed crop height of 0.12 m, a fixed surface resistance of 70 s m^{-1} and an albedo of 0.23."

The reference surface closely resembles an extensive surface of green grass of uniform height, actively growing, completely shading the ground and with adequate water. The requirements that the grass surface should be extensive and uniform result from the assumption that all fluxes are one-dimensional upwards.

The FAO Penman-Monteith method is selected as the method by which the evapotranspiration of this reference surface (ET_0) can be unambiguously determined, and as the method which provides consistent ET_0 values in all regions and climates.

FAO Penman-Monteith Equation

Equation

A consultation of experts and researchers was organised by FAO in May 1990, in collaboration with the International Commission for Irrigation and Drainage and with the World Meteorological Organisation, to review the FAO methodologies on crop water requirements and to advise on the revision and update of procedures.

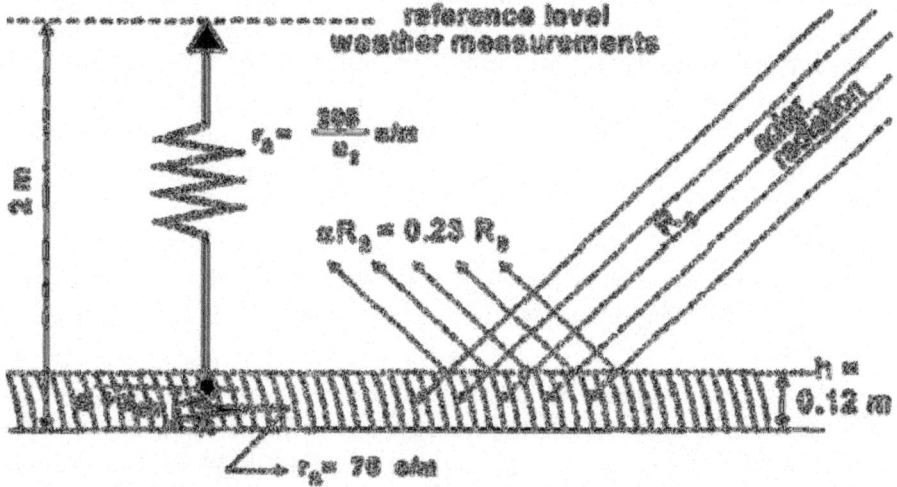

Figure: Characteristics of the hypothetical reference crop

The panel of experts recommended the adoption of the Penman-Monteith combination method as a new standard for reference evapotranspiration and advised on procedures for calculation of the various parameters. By defining the reference crop as a hypothetical crop with an assumed height of 0.12 m having a surface resistance of 70 s m[-1] and an albedo of 0.23, closely resembling the evaporation of an extension surface of green grass of uniform height, actively growing and adequately watered, the FAO Penman-Monteith method was developed. The method overcomes shortcomings of the previous FAO Penman method and provides values more consistent with actual crop water use data worldwide. From the original Penman-Monteith equation (Equation 3) and the equations of the aerodynamic (Equation 4) and surface resistance (Equation 5), the FAO Penman-Monteith method to estimate ET_0 can be derived:

$$ET_0 = \frac{0.408\Delta(R_n - G) + \gamma\dfrac{900}{T+273}u_2(e_s - e_a)}{\Delta + \gamma(1 + 0.34u_2)}$$

where

ET_o reference evapotranspiration [mm day^{-1}],

R_n net radiation at the crop surface [MJ m^{-2} day^{-1}],

G soil heat flux density [MJ m^{-2} day^{-1}],

T mean daily air temperature at 2 m height [°C],

u_2 wind speed at 2 m height [m s^{-1}],

e_s saturation vapour pressure [kPa],

e_a actual vapour pressure [kPa],

$e_s - e_a$ saturation vapour pressure deficit [kPa],

D slope vapour pressure curve [kPa °C^{-1}],

g psychrometric constant [kPa °C^{-1}].

The reference evapotranspiration, ET_o, provides a standard to which:

- evapotranspiration at different periods of the year or in other regions can be compared;
- evapotranspiration of other crops can be related.

The equation uses standard climatological records of solar radiation (sunshine), air temperature, humidity and wind speed. To ensure the integrity of computations, the weather measurements should be made at 2 m (or converted to that height) above an extensive surface of green grass, shading the ground and not short of water.

No weather-based evapotranspiration equation can be expected to predict evapotranspiration perfectly under every climatic situation due to simplification in formulation and errors in data measurement. It is probable that precision instruments under excellent environmental and biological management conditions will show the FAO Penman-Monteith equation to deviate at times from true measurements of grass ET_o. However, the Expert Consultation agreed to use the hypothetical reference definition of the FAO Penman-Monteith equation as the definition for grass ET_o when deriving and expressing crop coefficients.

It is important, when comparing the FAO Penman-Monteith equation to ET_o measurements, that the full Penman-Monteith equation (Equation 3) and associated equations for r_a and r_s (Equations 4 and 5) be used to enable accounting for variation in ET due to variation in height of the grass measured. Variations in measurement height can significantly change LAI, d and z_{om} and the corresponding ET_o

measurement and predicted value. When evaluating results, it should be noted that local environmental and management factors, such as watering frequency, also affect ET_o observations.

The FAO Penman-Monteith equation is a close, simple representation of the physical and physiological factors governing the evapotranspiration process. By using the FAO Penman-Monteith definition for ET_o, one may calculate crop coefficients at research sites by relating the measured crop evapotranspiration (ET_c) with the calculated ET_o, i.e., $K_c = ET_c/ET_o$. In the crop coefficient approach, differences in the crop canopy and aerodynamic resistance relative to the hypothetical reference crop are accounted for within the crop coefficient. The K_c factor serves as an aggregation of the physical and physiological differences between crops and the reference definition.

Data

Apart from the site location, the FAO Penman-Monteith equation requires air temperature, humidity, radiation and wind speed data for daily, weekly, ten-day or monthly calculations. The computation of all data required for the calculation of the reference evapotranspiration. It is important to verify the units in which the weather data are reported.

Location

Altitude above sea level (m) and latitude (degrees north or south) of the location should be specified. These data are needed to adjust some weather parameters for the local average value of atmospheric pressure (a function of the site elevation above mean sea level) and to compute extraterrestrial radiation (R_a) and, in some cases, daylight hours (N). In the calculation procedures for R_a and N, the latitude is expressed in radian (i.e., decimal degrees times p/180).

A positive value is used for the northern hemisphere and a negative value for the southern hemisphere.

Temperature

The (average) daily maximum and minimum air temperatures in degrees Celsius (°C) are required. Where only (average) mean daily temperatures are available, the calculations can still be executed but some underestimation of ET_o will probably occur due to the non-linearity of the saturation vapour pressure - temperature relationship. Using mean air temperature instead of maximum and minimum air temperatures yields a lower saturation vapour pressure e_s, and hence a lower vapour pressure difference (e_s - e_a), and a lower reference evapotranspiration estimate.

Humidity

The (average) daily actual vapour pressure, e_a, in kilopascals (kPa) is required. The actual vapour pressure, where not available, can be derived from maximum and minimum relative humidity (%), psychrometric data (dry and wet bulb temperatures in °C) or dewpoint temperature (°C) according to the procedures outline.

Radiation

The (average) daily net radiation expressed in megajoules per square metre per day (MJ m^{-2} day^{-1}) is required. These data are not commonly available but can be derived from the (average) shortwave radiation measured with a pyranometre or from the (average) daily actual duration of bright sunshine (hours per day) measured with a (Campbell-Stokes) sunshine recorder.

Wind Speed

The (average) daily wind speed in metres per second (m s^{-1}) measured at 2 m above the ground level is required. It is important to verify the height at which wind speed is measured, as wind speeds measured at different heights above the soil surface differ.

Missing Climatic Data

Situations might occur where data for some weather variables are missing. The use of an alternative ET_0 calculation procedure, requiring only limited meteorological parameters, should generally be avoided. It is recommended that one calculate ET_0 using the standard FAO Penman-Monteith method after resolving the specific problem of the missing data. Procedures for estimating missing climatic data are outline. Differences between ET_0 values obtained with the FAO Penman-Monteith equation with, on the one hand, a limited data set and, on the other hand, a full data set, are expected to be smaller than or of similar magnitude to the differences resulting from the use of an alternative ET_0 equation.

Even where the data set contains only maximum and minimum air temperature it is still possible to obtain reasonable estimates of ten-day or monthly ET_0 with the FAO Penman-Monteith equation. As outline radiation data can be derived from the air temperature difference, or, along with wind speed and humidity data, can be imported from a nearby weather station. Humidity data can also be estimated from daily minimum air temperature. After evaluating the validity of the use of data from another station, ten-day or monthly estimates of ET_o can be calculated.

The procedures for estimating missing data should be validated at the regional level. This can be done for weather stations with full data sets by comparing ET_o calculated with full and with limited data sets. The ratio should be close to one. Where the ratio deviates significantly from one, the ratio can be used as a correction factor for estimates made with the limited data set. Where the standard error of estimate exceeds 20% of the mean ET_o, a sensitivity analysis should be performed to determine causes (and limits) for the method utilised to import the missing data. A validation should be completed for each month and variable, for the monthly as well as for the daily estimates.

Chapter 6

The Mitigation of Drought Stress

Irrigation, where available, is the major means for combating drought conditions. It is a prime approach to the intensification of agriculture and the generation of stable income. The development of irrigation depends on various environmental, economical and social factors on both the macro and micro scales.

There are hazards in irrigation if practiced indiscriminately, such as soil erosion, soil salination, soil leaching and soil disease infection. Irrigation as such is not an important topic in this site. Links to irrigation sites throughout this section should provide additional information.

General Crop Irrigation Guidelines

The key to planning irrigation system and scheduling is knowledge of the crop, the soil properties and the potential evapotranspiration (PET) of the specific crop at the site. This information can also be used to estimate dryland crop water use and deficit at any given time during the crop cycle, which is actually an index of crop drought stress.

The Penman-Monteith potential evapotranspiration equation is recommended by the FAO as the standard method for estimating reference and crop evapotranspiration. The new method has been proved to have a global validity as a standardized reference for grass evapotranspiration and it has been recognised by both the International Commission for Irrigation and Drainage and by the World Meteorological Organisation.

The (FAO) Penman-Monteith method was developed by defining the reference crop as a hypothetical crop with an assumed height of 0.12 m having a surface resistance of 70 s m^{-1} and an albedo of 0.23, closely resembling the evaporation of an extensive surface of green grass of uniform height, actively growing and adequately watered.

Further educational information and guidelines on the applications of the Penman-Monteith method and the general approaches to prediction of crop water requirements and is provided by:

The FAO Methodology for Crop Water Requirements; FAO CROPWAT, is a downloadable software to carry out standard calculations for the design and management of irrigation. More recently FAO released AQUACROP – a model to simulate yield response to water.

Irrigation to Control Drought in Various Crops

Deficit (or supplemental) irrigation is the more common irrigation practice for crops not designated *a priori* for fully irrigated conditions and maximised yield. Supplementary irrigation is a practice dictated by constraints, which can be derived from the limited availability of water, irrigation equipment, the cost of water, or other economical and technical constraints. With supplemental irrigation the amount of water applied to the crop in irrigation is well below the full requirement of the crop. The water-use efficiency of supplemental irrigation is generally high if applied logically. It can be applied to save the crop in case of unexpected drought or as a planned practice to supplement the expected total seasonal rainfall. The practice may vary extensively with crop and region. In many environments, and especially the Mediterranean region, if only a single supplementary irrigation is given it is usually more effective if applied pre-planting. As such the crop enters the season with a stored supply, which can insure growth despite unexpected transient rainfall fluctuations. An example for using AQUACROP for planning deficit irrigation in cotton is available here.

Managing the Dryland Crop Environment

Modern dryland farming is a system of low inputs combined with soil and water conservation practices and risk reducing strategies. The system can be sustainable if practiced properly. Water shortage is the main limiting factor, but successful dryland systems also maintain reasonable practices to eliminate other limiting factors (poor nutrient status, weeds, biotic stresses, etc'), which can reduce the effectiveness by which the crop uses the limited moisture. However, as water shortage *a priori* dictates a limit on yield, all other inputs must be carefully adjusted downwards to fit the expected low economic return.

The most advanced systems have been developed in the Great Plains of the USA, Canada and Australia, while traditional systems employed in Asia and the Middle East also offer important insights. In the USA, the lesson learned during the "dustbowl" years in the early 1930's prompted extensive legistration and investments in developing sustainable dryland farming systems. These systems and the associated technological progress such as plant breeding, brought about an increase in mean winter wheat yield from 0.5 ton ha^{-2} in 1930 to about 2 ton ha^{-2} in 1980.

In Southern Australia the "ley farmimg" system was developed in the 1920's and adopted widely in the late 1940's. The system involves a rotation between a self-seeding legume grown for several years and wheat. The farmer grows wheat and raises sheep while the legume serves to sustain soil fertility (mainly nitrogen). This system has become less popular in recent years with the increase in economic pressures and other considerations.

The lesson learned from the American and Australian experience is that the development of a sustainable dryland farming system involves the following principles, not necessarily in their order of importance:

1. Improved soil and water conservation practices and the associated reduced tillage systems.
2. Optimisation of the fit between crop growth cycle and the available moisture.
3. Weed control
4. Soil fertility management.
5. Optimised plant population density and spatial arrangement of plants with respect to the expected soil moisture regime.
6. Control of soil biotic stress factors that reduce root development.
7. Improved forage/livestock/grains integration and rotation.
8. Avoidance of mono cropping and enhancement of crop diversification.
9. The increase of precipitation by cloud seeding, as an ongoing experiment.

Some of the above principles of the dryland farming system constitute general knowledge in agronomy, which can be explored in our Web Resources page as well as in standard agronomy textbooks and other publications (*e.g.* Drought Management of Farmland - by Joan Sydney Whitmore, 2000, 360 pp., Springer, SBN 0792359984). Here only several topics will be touched upon.

Soil and Water Conservation

Fallow and Conservation Tillage

The fallow system is designed to conserve soil moisture from one season to another or from one year to the other, depending on climate and crop. Increasing storage of soil moisture by the fallow system with or without conservation tillage is standard agricultural practice in dryland farming. The benefit of fallow and conservation tillage in terms of increasing available soil moisture to the crop depends on soil water-holding capacity, climate, topography and management practices. Fallow efficiency, in terms of percent increase in soil moisture availability to the crop measured at planting date normally ranges from about 5% to 30%. While these amounts are not impressive they can make a difference between crop failure and success. The fallow carries additional benefits such as improved soil nutrients availability and the eradication of certain soil-born pests, such as nematodes.

Conservation tillage involves the principle of minimised tillage operations to conserve soil structure and to maintain ground cover by mulch, such as stubble. These practices reduce water runoff and increase soil infiltration. Conservation tillage has become the cornerstone of dryland systems in certain regions of the USA, Canada and other regions. It has been re-demonstrated in dryland wheat experiments carried out in Southern Israel (#4407). While the benefits of conservation tillage are well-documented it has also been noted that crop residues under this system may promote certain crop diseases. To obtain some real impressions on the subject spend an hour in a farmers' meeting in California.

Deep tillage is a system to overcome hardpan, very high bulk density and compacted soils. It can be performed by deep plowing or deep ripping. Deep plowing involves actual plowing to depth which is an expensive operation. It is uncommon in dryland farming. Deep ripping is less expensive and often used in crop production. The important consideration in deep ripping is to operate at the correct depth in order to break the hardpan, no less and no more.

Furrow dikes and Soil Pitting

These techniques constitute a field surface tillage manipulation to minimise runoff away from the field.

Furrow dikes are furrows, which are divided into short basins by small dikes. This is achieved by special equipment. The system is very amenable to row crops such as cotton, corn and sorghum and it can be integrated with or without furrow irrigation. It is generally considered effective for increasing rainfall capture and raising dryland yield where annual rainfall ranges between 500 and 800 mm.

Soil pitting (left side photograph) involves the formation of small depressions at close proximity to reduce runoff from rainstorms. The crop is planted over this modified surface. Experiments performed with wheat in nine farmer demonstration plots in Southern Israel during 1988 showed that pitting increased yield by an average of 7.5% at a mean yield of 3.25 ton/ha. Unlike furrow dikes these system is not limited to row crops.

For further online information on dryland farming and its research visit the following sites and the links therein:

- Soil and Water conservation in Semi Arid Areas – (FAO Manual, 1987)
- USDA-ARS Bushland Texas Experiment Station – a distinguished centre of excellence in dryland conservation research
- Soil, water and reclamation publications —From Alberta, Canada.

Water Harvesting/Spreading

This is a broad term to describe various methods to collect runoff from large contributing areas and concentrate it for use in smaller crop area. This is an ancient practice already adopted by Nabatian desert settlements in the Middle East several centuries A.D. The photo on left represents a view from the ancient city of Avdant in the Negev region of Israel. In the front there are several ancient water spreading plots while at the back is a modern experimental farm (set-up by Prof. Evenari) seeking to evaluate the effectiveness of these systems in sustaining agriculture with around 100mm of annual rainfall.

Presently, the basic water harvesting systems involve an external contributing area to induce runoff. This area is physically or chemically treated for maximising runoff. The water is diverted into a receiving area comprising of cultivated plots, individual trees or small terraces. The contributing area may lie in the agricultural field (a system sometimes referred to as "conservation bench terrace") or outside the field in the natural watershed system. In the Avdat photo the small valley is a water-shed system experiencing flash flood once or twice a year. The size ratio between the contributing and the receiving areas is determined by the expected rainfall events, crop water requirements, soil characteristics and topography. The resulting yield increase in the receiving (crop) area is proportional to the amount of water gained.

Diversification of Farming

Diversification of farming is an ancient but an effective approach to reduce the risk associated with farming in unpredictable environments. Reduced diversification to the extent of mono-cropping is possible only with a high level of control over the crop environmental conditions. Such control method (irrigation, chemical pest control, etc') are among the main reasons for the more recent environmental quality problems found to be associated with mono-cropping.

Diversification of cropping to reduce risk is especially important under dryland conditions. It is achieved on several levels, as described by Pandey et al (#4194) for the case of traditional rain-fed rice in Eastern India.

1. Spatial diversification of fields. The farmer's land is divided into several fields or plots which may differ in their topography, soil and hydraulic properties. Some fields may be prone to flooding while others do not hold water. Certain fields may be on a warmer slope while others on a cooler one. The different field conditions allow to achieve a better fit between the crop and the environment and to reduce the general probability of stress affecting the farmer.

2. Crop diversification is an important feature of traditional farming. It takes an advantage of the generally low correlation between crops in performance when grown in a single stress environment. Crops differ in their response to a given environment and this difference is used to reduce the risk associated with growing one crop. "Mixed cropping" or "intercropping" is an example of a traditional and a successful approach to crop diversification on a single parcel of land, where two or more crops are grown together in various possible configurations. If for some reason only one crop is grown, a certain (though lower) level of risk reduction can be achieved by varietal diversification. Planting of several crop varieties offer a better probability for reducing loss due to environmental stress, as compared with growing one variety only. For environmental stress conditions varietal diversification is based mainly on differential phenology, primarily flowering date. A typical example is a transient frost or heat wave that is likely

to occur around flowering time of the specific crop. Damage reduction can be achieved when the crop is sown to several varieties of different flowering dates.

3. Temporal diversification may achieve the same result as varietal diversification, when phenology is concerned. The purpose of setting a distinct planting date is to optimise crop development with respect to seasonal climate, mainly rainfall in rain-fed agriculture. Ideally the crop is planted at the beginning of the rainy season, rainfall peaks when crop evapotranspiration peaks and it terminates just before harvest time. When such conditions are reasonable predictable, planting date can be set to optimise production. Where the timing of rainfall is very unpredictable, adopting more than one planting date for the given crop can reduce the risk involved with untimely rainfall and a given planting date.

Cloud Seeding

Cloud seeding is a form of weather modification attempt. The process of cloud seeding involves deposition of cloud condensation nuclei (CCN) into a specific region of the cloud. Seeding may be achieved from above or through the clouds by aircraft, and from below where CCN are carried into the cloud by updrafts. With either method, the CCN must reach the super cooled cloud region, where water molecules remain unfrozen at temperatures below ^0C.

Experiments in cloud seeding have been performed for the last 60 years. The results and benefits of this practice are still under debate.

Information is available in the report on 'Weather Modification by Cloud Seeding-A Status Report 1989-1997 by William R. Cotton, Colorado State University; and at the Oklahoma Weather Modification Demonstration Program.

Mitigation of Drought Stress by Crop Plant Resistance

The Nature of Drought Resistance

Drought Resistance and Crop Yield : Crop plant breeding for drought resistance has long been part of the breeding process in most crops that have been or are being grown under dryland conditions. During the period of the pre-scientific agriculture the genetic improvement of plant adaptation to dry conditions was simply attained by repeatedly selecting plants that appeared to do well when drought stress occurred. As a result of many generations of selection by generations of farmers we now encounter such materials, which are defined as "landraces" of the crop. Such landraces were shown to posses distinct drought resistance traits. Later, as scientific agriculture developed and following the emergence of Mendelian genetics, elaborate biometrical and statistical methods for quantitative genetics analysis were developed to enable selection for yield and yield stability more effectively and efficiently. An important factor of yield stability is coping with drought and other abiotic plant stresses. Subsequently, yield-based selection programs were augmented by observing plants under carefully managed stress environments, followed by the development of physiological selection criteria for stress resistance. More recently, molecular methods, such as marker-assisted selection are being adopted to facilitate more efficient selection for distinct components of abiotic and biotic stress resistance. Finally, biotechnology is experimenting with genetic transformation, which is in the process of being applied as an additional solution to breeding for drought resistance.

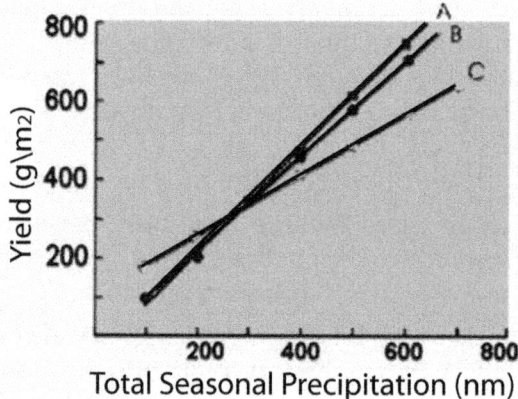

Figure: The association between yield and total seasonal precipitation for 3 different wheat cultivars.

Looking at crop drought resistance from a botanical perspective it must be realised at the onset that there is a vast difference between drought resistance in natural vegetation and in crop plants. Natural vegetation has evolved to conserve the species. Henceforth, plant survival and the capacity to produce at least one seed per life cycle despite stress is the most powerful component of natural selection. On the other hand, drought resistance in modern agriculture requires sustaining economically viable plant production despite stress. Crop survival is of a lesser consequence to economical farming. On the other hand, plant survival can be a critical factor in subsistence agriculture, where the ability of a crop to survive drought and produce some yield at all may translate into a difference between famine and livelihood. Breeding for drought resistance is therefore very tightly linked to the target environment of the crop, not only with respect to its physical and chemical features but also its social grounds. The recent developments in GIS technology are extremely important as a tool for defining a target environment for the breeding program. GIS has been discussed under impact of stress above.

For a variety of reasons there is a general trade-off between a genetically high yield potential and drought resistance. At the same time there is a yield advantage under drought stress brought about by a high yield potential, to a limit.

Wheat cultivar C is different from A and B in that it has a lower yield potential (yield at high moisture conditions) but as moisture becoming deficient C turns out to be superior to A and B. In terms of yield, C may be defined as drought resistant while cultivars A and B are of high yield potential but are relatively drought susceptible. The "crossover" where the advantage of C over A and B under stress begins to be expressed is at about 300 mm or at a yield level of about 300 g m^{-2}. Hence, drought resistance of C is expressed only when stress is sever (<300 mm). Still, it is extremely important to realise that the high yielding cultivars A and B are superior to the drought resistant C when drought stress is moderate (e.g. at 400 to 500 mm). A high yield potential therefore ascribes an advantage under moderate stress conditions. By definition drought resistant cultivars have lower yield potential. Cases where drought resistance has been improved together with yield potential exist but they are very rare and exceptional and cannot be used to indicate a general rule. With the available evidence the rule seems also implies that breeding for real drought resistance is not required if yield in the target environment is not reduced (schematically) to below 300-400 g m^{-2}. On the other hand, real drought resistance cannot be field- tested or evaluated if yield is above around 300-400-g m^{-2}, schematically. Consider the principle not the actual numbers.

The Components of Drought Resistance

Drought resistance in crop plants is conditioned by two major pathways: Dehydration avoidance and dehydration tolerance. Dehydration avoidance is the capacity to avoid plant tissues and cells dehydration under drought stress. Dehydration tolerance is the capacity to sustain function when the plant is dehydrated.

As discussed above under the "Impact of Stress", moisture stress signals certain stress responsive genes, which are responsible for a chain of events, expressed at various levels of plant organisation. It has been assumed almost axiomatically that stress responsive genes are involved in adaptation; henceforth that they are 'stress adaptive'. It was later realised that not every stress responsive gene is necessarily adaptive in terms of drought resistance or survival.

Irrespective of the role and function of stress adaptive genes in plant drought resistance, it should be recognised that not only stress adaptive genes determine plant performance under drought stress. Genes that are expressed irrespective of the environment also condition plant function and performance under stress. These genes are expressed constitutively. An example for a constitutive (non-adaptive) plant trait that may control drought resistance is potential root size. Stress and soil conditions can affect root size in several ways but potentially a deep rooted genotype will maintain its advantage over a potentially shallow rooted genotype under conditions of deep soil moisture. For this difference to be expressed plants do not have to be subjected to drought stress conditions. On the other hand, stress-adaptive traits will be expressed only when plants are subjected to drought stress.

Dehydration Avoidance

Plant Development and Size

Plant size as expressed mainly in terms of single plant leaf area or leaf area index (LAI) has a major control over water-use, as explain under Impact of Stress. Small plants and reduced leaf area are generally conducive to low productivity while they limit water use. Botanists have long recognised small plants bearing small leaves as typical ecotypes of xeric environments. While such plants withstand drought very well their growth rate and biomass are relatively low.

In the domain of plant breeding, cultivars developed for dryland conditions by selecting mainly for yield under such conditions often result in plants of moderate size and water-use. For example this can be seen in dryland temperate cereals, which tend to have moderate tillering. On the other hand researchers in the CSIRO Australia have concluded (#3903) that early plant (and seedling) vigour are important traits for dry conditions. The reason is in the rapid ground cover achieved and the subsequent decrease in water loss by direct soil evaporation at this stage. However, other benefits for seedling vigour were also noted, such as the nitrogen status of the plant (#7059). Early flowering which determines 'drought escape' generally involves a reduction in plant size and leaf area leading to reduced water-use. Small plants and small leaf area is a decisive link between improved drought resistance and lower potential yield.

The Root

The most important control of plant water status is with the root, whereas roots are the main engine for meeting transpirational demand. Two major dimensions describe the root: root depth and root-length density. The more practically important dimension for most breeding scenarios is root depth, which facilitate deep soil moisture extraction where such moisture is available. It is a primary component of drought resistance. The development of lateral roots at very shallow soil depth may have a role in capturing small amount of intermittent rainfall.

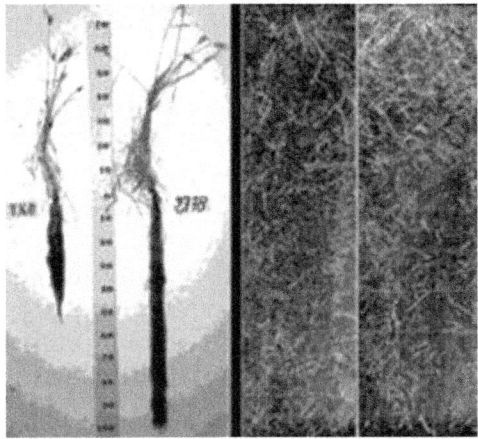

Figure: Left panel: hydroponically grown roots of two wheat cultivars differing in root length. Right panel: roots of two sorghum cultivars in soil in a root observation box, differing in root-length density.

Root depth in the cereals is generally associated with a small number of main thick axes. Such fibrous root system is often seen in upland rice, which has a deeper root system, in contrast to lowland rice with the shallower roots. The control of root growth is not only in the root. In the cereals, tillering is associated with production of new crown roots from each developing tiller. Such profuse rooting is at the expense of the growth of existing roots, deeper into soil. Hence, limited tillering in cereals and grasses has been repeatedly observed to be associated with relatively deeper root extension.

In certain soils a hardpan can limit deep root growth and the capacity for hardpan penetration by roots becomes a critical factor in drought resistance. The factors, which may support axial root force and hardpan penetration, are not known and most research in this area has been performed mostly with seedlings. In mature plants the penetration of hardpan by roots seems to be better in plants that constitutively develop fibrous and thick roots.

Many drought environments present a situation where rainfall is low and soil depth that contains moisture is permanently shallow. For example, in many of the drier Mediterranean wheat-growing regions the wetted soil depth of around 60-80 cm is shallower than the maximum root depth of wheat (>100 cm). Under such conditions a deep root is not an issue. Other plant factors may then become far more important in the control and use of the limited soil moisture, such as shoot developmental characteristics (e.g. leaf area development or growth duration), osmotic adjustment, leaf surface properties, etc.' Greater root length density will allow to extract more moisture from a given soil volume which in certain cases should provide several more days before wilting.

Another scenario of seasonal soil moisture status is where the crop is grown on stored soil moisture and there is little effective rainfall during the growing season. Under such conditions the main consideration is to manage seasonal soil moisture use such that sufficient moisture will remain for carrying the crop to maturity. It is to be expected that with the available moisture the crop might grow luxuriously leading to a large leaf area and an even greater water requirement towards the latter part of the season. Hence, short growth duration, small leaf area and perhaps a higher root hydraulic resistance can achieve the control of seasonal water use. The last option has been researched at the CSIRO Australia and an increase wheat root hydraulic resistance was effectively attained by selection for smaller xylem element diametre. It is not known whether this approach has found its way into actual application in wheat breeding

Whatever may be the constitutive form and function of roots, the environment can modify the root in a pronounced way. Offcourse, soil conditions in terms of topsoil moisture and deep soil hardness alter root growth and depth. Drought stress generally inhibit total root mass (while it can modify its distribution). Root-length density may locally increase in wet regions in the soil while it will decrease in the drying parts. As soil moisture deficit develop throughout the profile, the proportion of dry to wet soil increase so that the proportion of dead to live roots increase. There is hardly evidence to show that total root mass increase with drought stress. The shoot/root mass ratios consistently decrease under drought stress, which is a universal expression of adaptation. The ratio changes mainly due to the reduction in shoot mass. The root system is highly dynamic and as long as it is not senesced or diseased it is capable of regrowth from meristems in the root axes and meristems in the root crown (in cereals and grasses). The renewal of root branching into wet soil immediately after rainfall is considered as an important factor in plant recovery from drought stress. Root hairs are considered an important component of root length density and the capacity for soil moisture extraction via improved contact with the soil.

Roots are a major target and a candidate for marker assisted selection (MAS) for the apparent reason that phenotypic selection for root traits is a slow and hard work in large population. Still, practical results from MAS for root traits (*e.g.* 8095) are still limited.

Plant Surface

Plant surface structure, form and composition carry a major impact on the plant interaction with the environment. Plant surface absorbs solar energy part of which is used for photosynthesis and most of which must be dissipated. Energy is dissipated by reflection, emission and the dissipation of latent energy by transpiration. Plant surface structure determines the reflective properties of the leaves and their resistance to transpiration. Leaf resistance to transpiration is offcourse largely determined by stomatal activity. However, plant surface structure determines the hydraulics of leaf surface and the boundary layer conditions, which affect the rate of water removal from the leaf surface, upon transpiration. Therefore plant surface help to avoid dehydration by two channels: improved reflectance of incoming g radiation (*i.e.* decreasing net radiation) and by improved cuticular hydraulic resistance.

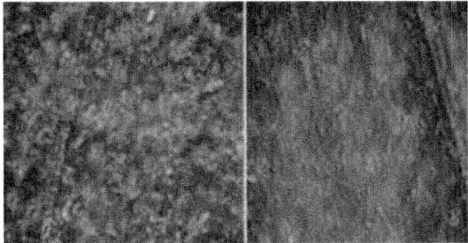

Figure: Sorghum leaf epicuticular wax by the scanning electron microscope; left normal (Bm genotype); right low wax (bm genotype).

Figure: Leaf pubescence in the wild plant *Solanum elaeagnifolium* (Silverleaf) (right) as compared with cotton (left).

After the stomata, the secondary site for water loss by transpiration is the cuticle. The hydraulic permeability of the cuticle is determined by the wax embedded in the cuticle matrix as well as by the wax deposited over the cuticle. High cuticular permeability not only affects non-stomatal transpiration pathway but it may also directly affect water loss from guard cells and therefore their water status and stomatal aperture. An example of a difference in epicuticular wax load between two sorghum genotypes. The lower wax (*bm*) genotype had far greater total leaf transpiration than the *Bm* genotype.

Epicuticular wax is deposited in different forms and structures, most likely as a function of its composition. The environment also affects the density of epicuticular wax. Conditions of water stress, high tem-perature, and high radiation increase its density. The full genetic potential for wax deposition is therefore best evaluated in plants subjected to stress.

In practical terms, the quantitative effect of wax on transpiration is finite, and for a given plant, the increase in epicuticular wax load beyond a given threshold would not reduce transpiration. Sorghum typically represents relatively high potential epicuticular wax deposition while rice represents species that lack in this respect, as estimated by quantifying epicuticular wax and by rate of cuticular transpiration. Hence, there is a potential for improving drought resistance in rice by genetically increasing epicuticular wax load.

The shape and angles of the cuticular wax deposits of may affect the spectral properties of the leaf. Thus, for example, the glaucous appearance of some wheat genotypes is determined by the structural properties of the wax deposits. Increased glucousness was found to result in an increase in leaf reflectance of wheat and sorghum within the spectrum range of at least 400 to 700 nm and possibly also at the

UV-B. This increase in reflectance may result in a reduction in net radiation and leaf temperatures in glaucous genotypes.

Leaf pubescence is a common feature in xerophytic plants as well as in some crop plants, such as soybean. Generally it increases reflectance from the leaf within the range of 400 to 700 nm and sometimes up to 900 nm, resulting in lower leaf temperatures under high irradiance. It is sometimes argued that the increased reflectance in the photosynthetically active waveband would reduce photosynthesis under non-stress conditions. Under conditions of stress, there is a trade-off between the effect of pubescence towards the reduced stress load and its possible effect on photosynthesis.

Increased leaf pubescence may increase the leaf boundary-layer resistance by up to 50%. However, it has been argued that this should carry a relatively small effect on water and CO_2 exchange, as compared with the effect of pubescence on the radiative properties of the leaf.

Leaf colour can affect the thermal properties of the leaf. In both wheat and barley there are 'yellow leaf' cultivars, which have about a third less chlorophyll than the 'normal' ones. The 'yellow' cultivars tend to perform relatively better under drought stress as compared with the normal green. Yellow leaves are more reflective and their temperature is relatively lower than that of green ones. Beyond this difference in reflective properties, the low chlorophyll lines seem to sustain lower injury to the photosystem under conditions of high irradiance and water deficit (#3817).

Osmotic Adjustment (OA)

Figure: Differential response to drought stress of a high OA cultivar (left) and a low OA cultivar (right) of wheat.

When water deficit develop various solutes accumulate in cells and subsequently tissue osmotic potential is reduced. Tissue osmotic potential can be reduced merely by the concentration of cellular solution due to water loss. This is not OA. OA is derived from the net increase in cellular osmolality caused by the accumulation of solutes such as various ions (mainly potassium), sugars, poly-sugars (*e.g.* fructan), amino acids (*e.g.* proline), glycinebetaine, etc.' Recently constitutive accumulation of natural solutes (*e.g.* glycinebetaine) and exotic solutes (*e.g.* mannitol) were engineered and tested for functionality in model plants such as tobacco. OA occurs when cellular water deficit exceeds a certain threshold, which is not universally determined. Nor has the exact signalling for OA been resolved.

OA is a slow process requiring time, and very rapid desiccation in experiments or even in the field may not allow for OA. Ideally the rate of desiccation should not be greater than about 0.1 MPa day^{-1}. Practically, it should take around 2 weeks from fully hydrated state to wilting on order for the full capacity and impact of OA to be expressed in whole plant.

The rate of OA varies greatly among species and cultivars. A minimal rate of OA, which can be considered as effective, is about 0.3 MPa and rates of up to 1.5 to 2.0 MPa were observed in certain cereal cultivars. Some crop plants generally tend to be better at OA than others with cowpea and maize generally having lower rates while *indica* rice, sorghum and wheat tend to express higher rates.

OA is probably one of the most crucial components of adaptation to drought stress. It help maintain cellular turgor at a given leaf water potential and thus delay wilting. OA enables to sustain growth and productivity at lower plant water status. Irrespective of the effect on turgor maintenance, the accumulated solutes protect cellular proteins, various enzymes, cellular organelles, and cellular membranes against desiccation injury.

Hence, cell and tissues may continue to function despite the progressing desiccation. This is why the accumulated osmotic solutes are sometimes defined as "protectants". One consequence of OA at the whole plant level is the continued growth of roots and the extraction of deeper soil moisture. Finally, OA is crucial for the conservation of meristem viability under desiccation towards the recovery of function upon dehydration. OA in different cultivars of wheat, sorghum, various pulses and brassicas has been shown to be well associated with biomass and/or yield under drought stress.

Upon rehydration the various solutes are recycled and metabolised to the extent that the accumulated sugars, for example, are considered as an important energy resource for recovery growth.

Extensive genetic engineering efforts are being made to use the phenomenon of OA in the design of stress resistant plants (e.g. #4635, #5897). Most experiments involve transgenic model plants that were modified to constitutively express the accumulation of osmolytes. Such transgenics that accumulate glycinebetaine, D-ononitol, mannitol, and trehalose gave positive or inconclusive results with respect to stress resistance, and work in this area is developing rapidly.

Non-Senescence (Delayed Senescence or "Staygreen")

Figure: "Stay green" (left) and "normal" (right) cultivars of sorghum under post-flowering stress

Plant senescence is a genetically programmed process, accelerated by environmental stress such as drought, heat, and nitrogen deficiency. The primary expression of leaf senescence is the breakdown of chlorophyll and the subsequent collapse of photosynthesis. Leaf greenness as measured by chlorophyll content or by leaf reflectance properties (using the Minolta chlorophyll metre for example) is becoming an acceptable estimate of senescence (and leaf nitrogen status). In various crops certain genotypes were identified as expressing delayed senescence or non-senscent or stay-green phenotype (#4440). These genotypes generally sustain leaf greenness and photosynthesis for a longer time and consequently tend to yield more. Since drought stress accelerates senescence, stay-green (SG) genotypes are important in sustaining green leaf area under stress.

SG does not present a uniform expression across different crop plants. In sorghum for example SG can be associated with high stem soluble carbohydrate content and greater resistance to lodging caused by stem 'charcoal rot'. In sorghum and millet at least, SG genotypes sustain higher RWC under stress as compared with normal ones. This is why SG is discussed under 'dehydration avoidance'. Maintenance of RWC is not necessarily an expected result of delayed chlorophyll loss or delayed leaf protein breakdown. Furthermore, certain SG genotypes of sorghum are expressed better when exposed to drought stress. Hence, the phenotypic selection of SG in sorghum (and perhaps other crops) is more effective under post-flowering drought stress.

SG is at least partly regulated by endogenous plant hormones, whereas in certain cases an increase in kinetin in leaves promoted SG. In other cases SG was associated with decrease in plant ethylene content. Such hormonal regulation can involve both nitrogen and water status of leaves.

The expression of SG and plant senescence in general can be markedly influenced by intra-plant interactions which involve assimilate partitioning and endogenous hormonal balance. A simple exercise to obtain a SG phenotype in grain producing crops is by detaching the inflorescence at flowering. Grain set and grain growth generally enhance leaf senescence by enhancing carbohydrate and nitrogen export from leaves into the grain. Very low yielding or partially sterile plants may present some delay in senescence when subjected to drought stress during grain filling.

There are ongoing attempts to achieve genetic transformation of SG trait by either promotion of endogenous kinetin or by antisense suppression of ethylene. QTLs for SG are being identified in several crops and marker assisted selection for the trait is possible in sorghum and probably other crops in the future.

Dehydration Tolerance

Effect of stress kinetics on differential gene expression of immature ears of maize. Plants were grown in buckets where drought stress reached the point of null photosynthesis (Pn) in 5 days. Plants were grown in the field where the same state of stress was reached after >5 weeks (From Barker et al., 2005).

Stress Phenotyping	Stress Kinetics	% genes responding
Stress in a large pot	Rapid (5 days)	27
Stress in the field	Slow (4 weeks)	2

Cellular and molecular adaptive processes in response to water deficit do not occur until a certain level of water deficit has been reached. Cellular and molecular adaptive responses serve one or more of the following major functions: (a) reduce whole plant growth in order to reduce plant water-use; (b) reduce the rate of cellular water loss and retain cellular hydration; and (c) protect various cellular structures and functions as cells desiccate.

With modern molecular research tools it becomes fairly straightforward to reveal hundreds of genes that are up regulated or down regulated in response to plant tissue water loss. However, research into the function of most of these genes is not as developed. Subsequently the exact function at the whole plant level of the found gene responses to cellular water deficit is not well understood to the extent that they can be used in plant breeding. However, there is slow progress in this area as can be seen in our 'Biotech Issues' files. In terms of application to plant breeding dehydration tolerance is the capacity to function in a dehydrated state which often (but not always) means the involvement of stress responsive and adaptive genes. Most of the information that is relevant for application to breeding is derived from whole plant physiological studies while some rudimental information comes from genomics.

Plant physiology always cautioned that the evaluation of plant response to drought stress and the evaluation of plant adaptation require sufficient time under stress. Adaptive plant responses to drought stress do not only depend on the level of tissue desiccation but also on its rate (*e.g.* #3418). It was well established that fast or slow desiccation may have totally different impact on results in terms of adaptation. Very rapid desiccation often exercised in laboratory experiments is totally irrelevant even though statistically significant results can be obtained. Tissue desiccation under natural conditions is slow. Confirmation of this axiom is now received from a gene expression study in maize as presented in Table, which speaks for itself.

Stem Reserve Utilisation

The current source of carbon for grain filling is assimilation by the light intercepting viable green leaf area. This source is normally diminishing due to natural senescence and the effect of various stresses. At the same time the demand by the growing kernel is increasing, in addition to the demand posed by maintenance respiration of the live plant biomass. When the demand by the grain sink is not fully supplied by the source of current assimilation, plant reserves can provide the balance.

Small grains and cereal stems store carbohydrates in the form of glucose, fructose, sucrose, fructans or starch. Total storage in cereal plant roots or leaves is relatively small to that in the stem (including leaf sheaths). This storage is commonly analysed as total non-structural carbohydrates (TNC) or water-soluble carbohydrates (WSC) (#2722) and it is available for translocation to the grain.

Usage of stem reserves depends on the available storage and the rate and duration of mobilisation of storage to the grain. The size of the storage strongly depends on favourable growing conditions before anthesis and genotype. Developmentally, potential stem storage as a sink will also be determined by stem length and stem weight density (stem dry weight per unit stem length). Stem length, as affected by the height genes is important in affecting stem reserve storage, as demonstrated in sorghum (#3561).

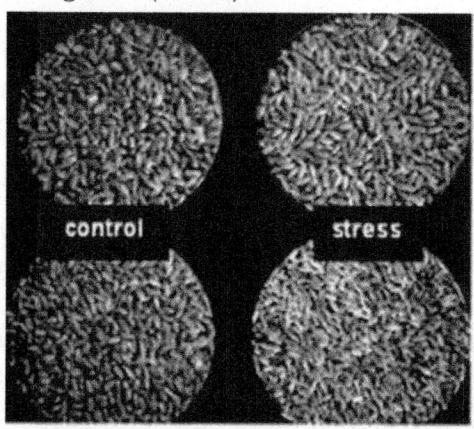

Figure: Grain of two wheat cultivars subjected to sever drought stress during grain filling (right). Top: cultivar with superior capacity for stem reserve utilisation; bottom: normal cultivar. Note the shriveled grain under stress in the latter.

Storage by about one third as a consequence of a reduction in stem length. This may be one of the reasons for the recognised greater drought susceptibility of the dwarf high yielding wheat cultivars.

Stem reserve mobilisation or the percentage of stem reserves in total grain mass is affected by sink size, by the environment and by cultivar. It is not surprising that different published estimates of the percentage of grain yield that is accounted by stem reserves range from 9 to 100%.

The demand by the grain yield sink is a primary factor in determining stem reserve mobilisation. When degraining reduced sink size, more reserves were stored in the stem, as compared with intact ears. The availability of storage at grain filling does not necessarily assure mobilisation. There are cases on record where despite stress conditions the available storage was not utilised. This may be traced to problems in enzymatic conversion of storage to transportable constituents or sometimes inhibition of sink processes. For example, when heat stress occurs starch synthesis in the wheat grain might be inhibited by a thermolabile enzyme (such as soluble starch synthase) and available stem reserves would not be in demand.

The reduction in current assimilation during grain filling, under different stresses, will induce greater stem reserve mobilisation to and utilisation by the grain. What is important is the reduction in assimilation and not the nature of stress causing the reduction. Thus, stem reserve mobilisation is a solid source of carbon for grain filling under any stress, which would inhibit current photosynthesis, including biotic stresses such as late developed leaf diseases. Tolerance to *Septoria* leaf blotch in wheat is expressed in sustained grain filling under sever epiphytotic. It has been demonstrated that mobilised stem reserve is a major component of Septoria tolerance in wheat (#2659).

The full potential for stem reserve utilisation of a cereal cultivar can be experimentally assessed by growing plants under favourable conditions and then detaching all leaf blades and shading the inflorescence at the onset of grain filling. Grain weight per inflorescence in such treated plants as compared with controls provide a reliable estimate. It appears that wheat genotypes differ in their capacity to store stem reserves. Cultivars that have this high capacity must also possess relatively long grain filling period in order to allow sufficient time for reserves to be mobilised into the grain.

A possible "penalty" for high stem reserve utilisation capacity is accelerated shoot senescence, due to the export of storage into the grain. Thus, it seems that the two factors cannot be recombined and a breeder must opt for either stem reserve mobilisation or delayed senescence trait as mechanisms supporting grain filling under stress.

Cellular Membranes

The fluid mosaic model of the cellular membranes (CM) describes the membrane as a bi-layer of phospholipids and glycolipids studded and spanned by proteins partially or fully solvated by the lipid matrix. CM is central site for various cellular functions, especially those associated with membrane bound enzymes and transport of water and solutes. The function and role of the CM under extreme temperature stress is somewhat clearer than with drought stress.

The phospholipids in terms of their quantity and composition are generally considered as the more crucial components of cellular membrane stability under drought stress.

The most notable factor in cellular membrane function under desiccation is that plant exposure to slow desiccation or to other stresses, typically extreme temperature stress signal a hardening ("acclimation") effect that is expressed in increased membrane stability under desiccation stress. While it is not clear how cellular membrane stability under stress is translated into a yield advantage in stress affected crops, such relationship has been indicated in several studies (*e.g.* #2676, #2985).

Water passage through both the plasma membrane and the tonoplast is crucial to cell life and specific proteins inserted in the membrane largely regulate it. These "water-Channel" proteins, also termed "aquaporins", respond to various signals and "molecular switches". These pores are highly selective to water and they play an import role in cellular water relations in response to plant water deficit and osmotic stress. For example, maize root aquaporins were found to be stimulated by water deficit, resulting in improved water transport. The study of aquaporins and their function is now at the forefront of research on cell water relations. Greater understanding of aquaporin role in drought response and adaptation is expected in the near future.

Antioxidation

Oxidative Stress is a general term used to describe a state of damage caused by reactive oxygen species (ROS). ROS, such as free radicals and peroxides, represent a class of molecules that are derived from the metabolism of oxygen. There are many different sources of ROS that can cause oxidative damage to an organism. Most come from endogenous sources as by-products of normal and essential reactions, such as energy generation from mitochondria or the detoxification reactions. Free radicals are unstable because they have unpaired electrons in their molecular structure. This causes them to react almost instantly with any substance in their vicinity. Free radicals destroy cellular membranes, enzymes and DNA.

Antioxidants are active substances naturally occurring in all organisms which detoxify free radicals. These are for example superoxide dismutase (SOD), catalase, glutathione reductase or ascorbate peroxidase. SOD, for example, converts the O_2° to H_2O_2 and Catalase converts H_2O_2 to molecular oxygen (O_2).

Drought, as well as other stresses cause oxidative stress in plants and antioxidant abundance and activity is important for the protection of metabolism under stress. When various studies are reviewed it is unclear whether the genetic enhancement of antioxidant production in plants beyond the natural level is indeed required to alleviate drought stress at the whole plant level and whether the naturally occurring active antioxidants are not sufficient to protect the plant. It has been shown that drought induced oxidative stress related genes and that this was associated with increased levels of various antioxidants in plants.

The most important information in this respect is coming from the developing work with transgenic plants, which over express antioxidant production. When these studies are taken as a whole, no clear conclusions can be made yet with respect to the advantage in the overproduction of antioxidants to improve plant production under drought stress. More information is available on this site under "The Stresses".

Stress Proteins and Chaperons

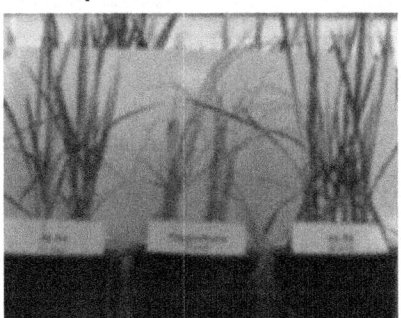

Figure: Rice transgenic plants over expressing the HVA1 barley embryo LEA protein and subjected to drought stress. The middle pot is the 'wild type' (control) plant. These transgenics were developed by Prof. R. Wu and associates at Cornell University. The photograph was taken from a study by Dr. H.T. Nguyen at Texas Tech University.

Stress proteins is a large group of different proteins induced by different environmental and biotic stress in various organisms ranging from prokaryotes to man.

A group of relatively small molecular weight proteins is developmentally regulated in growing seed such as that of barley. Their accumulation during embryo development has a role in protecting the embryo as the seed matures and desiccates during maturation (typically to about 10% water content). These are defined as 'late embryogenesis abundant' (LEA) proteins. Further research found that LEA proteins consist of a family, including several similar

proteins such as "dehydrins". These are not limited to seed embryo and they can be induced by drought stress in various plant tissues. Some are ABA responsive while others are not. Additional information especially on LEA protein and desiccation tolerance in seed is available on line.

Work with transgenic plants indicated that the LEA family of stress protein might have a role in drought and osmotic stress resistance. Their exact function is not clear but it may involve osmotic adjustment or protection of cellular membranes or organelles during desiccation.

They may also act as molecular chaperons and in that respect they are very similar to low molecular weight heat shock proteins (HSP) (#3741). In this role they may conserve protein structure during stress. The LEA family of proteins may carry an important potential for enhancing stress resistance.

ABA (Abscisic Acid) Accumulation and Its Consequences

ABA accumulates in various plant parts subjected to desiccation. ABA responsive genes are often assumed to be stress adaptive. The rational is that if plants under stress consistently respond by producing ABA in leaves and root, then ABA must be important for coping with stress.

The most prominent effect of ABA accumulation in the plant is stomatal closure and reduced transpiration. However, a review of the literature indicates that ABA has numerous and critical negative effects on plants especially when crop productivity is considered.

Table 1. Effects of ABA on plant processes involved with growth and reproduction

Growth	
General growth	Inhibition
Cell division	Decrease
Cell expansion	Decrease
Leaf initiation	Inhibition
Germination	Decrease
Root growth	Increase
Tillering	Decrease
Dormancy	Improved

Reproduction	
Flowering (annuals)	Advance
Flower induction	Inhibition
Flower abscission	Increase
Pollen viability	Decrease
Seed set	Decrease

Figure: Wheat seedlings grown in vermiculite and severely desiccated after which they were irrigated. The seedling on the right received 0.1 μmol of ABA in the irrigation water before the onset of stress. Control seedlings on the left.

Genotypes of wheat that were selected for a high capacity for ABA accumulation under drought stress were found to be no better or even worse than the normal ones in terms of function and yield under drought stress. Selection for low leaf ABA content in maize was correlated with reduced yield under conditions of limited water supply (#5393).

On the other hand ABA may have an important role in regulating an orderly shutdown of plant functions towards a state of dormancy, as the case is for the maturing seed. Dormancy is essential for surviving extreme plant desiccation. ABA mediated dormancy is crucial for attaining freezing tolerance, which involves cellular desiccation. The value of plant survival under sever desiccation depends on the agricultural ecosystem concerned. It can be important in subsistence agricultural where plant recovery from sever drought stress can provide some growth and production. Plant survival is of lesser consequence to commercial crop production as practiced in developed countries. Even the ability of seedling survival and recovery after a prolonged drought in commercial wheat production does not carry great impact when re-seeding the crop is a technically and an economically viable option. A commercial crop based on recovered seedlings is likely to be inferior to that grown from newly planted seed. The present knowledge on ABA and it role in plant adaptation to drought stress does not allow yet to formulate a breeding strategy with respect to ABA.

Drought Escape and Plant Phenology

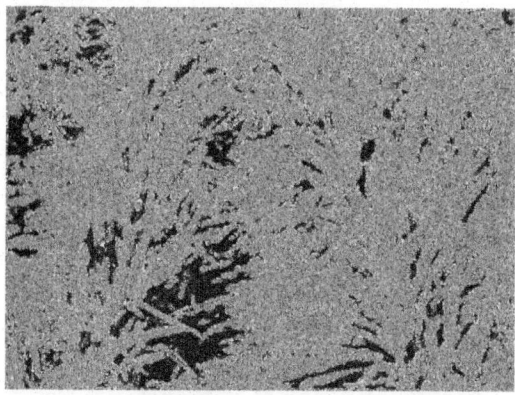

Figure: Early flowering (left) and late flowering (right) sorghum cultivars under late-season drought stress. The late cultivar will not flower at all due to stress.

Short growth duration (generally defined by early flowering) constitutes an important attribute of 'drought escape', especially for conditions of a late-season drought stress. On the other hand, longer growth duration is often associated with high yield potential. Consequently, using drought escape as a solution may involve a cost in terms of reduced yield potential. This is serious, especially when the moisture environment is unpredictable and may vary to a large extent between years. The more predictable the environment is, the easier it becomes to optimise phenology. The unpredictability of the environment may reach a state where short growth duration is a drawback, especially in indeterminate plants that offer a potential for regrowth and productivity upon recovery. A longer growth duration in both determinate and indeterminate plants would improve the probability for regrowth upon recovery simply because, on the same calendar day, late-maturing genotypes are younger than early ones and younger plants recover better. The final decision on the optimum growth duration has, of course, to consider additional factors, such as late-season disease and insect pressure or periods of frost.

Early maturity leads to reduced, total seasonal evapotranspiration simply because of the shorter time in the field. However, as growth duration is genetically linked with leaf number, early genotypes tend to have a small transpiring leaf-area index. Thus, early genotypes show reduced evapotranspiration during most growth stages, up to the point where a full ground cover is achieved. At most growth stages, root-length density and total root length per plant is generally greater in a late than in an early cultivar. This should be reflected in an advantage for the late genotype under conditions where extensive rooting is required.

A phenological feature specific to maize is the timing of anthesis with respect to silking, defined as anthesis-to-silking interval (ASI). Evidently a short interval is desirable whereas a large interval results in poor pollination. The maize program at CIMMYT dedicated many years of work to research the trait and explore its significance in tropical maize breeding for stress environments (#). Maize germplasm can vary for ASI irrespective of the effect of stress; a short ASI is a universally important trait for maize production. However, stress, and especially drought during the reproductive stage may extend ASI and thus reduce yield. Maize genotypes may vary in ASI under drought stress from few days up to a month or more. The effect on yield of change in ASI between null and 10 days is exponential. Selection for short ASI under drought stress proved to be the most effective approach to improve drought resistance of tropical maize. QTLs (quantitative trait loci) controlling ASI were located and marker-assisted selection is possible.

Different crop plants may advance (*e.g.* wheat) or delay (*e.g.* rice) their flowering when stress occurs before flowering. The rate of delay is a function of plant water deficit and probably also ABA signalling. The rate of change in flowering time under stress can be taken as an index of genotypic rate of stress in the field.

Water-use Efficiency (WUE)

WUE is not a component of drought resistance but the term implies greater production for a given amount of limited water. Namely "more crop per drop". This is not necessarily the case. As the following discussion will clarify, high WUE result (in most cases) from "less drop per crop".

WUE was originally developed by agriculture engineers as a ratio between yield and irrigation water in order to assess returns for irrigation input and cost. WUE is an important yardstick to measure irrigation efficiency. The WUE term was later adopted by soil scientists and agronomists for a wider use in agronomy, including dryland-rainfed crop production. Physiologists found the term useful also at the leaf level in studies of gas exchange where WUE (*i.e.* "transpiration ratio") is defined as the ratio of carbon fixation to transpiration. WUE can therefore be used at various levels of the crop, from the single leaf to the field.

Studies of water use efficiency at the whole plant and field level were cumbersome due to the work load and costs involved in assessing whole plant or crop water use, especially when large plant populations in plant breeding were considered. The breakthrough came with the development of better understanding of stomatal dynamics, gas exchange and photosystem function, leading to the carbon isotope discrimination (delta) assay as a heritable marker for WUE at the whole plant level (Farquhar et al. 1989; Hall et al. 1994). The reader is referred to these papers for details on the theory and the method (which is not cheap). In the majority of cases low carbon isotope discrimination (low delta) as measured in the grain or the leaves was found to be well correlated with high WUE across variable genetic materials and vice versa, with very few exceptions where delta was not associated with WUE.

An important contribution of the carbon isotope discrimination method was that it enhanced research on WUE and provided extensive data on the subject especially in the context of breeding and genetic diversity. At the same time the large volume of published information on delta, WUE and their implications towards selection for water limited environments created some confusion in the plant breeding community. Confusion was largely created by the fact that the relations between delta (WUE) and yield were sometimes positive and sometimes negative, depending on the crop growing conditions. It therefore appears that WUE as a target in breeding for water-limited environments is obscure if not constantly moving. Plant breeders discussing carbon isotope discrimination and WUE expressed confusion on two primary questions: (1) under what environmental conditions selection for carbon isotope discrimination is expected to result in yield gain, and (2) which direction should selection be made, high (low delta) or low (high delta) WUE.

Once a breeder can resolve the question for what delta value he should select for under the drought conditions of interest, the second question is if what he really requires is a genotype expressing high WUE under drought stress. WUE is often equated in a simplistic manner with drought resistance without considering the fact that it is a ratio between two physiological (photosynthesis and transpiration) or agronomic (yield and crop water use) variables. As a ratio it is often susceptible to misinterpretation, especially when the dynamics of the nominator and the denominator are ignored. A discussion of WUE in the context of plant breeding for plant production under water limited environments is presented by Blum (2005). However, a second paper (Blum 2009) provides further insight into WUE in breeding and explains the source of the confusion about WUE in breeding and why it is an ambiguous selection criterion for yield in most water limited environments. This review suggests that the target of plant breeding for water limited environments is effective use of water (EUW) rather than WUE.

Photosynthetic Systems and Water-use Efficiency

Plant science is still seeking ways to genetically increase productivity for a given unit of water-use under drought stress. The key is in photosynthesis. The C_4 photosynthetic metabolism as compared with the more widely common C_3 type photosynthetic metabolism is intimately associated with superior productivity at given water-use. The C_4 pathway of photosynthesis as found in maize, sorghum, pearl millet, and various forage grasses is essentially a pumping mechanism that moves CO_2 from the mesophyll cells and causes high CO_2 concentrations in the specific biochemically active vascular-bundle sheath cells. This mechanism goes hand in hand with certain anatomical and morphological features of the C_4 plant ("Kranz leaf anatomy") that are inseparable from the system as a whole. The CO_2-concentrating mechanism results in a high utilisation efficiency of low intercellular CO_2 concentrations. This is due to the PEP carboxylase enzyme in the C_4 plant, which unlike RuBP carboxylase is insensitive to atmospheric O_2 concentrations. Atmospheric O_2 concentrations are strongly inhibitive to CO_2 uptake in C_3 plants where CO_2 is fixed directly by RuBP carboxylase. In C_4 plants CO_2 fixation is carried out in the bundle-sheath cells using CO_2 from decarboxylated C_4 acids in the mesophyll cells. This sequence results in sufficiently high CO_2 concentration maintained at the bundle sheath cell. The efficiency of the CO_2 fixation pathway in the C_4 plant bears significance toward its transpiration-ratio. For a given rate of transpiration, photosynthesis is greater in C_4 than in C_3 plants. This advantage is also translated into a greater plant or crop WUE. It is not, however, necessarily related to drought resistance. However, under well-watered conditions, such as with irrigation, the greater WUE of the C_4 plant is most likely translated into better economic returns on the cost of irrigation. The normal WUE (for grain yield) of supplemental irrigation in grain sorghum (C_4) is about 20 kg mm^{-1} ha^{-1}, as compared with 10 kg mm^{-1} ha^{-1} in wheat (C_3).

Whereas WUE is often confused with drought resistance it is very important to take note of a comparative study of C_4 and C_3 Panicoid grasses (#10259). It concluded that declining C_4 photosynthesis with water deficit was mainly a consequence of metabolic limitations to CO_2 assimilation, whereas, in the C_3 species, stomatal limitations had a prevailing role in the drought-induced decrease in photosynthesis. The drought-sensitive metabolism of the C_4 plants could explain the observed slower recovery of photosynthesis on re-watering, in comparison with C_3 plants which recovered a greater proportion of photosynthesis through increased stomatal conductance. Therefore, within the Panicoid grasses, the high WUE C_4 species are metabolically more sensitive to drought than the lower WUE C_3 species and recover more slowly from drought.

Plant science is attempting to improve yield of C_3 plant such as rice by converting their biochemistry to C_4. Less ambitious but more closely at hand is the possibility of improving C_3 leaf internal, or mesophyll, conductance to CO_2, leading to greater leaf productivity per unit transpiration (#10465).

It is not uncommon to come across opinions that drought resistance is "very complex" or "confusing" or "difficult". However, while drought resistance is not simple in terms of its physiological nature, it is conceptually simple if one accounts for two main considerations.

Firstly, besides adaptive traits drought resistance is strongly dependant on plant constitutive traits that are not necessarily induced by stress and do not require stress for their expression.

Secondly, the most important factor of drought resistance in crop production and very possibly also in natural vegetation is the ability of the plant to maintain high water status or high turgidity. This would allow sustaining function better as environmental stress increases. Various traits affect the capacity to maintain high water status and turgor. Depending on the drought stress profile, the most effective traits in terms of agronomic value are growth duration, plant size, root depth, and osmotic adjustment.

The capacity to sustain plant function at low plant water status is a rare occurrence in crops. Research shows that genotypic differences in crop plant function under drought stress can often be accounted for by respective differences in plant water status under stress rather than by true difference in function at low water status.

Breeding for Drought Resistance

Some Principles

The primary difficulty and the most important task in planning a breeding program for the improvement of drought resistance is the formulation of the drought resistant ideotype with respect to the target of the breeding program. This involves an educated logical integration of most of the information discussed on these web pages and their links, applied to a well-understood and defined target environment. The primary issue is the decision on the important phenological, developmental and adaptive traits that would be most effective in supporting production or survival under drought stress, depending on the agro-ecological, social and economic conditions of the target environment. The level of funding and intellectual support for the specific breeding program will determine whether the ideotype is likely to be attainable.

Conventional breeding for general yield improvement relies very strongly on selection for yield and its components as a main approach. Conventional breeding for drought resistance supplements selection for yield by selection for developmental and physiological attributes, which may require physiological measurements in breeding populations. Physiological methodology is generally slow and meticulous and it does not allow to measure and screen large plant populations. In most cases, indirect or rapid methods were developed as screening aids to replace the slow physiological methods. While this resulted in reduced accuracy of the measurement, it still allows partitioning the population into the desirable subpopulations. This is sufficient in the eye of the breeder who is not interested in outmost accuracy of measurements but rather in being able to reduce the population by excluding the least appropriate phenotypes.

The flaw in conventional breeding is that the breeder can identify the genotype only by measuring the phenotype. The efficiency of this approach depends on many factors, including inheritance of traits, environmental effects, measurement error and more. For certain traits, such as root depth, phenotypic measurements in very large breeding populations are technically impractical. Marker assisted selection (MAS) is a molecular technique which allows to select the desirable genotype without actually measuring the phenotype. Read more on the basics of MAS and potential application to drought resistance breeding. A general example for a case of upland rainfed rice in South-East Asia is briefly discussed below, as a very simplified demonstrative exercise. In most upland/rainfed environments where rice may be grown soils generally contain moisture at depth. Rice is known for its shallow roots. The development of deep roots is therefore crucial for upland/rainfed rice as a major trait for sustaining plant water status. Deep roots are associated with a limited number of main root axes and limited tillering. Limited tillering is associated with reduced canopy area, which should moderate water-use. On the other hand a sparse canopy does not allow good competition with weeds, where weeds are a serious problem. In some rice soils roots will not grow and penetrate the soil due to the existence of a hardpan. The capacity for hardpan penetration by roots is important and it can be improved. Molecular marker assisted selection (MAS) for these root traits in rice is possible.

Rice is characterised by very high leaf cuticular conductance, leading to a poor control over transpiration. Improved cuticular resistance by increasing epicuticular wax deposition is a reasonable breeding target, in addition to root traits.

Rice is especially sensitive to drought stress during flowering resulting in high rate of floret sterility. A complete understanding of the exact causes of drought-affected sterility is not at hand yet. It is however clear that increased cuticular resistance of spikelet surfaces would moderate panicle desiccation. Higher rate of osmotic adjustment and the accumulation of protective solutes should protect floral part against desiccation. MAS for osmotic adjustment in rice might be possible. Drought stress at the reproductive stage is a common and a major problem in these ecosystems, therefore phenotyping for panicle fertility is important. Early flowering might ascribe an advantage, pending consideration of other factors involved with early flowering (reduced yield potential, etc'). Delayed senescence may be important in other crops. Information on the value of this trait in rice under drought stress during grain filling is not sufficient. Stem reserves are an important factor in sustaining rice grain filling under drought stress (e.g. #5351).

In conclusion, there are potentially important traits for improving drought resistance in upland rice, such as root traits, cuticular resistance, osmotic adjustment and probably stem reserve utilisation for grain filling. MAS can be used for some of these traits. A managed stress environment for the control of drought phenotyping at the reproductive stage would be an essential requirement for such a breeding program.

The Managed Stress Environment

While the field in the target stress environment is the primary goal of the breeding program, paradoxically, it is often inappropriate for selection work. Besides the amplified spatial field variability when water is limited, stress is also variable from year to year. The water regime can be too sever in one year, causing complete loss of breeding materials on one hand or too favourable to constitute any stress pressure in another year. Drought stress in different seasons can also occur in different growth stages. Stress in the target field environment is typically inconsistent, causing reduced efficiency in the overall selection program. It may be argued that this variability is an inherent problem to be addressed in breeding. While this may be true, selection becomes very ineffective if it is practice under such a variable protocol. For example, if drought resistance is to be improved at two different growth stages, it must be logically addressed separately for each different stage, followed by recombination. It follows therefore that the field-screening environment must be managed for stress intensity and timing to a level that can result in a consistent selection pressure from one cycle to the next. This is roughly analogous to the use of controlled disease infection in the selection for disease resistance.

Controlled drought stress implies the appropriate duration and severity of stress at the appropriate plant growth stage. Controlling drought stress in the greenhouse or the growth chamber is relatively straightforward. In the field, however various means are required to achieve control by eliminating rainfall on one hand and by providing irrigation on the other. The ideal field selection site for drought resistance would be in a desert environment with a minimal amount of rainfall, where almost any crop water regime can be designed by irrigation. While this may not always be possible it is the conceptual basis of the managed stress environment. It follows that most breeding programs which have a component for drought resistance must develop a special phenotyping site where stress can be managed to a reasonable extent. Alternatively, certain natural drought stress conditions may be quite repeatable from year to year or very easy to manage by irrigation. This is the case for crops grown exclusively on stored soil moisture from previous season precipitation. This stress scenario is found for example in the Mediterranean summer crops or the "rabi" season in parts of India.

When terminal stress (stress at the final reproductive growth stages) is considered, a delay in planting in most cases would put this stage in a dry season. Another possibility is to grow the population during a dry off-season if climate and biotic factors allow it. This approach was very successful with upland rice breeding at IRRI in the Philippines. Since growing plants in the dry offseason might expose them to somewhat different climatic conditions, an offseason nursery should be used mainly for recording results on drought resistance responses but actual selection should be performed with the same (duplicate) materials during the normal season. Exceptions are noted where selection for yield under stress in an offseason stress nursery was effective in gaining progress for drought resistance. The corollary is to understand the climatic and biotic factors which might affect plant growth and yield in the offseason nursery.

When a managed field site is impossible to achieve, the next option is the rainout shelter. A complete discussion of this option is available here.

Figure: Forage sorghum breeding materials (tall plants) grown on a line-source irrigation system. Source is indicated (arrow) and the growth of plants is seen reduced perpendicular to the source.

Where rainfall is limited in the natural field selection environment, such as dry season in the tropics or the summer season in the Mediterranean, managed stress environments can be designed by irrigation control. Options range from having a stress and non-stress environments side by side to the *line-source irrigation system*. This system is based on the fact that any sprinkler irrigation system spreads water in a gradient where the maximum amount is discharged at the source with a diminishing amount away from the source. Hence the amount of water available to the plants decreases perpendicular to the sprinkler irrigation line. Breeding materials can be planted in long plots or rows perpendicular to the line and be subjected to an increasing drought stress away from the line. Observations on plant response along a water supply gradient within each genotype can be very effective for revealing resistant materials. The 'Mixed-procedure' application (SAS) can perform the statistical analysis of data from a line-source irrigation system, if needed.

Variability of the Field Environment

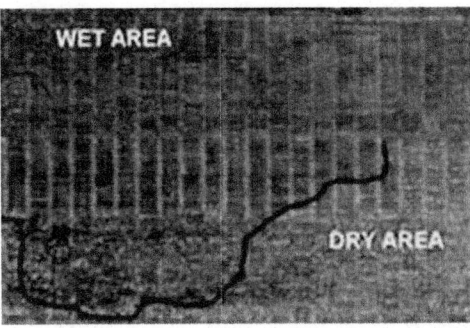

Figure: Aerial photograph of a sorghum breeding nursery under dryland conditions in a farmer field.

Any book on field experimentation deals extensively with the issue of spatial variability in the field, which is a major problem requiring detailed statistical design and analysis of field experiments. The field is variable in terms of topography and soil characteristics. Soil characteristics vary in all dimensions. This variability is amplified when the water regime is concerned and especially when water is lacking. Demonstrates field variability for soil moisture seen in a sorghum breeding nursery subjected to drought stress. Plots that are generally above the drawn line are situated on wetter soil and therefore appear more colour saturated (having higher leaf water status) as compared with more desiccated plots below the line.

Whichever methods and precautions are deployed to handle field variability as a generator of experimental error, these become especially critical in experiments involving water deficit. While suitable field topography (flat with a slight homogenous slope) can be identified, it is extremely difficult to estimate soil variability at the site with respect to its water characteristics, just by using soil tests. It is therefore highly recommended to perform a field homogeneity test by growing a homogenous commercial crop in the candidate field before choosing it for screening work. The crop should be water stressed and then observed for variability in plant development. Photogrammetric methods can also be applied for this purpose. Machinery that surveys the field by measuring soil electroconductivity is becoming a popular method after its use in precision agriculture application.

Yield as a Selection Criterion for Drought Resistance

The issue of selection for yield and the impact of the environment on genetic gains from selection and selection efficiency are under continuous debate as a central issue in conventional plant breeding. The comparative yield performance of two genotypes with respect to one another can vary from one environment to the other and this is basically defined as genotype by environment interaction (GxE) for yield. Generally, the ideotype preferred by most breeders and breeding textbooks is one that expresses minimal GxE and its yield is "stable" across all environments. The question is the spectrum of environmental diversity across which one variety can be stable.

However, while yield under stress is the target of the breeding program, selection for yield under stress is generally inefficient. Yield is a complex trait that is basically not directly inherited. It is the various developmental and physiological processes which make up yield that are inherited. Therefore the heritability of yield is generally not high and it becomes especially low under stress. It has been the general and repeatable observation of breeders that using yield as a selection index under stress to improve drought resistance is not very efficient, with exceptions. To compensate for the low efficiency breeders screen very large populations with the expectation that the "numbers game" will allow to identify the desirable genotype. While this approach has been successful, it is highly expensive. The use of molecular markers to tag and select for certain yield related quantitative trait loci (QTLs) can increase the efficiency of selection for yield, but again, less effectively under stress.

There are however indications that when the breeding population contains effective genes for drought resistance (say, segregation for deep roots in upland rice population) the efficiency of selection for yield under drought stress can be increased, provided all precautions are taken to minimise spatial variability at the selection site.

If selection for yield under stress is practiced, a positive GxE for the specific drought stress conditions is a strong indicator of resistance. There are many statistical models and methods that estimate GxE in different contexts and accuracies. In most cases of a planned breeding program for drought resistance the evaluation of GxE simply requires a comparison of yield performance under stress and non-stress (fully irrigated) conditions. The comparison of genotypic performance between the two environments can be simply evaluated in yield under stress as percent of yield under non-stress. Alternatively, the 'Fischer and Maurer stress resistance index' (RI) can be computed as:

$$RI = (Gs/Gi)/(Ms/Mi)$$

Where genotype yield under stress (Gs) and non-stress (Gi) is normalised for mean yield of all genotypes under stress (Ms) and non-stress (Mi). Values above 1 indicate a relative resistance as compared with the mean of the population.

A major impediment in comparing genotypic response to a managed water stress environment is the variation among genotypes in their phenology. With such variation, different genotypes may be water-stressed at different growth stages.

The solution is to divide the population into several phenology sub-populations and compare the effect of stress only within sub-populations of similar phenology.

Alternatively, staggered planting dates can be attempted where the earlier materials are planted later as compared with the late flowering materials. However, in most breeding programs, when tests for drought resistance are performed in the field at more advanced generations (*e.g.* =>F_4), the population of lines does not express large variability for phenology.

Selection for Drought Resistance by Developmental Traits

Selection for most plant developmental traits involving drought resistance are common to general breeding practices, such as the case for phenology, anthesis to silking interval (ASI) in maize, tillering, plant size, etc'. Two unique developmental features are discussed here: roots and stem reserve utilisation.

Roots

Figure: A glass-panelled in-ground root box lysimetre installation allowing measurement of roots and transpiration. (Texas A&M Research Centre at Temple, TX).

Figure: Wheat grown in soil-filled tubes. Inset: roots as they appear in the opened tube.

Root size and development is a crucial parameter in most selection programs for drought resistance. Detailed measurements of roots or even rough screening techniques for roots are generally labourious. Probably the most practical way to select for deep and effective roots is to judge their performance by observing the shoot performance under drought stress. Much has been published on root measurement techniques, including books.

Very detailed root measurements can be performed with special growing containers and installations where roots can be observed *in situ* though a glass panel. These installations are defined as rhizotrons and they may take various forms such as individual glass-panelled soil-filled root boxes set in the ground.

By weighing these root boxes it is also possible to estimate plant transpiration and relate it to shoot and root development. Rhizotrons and lysimetres are important tools for root research and for studying a few cultivars but they are not amenable to large scale screening work.

The simplest method for a direct selection for root length involves growing single plants in soil filled disposable polyethylene tubes (used in polyethylene bags production) and then washing the root out of the soil at the time of measurement (usually at flowering). Alternatively plants can be grown in re-useable soil filled PVC tubes (~10 cm in diametre) sawed longitudinally into two halves and taped together. Tubes can be opened at any time and destructive measurements are taken on roots *in situ* or after washing away the soil. The two methods can be combined into one, where PVC tubes are set in a trench and weighed for water-use measurements while being lifted periodically.

Root penetration capacity through a hardpan is phenotyped by challenging the root to penetrate a hard layer of paraffin wax at depth. The number of roots penetrating the wax layer is an estimate of root penetration capacity (#7965).

Stem Reserve Utilisation for Grain Filling

Figure: Wheat nursery rows with the centre one sprayed with KI to destroy chlorophyll and eliminate current photosynthesis at the onset of grain filling.

The capacity for stem reserve utilisation for grain filling when the photosynthetic source is completely inhibited by stress is estimated in selection work by destroying the photosynthetic source at the onset of grain filling and measuring grain filling with no current photosynthesis in comparison with normal plants. Spraying the plants with an oxidizing chemical such as magnesium chlorate or with potassium iodide (KI) destroys chlorophyll and the photosynthetic source. The chemical is applied by spraying the whole plant or just the leaf canopy. The treatment is applied at the onset of the exponential stage of grain filling, which is about two weeks after anthesis in the small grains. Work is therefore scheduled according to the different dates of anthesis of the different genotypes. Non-treated control plots are required. Since the capacity for stem reserve support of grain filling is measured by the difference in final kernel weight between treated and control plots of any given genotype, the control must be totally free of stress, especially drought or disease.

The Method has been Thoroughly Tested and Applied in Wheat Breeding

Methods can vary from purely physiological to indirect assessments that are useful mainly for selection purposes. The physiological methods were reviewed in previous pages. Here only methods for applied selection work are indicated.

Stress Symptoms

When plants reduce their water status and loose turgor under stress they display various and very distinct symptoms. Symptoms progress in proportion of plant water deficit and they can be visually scored for and used in selection. The most notable symptom is leaf wilting. Leaf Rolling is an expression of wilting in the cereals and it is being widely used in field selection work. It is visually scored (typically on a 0 to 5 scale) and used in selection for drought resistance in various crops such as rice, wheat, barley and sorghum. Other leaf stress symptoms include leaf desiccation ("firing"), leaf tip "burning", leaf "drooping" and leaf drop. General scores of plot appearance under stress is also used by experienced breeders who are well acquainted with the various responses of their crop to drought stress. Genotypes of delayed wilting or leaf rolling are preferred offcourse.

Stress symptoms are also expressed in flowering time if stress occurs before normal flowering time. A delay in inflorescence appearance or exertion is typical of rice or sorghum. An advance in flowering is seen in wheat. Assessing the rate of delay or advance requires a comparison between stress and non-stress plots.

Canopy Temperature

Selection methods were developed and applied to plant breeding following principles and techniques used in Remote Sensing in Agriculture. Most methods are based on the spectral response of leaves and its modification with plant response to the environment.

From the previous discussions it is indicated that canopy temperature is a function of transpirational cooling. As water deficit develops, canopy temperature differences among genotypes increase and plant water status becomes the main source of this variation. Canopy temperature was used to develop a crop water stress index as a tool for crop management. Canopy temperature measured under drought stress has become a most popular, fast and significant field screening method for plant water status under drought stress. Since dehydration avoidance is the major drought resistance mechanism in crop plants, canopy temperature is a most relevant screen for drought resistance. Relatively lower canopy temperatures under stress indicate a relatively better plant water status. Lower canopy temperatures were generally found to be correlated with relatively higher yield under stress across diverse genetic materials. Canopy temperature can be measured remotely with the infrared thermometre, provided the correct protocol is strictly followed. Since its initial development as a screening method for dehydration avoidance by Blum et al. (1982), infrared thermometry of plant canopies under drought stress has become a popular method in breeding and phenotyping drought resistance. Twenty five years later and in tune with some 30 reports verifying the utility of the method in different crops, Olivares-Villegas et al. (2007) summarised their exhaustive study with wheat as follows: "Field trials under different water regimes were conducted over 3 years in Mexico and under rainfed conditions in Australia. Under drought, canopy temperature was the single-most drought-adaptive trait contributing to a higher performance, highly heritable and consistently associated with yield phenotypically and genetically. Canopy temperature epitomizes a mechanism of dehydration avoidance expressed throughout the cycle and across latitudes, which can be utilised ... as an important predictor of yield performance under drought"

Infrared Photography

The spectral reflection from leaves spans over the visible range (400-700 nm) and beyond. Leaves reflect in the near infrared (800-900 nm) and this reflectance is affected by mesophyll cell turgor. When turgor is lost and cells reduce their volume, near infrared reflectance decreases. This reflectance can be recorded on infrared sensitive film. Images with low near infrared reflectance appear less colour saturated as compared with those that reflect more.

Therefore, low-altitude aerial infrared photography of a drought stressed nursery can be used to differentiate between nursery entries in their water status. The overflight must be done around solar noon under clear and calm conditions and when drought stress is well developed. Infrared aerial photography providers can supply the photograph on a special Kodak infrared transparency film in a 22x22 cm size. Individual entries can be well recognise under a magnifying glass and scored for their colour saturation. Alternatively the image can by analysed by densitometrical methods which would provide the colour densities in numerical values.

An example is given of an enlarged section of a larger aerial infrared colour photograph of a drought stressed sorghum nursery. The relatively drought stressed lines appear less colour saturated while the more saturated coloured rows are less stressed.

Infrared digital cameras are becoming available on the market but experience with these in breeding nurseries is not available at this time.

Selection for Drought Resistance by Assessing Plant Function

As discussed previously the comparative assessment of plant function in different genotypes under drought stress (dehydration tolerance) must be normalised for plant water status. Else, differences in function among genotypes can be ascribed to differences in water status and not necessarily to real difference in function at a given plant stress. This is a difficult requirement in selection work and especially under field conditions.

An effective approach for normalising measurements of function against plant water status is to measure plant water status at the time of function measurement. Then regress function on water status across all genotypes. Genotypes that deviate positively from the regression are resistant while those that deviate negatively are susceptible in terms of the specific function. Off course, these measurements cannot be performed on large breeding populations and they may be useful for screening potential parents for crosses or very advance lines.

Two examples of the more popular selection methods for function are given here, cell membrane stability (CMS) and chlorophyll fluorescence.

Cell Membrane Stability (CMS)

This method is based on the fact that stress cause injury to cellular membranes. This injury is expressed in leakage of various cellular solutes, including electrolytes. Electrolyte leakage can be easily measured by the electro conductivity of the medium in which the affected leaf sample is placed. The method is being used for assessing thermotolerance in heated leaf samples. Basically the method compares leakage from stress-affected with leakage from control samples, calculating the relative injury or stability (the inverse of injury).

When CMS is used as a dehydration tolerance trait it is estimated in leaves subjected to advanced stress, typical to RWC of around 50-70%. Samples are taken from stressed and non-stressed (control) leaves. Samples are also taken for estimating RWC as a measure of water status. The first case of such a study was in rice where QTLs for CMS under drought stress were identified (#4793).

Chlorophyll Fluorescence

The light harvesting system of photosystem II absorbs light reaching the leaf. This light energy can be used to drive photochemistry providing the chemical energy (in the form of ATP and NADPH) for CO_2 fixation in the Calvin cycle. PSII extracts electrons from water releasing oxygen in the process. The excessive light energy that is not used in photochemistry (for different reasons, including photosystem dysfunction due to stress) can be dissipated as heat or re-emitted as fluorescence. This fluorescence is being widely used in plant science to probe the activity and performance of the photosystem under a variety of stress conditions. In some case such as wheat and maize quantitative trait loci (QTLs) for chlorophyll fluorescence signals under given set of conditions were mapped and can be used in MAS.

A host of instruments and imaging systems were developed for analysing chlorophyll fluorescence. There are different levels of analysis, depending on the purpose of the study. A unique and detailed analytical probe defines as the JIP test has been developed by Prof. Strasser in Geneva. It allows a very comprehensive dissection and interpretation of the fluorescence phenomenon. This is mainly a research rather than a selection tool. Chlorophyll fluorescence is even entering the domain of remote sensing where vegetation function might be monitored from aerial platforms in the near future.

However, fast portable and simple instruments are needed for selection work and these are available from various commercial suppliers. It must be realised however that the measurement and interpretation of chlorophyll fluorescence signal, even with simple instrumentation require a complete understanding of the phenomenon, also referred to as the "Kautsky effect".

The Polyethylene Glycol Root Medium

It is practically impossible to control plant water status of different genotypes under drought stress for reasons discussed under Impact. The only method that can provide some control over root medium water status is the use of polyethylene glycol (PEG) 8000 in the root medium. PEG is an osmoticum providing given water potential for a given concentration in aqueous solution. When plants are grown in a hydroponics system with roots immersed in PEG, they must reduce their water potential in order to take up water. This however does not necessarily imply that different genotypes will reach the same leaf water status when all are exposed to the same PEG concentration in the root medium. Still, when morphological variations among genotypes are slight, such as in seedlings, similar leaf water status might be expressed in all.

The problem with PEG is that roots may take it up and slowly toxify the plant. While PEG of large molecular weight (*e.g.* 8000) is less of a problem in this respect, still many plant species are not quite suitable for the method, unless treated in a special protocol.

The Impact of Salinity Stress

According to the FAO Land and Plant Nutrition Management Service, over 6% of the world's land is affected by either salinity or sodicity. The term *salt-affected* refers to soils that are saline or sodic, and these cover over 400 million hectares, which is over 6% of the world land area. Much of the world's land is not cultivated, but a significant proportion of cultivated land is salt-affected. Of the current 230 million ha of irrigated land, 45 million ha are salt-affected (19.5 percent) and of the 1,500 million ha under dryland agriculture, 32 million are salt-affected to varying degrees (2.1 percent).

Table 1: Regional distribution of salt-affected soils, in million hectares

Regions	Total area	Saline soils		Sodic soils	
	Mha	Mha	%	Mha	%
Africa	1,899	39	2.0	34	1.8
Asia, the Pacific and Australia	3,107	195	6.3	249	8.0
Europe	2,011	7	0.3	73	3.6
Latin America	2,039	61	3.0	51	2.5
Near East	1,802	92	5.1	14	0.8
North America	1,924	5	0.2	15	0.8
Total	12,781	397	3.1%	434	3.4%

Source: FAO Land and Plant Nutrition Management Service

Salinity occurs through natural or human-induced processes that result in the accumulation of dissolved salts in the soil water to an extent that inhibits plant growth. Sodicity is a secondary result of salinity in clay soils, where leaching through either natural or human-induced processes has washed soluble salts into the subsoil, and left sodium bound to the negative charges of the clay.

A *saline soil* is defined as having a high concentration of soluble salts, high enough to affect plant growth. Salt concentration in a soil is measured in terms of its electrical conductivity, as described in the section below on measurements. The USDA Salinity Laboratory defines a saline soil as having an EC_e of 4 dS/m or more. EC_e is the electrical conductivity of the 'saturated paste extract', that is, of the solution extracted from a soil sample after being mixed with sufficient water to produce a saturated paste. However, may crops are affected by soil with an EC_e less than 4 dS/m. The moisture content of a drained soil at field capacity may be much lower than the water content of its saturated paste. Further, under dryland agriculture, the soil water content might drop to half of field capacity during the life of the crop. The actual salinity of a rain-fed field whose soil had an EC_e of 4 dS/m could be 8-12 dS/m. As described below, this would severely limit yield of most crops.

Types and Causes of Salinity

Natural or Primary Salinity

Primary salinity results from the accumulation of salts over long periods of time, through natural processes, in the soil or groundwater. It is caused by two natural processes.

The first is the weathering of parent materials containing soluble salts. Weathering processes break down rocks and release soluble salts of various types, mainly chlorides of sodium, calcium and magnesium, and to a lesser extent, sulphates and carbonates. Sodium chloride is the most soluble salt.

The second is the deposition of oceanic salt carried in wind and rain. 'Cyclic salts' are ocean salts carried inland by wind and deposited by rainfall, and are mainly sodium chloride. Rainwater contains from 6 to 50 mg/kg of salt, the concentration of salts decreasing with distance from the coast. If the concentration is 10 mg/kg, this would add 10 kg/ha of salt for each 100 mm of rainfall per year. Accumulation of this salt in the soil would be considerable over millennia. The amount of salt stored in the soil varies with the soil type, being low for sandy soils and high for soils contain a high percentage of clay minerals. It also varies inversely with average annual rainfall. For example, in Western Australia, the salt content of a 40 m profile ranges from 170 to 950 tonne/ha for rainfall averaging from 1000 mm to 600 mm per year.

The composition of rainwater varies greatly depending on prevailing winds and distance from the coast, and the table gives the composition of rainwater from a northern hemisphere source (Encyclopaedia Britannica). It is measured as mg/kg or ppm (parts per million). The composition of seawater is uniform around the globe, and is expressed as g/kg or ppt (parts per thousand). The electrical conductivity of rainwater is about 0.01 dS/m, and of seawater is 55 dS/m.

Table 2: Concentration of salts in rain and seawater.

Ion	Rainwater (local)		Seawater (global)	
	mg/kg(ppm)	(μmol/L)μM	g/kg(ppt)	(mmol/L)mM
Sodium (Na⁺)	2.0	86	10.8	470
Chloride (Cl⁻)	3.8	107	19.4	547
Sulfate (SO₄²⁻)	0.6	6	2.7	28
Magnesium (Mg²⁺)	0.3	11	1.3	53
Calcium (Ca²⁺)	0.1	2	0.4	10
Potassium (K⁺)	0.3	8	0.4	10
Total	7.0		35.0	

Secondary or Human-Induced Salinity

Secondary salinisation results from human activities that change the hydrologic balance of the soil between water applied (irrigation or rainfall) and water used by crops (transpiration). The most common causes are (i) land clearing and the replacement of perennial vegetation with annual crops, and (ii) irrigation schemes using salt-rich irrigation water or having insufficient drainage. Prior to human activities, in arid or semi-arid climates, the water used by natural vegetation was in balance with the rainfall, with the deep roots of native vegetation ensuring that the water tables were well below the surface. Clearing and irrigation changed this balance, so that rainfall on the one hand, and irrigation water on the other, provided more water than the crops could use. The excess water raises water table and mobilises salts previously stored in the subsoil and brings them up to the root zone. Plants use the water and leave the salt behind until the soil water becomes too salty for further water uptake by roots. The water table continues to rise, and when it comes close to the surface, water evaporates leaving salts behind on the surface and thus forming a 'salt scald'. The mobilised salt can also move laterally to water courses and increase their salinity.

Irrigated lands of the world in 1987 totalled 227 Mha. In many irrigated areas, the water table has risen due to excessive amounts of applied water coupled with poor drainage. In most of the irrigation projects located in semi-arid and arid areas, the problems of waterlogging and soil salinity have reached serious proportions even before the full potential of the irrigation project could be realised. Most of the irrigation systems of the world have caused secondary salinity, sodicity or waterlogging. Table below shows that the proportion of salt-affected irrigated land in various countries ranges from a minimum of 9% to a maximum of 34%, with a world average of 20%. Irrigated land is only 15% of total cultivated land, but as irrigated land has at least twice the productivity of rainfed land, it may produce one-third of the world's food.

Table 3: Global estimate of secondary salinisation in the world's irrigated lands

Country	Total Land Area Cropped	Area Irrigated		Area of Irrigated Land that is salt-affected	
	Mha	*Mha*	*%*	*Mha*	*%*
China	97	45	46	6.7	15
India	169	42	25	7.0	17
Soviet Union	233	21	9	3.7	18
United States	190	18	10	4.2	23
Pakistan	21	16	78	4.2	26
Iran	15	6	39	1.7	30
Thailand	20	4	20	0.4	10
Egypt	3	3	100	0.9	33
Australia	47	2	4	0.2	9
Argentina	36	2	5	0.6	34
South Africa	13	1	9	0.1	9
Subtotal	843	159	19	29.6	20
World	1,474	227	15	45.4	20

Source: Ghassemi et al. (1995) compiled from FAO data for 1987

Irrigation water adds appreciable amounts of salt, even with good quality irrigation water containing only 200-500 mg/kg of soluble salt. Irrigation water with a salt content of 500 mg/kg (i.e. 500 mg/L) contains 0.5 tonnes of salt per 1,000 m^3. Since crops require 6,000-10,000 m^3 of water per hectare each year, one hectare of land will receive 3-5 tonnes of salt. Because the amount of salt removed by crops is negligible, salt will accumulate in the root zone, and must be leached by supplying more water than is required by the crops. If drainage is not adequate, the excess water causes the water table to rise, mobilising salts which accumulate in the root zone. When the crop is unable to use all the applied water, waterlogging occurs.

Land clearing also changes the hydrological balance. In its natural state, native deep-rooted and perennial vegetation use almost all the rainwater that falls on the land. In arid or semi-arid climates the growth rate of the natural vegetation is limited by the availability of fresh rainwater. Salts will be flushed down by rain, and accumulate at the bottom of the root zone to the limiting concentration for roots to extract water, at approximately 50 dS/m. Clearing the deep-rooted native vegetation, and replacing it with shallow-rooted annual species that do not use all the rainfall, allows rainwater to escape below the roots, and 'recharge' the groundwater. Clearing of native vegetation for dryland agriculture can increase the rate of drainage by 100 times. In the Mallee Region of southern Australia, measurements show that drainage under native mallee vegetation is only 0.1 mm per year, but under annual crops it is 10 mm per year. This additional rainwater enters the groundwater or aquifers, causing the watertable to rise. In Australia, 2 Mha of land have been damaged by rising watertables due to land clearing, and another 15 Mha are at risk of salinisation by rising watertables over the next 50 years.

Soil Sodicity and Sub-soil Salinity

Sodic soils have a low concentration of soluble salts, but a high percent of exchangeable Na^+; that is, Na^+ forms a high percent of all cations bound to the negative charges on the clay particles that make up the soil complex. Sodicity is defined in terms of the threshold ESP (exchangable sodium percentage) that causes degradation of soil structure. The negatively charged clay particles are held together by divalent cations. When monovalent cations such as Na^+ displace the divalent cations on the soil complex, and the concentration of free soluble salts is low, the complex swells and the clay particles separate ('*disperse*'). The USDA Salinity Laboratory defines a sodic soil as having an ESP greater than 15, but in Australia it is considered sodic when the ESP is greater than 6. This lower threshold is due to Australian soils having a low content of other soluble cations, particularly Ca^{2+}, which help to stabilise clay colloids during leaching. If the concentration of soluble salts is sufficiently low, hydrolysis of the sodic clay will occur, creating a highly alkaline soil. *Alkaline soils* are a type of sodic soil with a high pH due to carbonate salts, and are defined as having an ESP of 15 or more with a pH of 8.5-10.

The process of sodicity is complex and occurs over a long period of time. Initially, salts that have accumulated within the soil profile, from either airborne deposition or mineral weathering, cause the clay fraction of the soil to become saturated with sodium. Subsequently, leaching of the profile, either by rainwater over prolonged periods, or by irrigation with fresh water, lowers the electrolyte concentration and the clay particles disperse. Further leaching washes the dispersed clay particles deeper into the profile where they block pores and hinder infiltration of water. The soil then is very slow to drain, and is readily waterlogged.

In semi-arid environments, soil profiles are commonly *saline/ sodic*, where the salt has accumulated due to the low permeability of the sodic subsoil. In theory, if sufficient salts accumulate, the threshold electrolyte concentration for flocculation will be exceeded and the clay will flocculate and take on pseudo-structure. However, given that permeability and leaching will then increase, the subsequent dilution of salts will cause colloids to disperse. Consequently, a quasi-steady state between flocculation and dispersion processes is maintained.

Saline/sodic soils are widespread in arid and semi-arid lands of the world, with a large component in Australia (over 250 million ha). Water infiltration is slow, and salts derived from rainfall or weathering reactions accumulate in saturated zones in the subsoil. The term '*transient salinity*' denotes the seasonal and spatial variation of salt accumulation in the root zone not influenced by groundwater processes and rising water table (Rengasamy, 2002). This transient salinity fluctuates in depth, due mainly to seasonal rainfall patterns. Transient salinity is extensive in many landscapes dominated by subsoil sodicity. Probably, two thirds of the agricultural area of Australia has a potential for transient salinity not associated with groundwater (Rengasamy, 2002).

Measuring Soil Salinity

Soil salinity is measured by its electrical conductivity. The SI unit of electrical conductivity (EC) is dS/m. The relationship to other units of conductivity, and to NaCl concentration (10 mM NaCl has an EC close to 1 dS/m). Originally the conductivity was measured in a saturated paste extract (EC_e), but the method is tedious as first the saturated paste has to be made, second the water needs to be extracted by a powerful vacuum pump, and third a very sandy soil does not make a saturated paste. A more convenient and universal method is a '1:5 extract'. The soil is dried or compressed, shaken with 5 g deionised water per 1 g soil, the probe of a hand-held conductivity metre is placed in the suspension, and the EC measured. The EC of the soil that was sampled can be calculated if its water content is known at the time of sampling. Alternatively, the measurement can be related to the field capacity of the soil using conversion factors. As a rough guide, a sandy soil will have a field capacity of 0.2 g water per g soil, and a clay soil about 0.4 g/g.

Table : Electrical conductivity (EC) of pure solutions at 20°C (dS/m).

The solutions represent those of salts found in soils or in seawater. Data from the Handbook of Physics and Chemistry (CRC Press, 55th editition, 1975). (Note that 1 dS/m = 1 mmho/cm).

Solution	EC (dS/m)
10 mM NaCl	1.0
100 mM NaCl	9.8
500 mM NaCl	42.2
10 mM KCl	1.2
10 mM CaCl2	1.8
10 mM MgCl2	1.6
50 mM MgCl2	8.1

The electrical conductivity of irrigation or river water is measured with the same hand-held conductivity metre as above, but is expressed in units 1000 times magnified, as channel or river water would normally have a very low concentration of salts. River water quality is often expressed as dS/cm (1000 x dS/m). Irrigation water quality is often expressed as total soluble salts, an international convention being that 1 dS/m is equivalent to 640 mg/L of mixed salts.

Soil salinity on a large scale is mapped with an electromagnetic (EM) conductivity metre. This instrument estimates the bulk electrical conductivity of the soil, which depends on the salinity of the soil solution, its water content, and the type and amount of clay in the soil. The output needs to be calibrated by chemical measurements of cores taken from the field. The ground EM conductivity metres consist of a small transmitter coil, energised with an alternative current. This current generates a primary magnetic field in the ground, which induces small currents which generate their own secondary magnetic field, which is smaller and proportional to the soil conductivity. A receiver coil close by measures the primary and secondary magnetic fields. The EM38 metre (Geonics Limited) is designed for agricultural surveys and measures to 1.5 m depth in the vertical mode and 0.75 m in the horizontal mode. An EM38 survey can be used to map the extent of subsoil salinity as well as discharge areas.

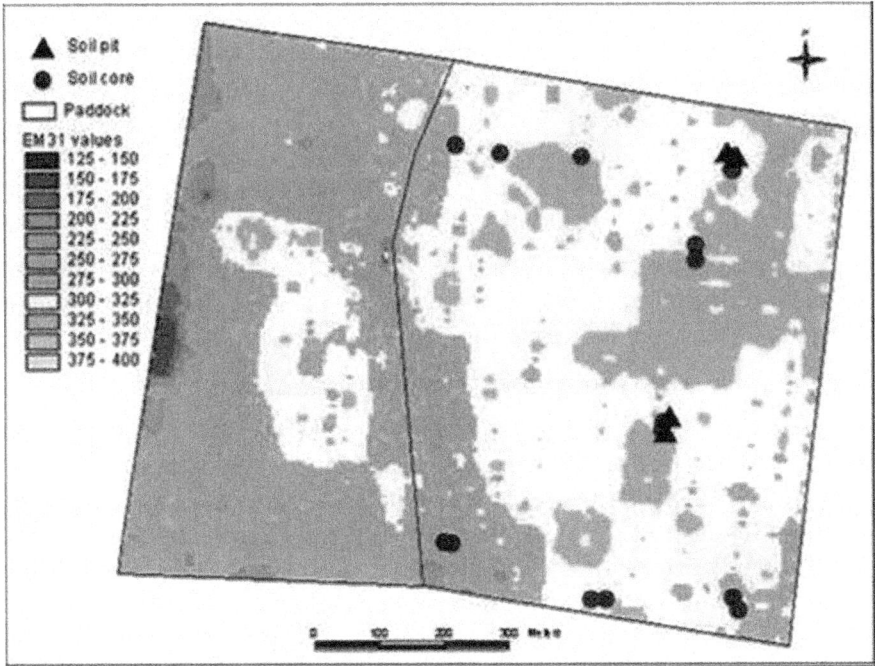

Figure: Example of an EM survey taken on a paddock basis at ground level. (Image courtesy of P. Rampant, Department of Primary Industry, Bendigo, Australia).

Airborne EM mapping is feasible now that global positioning systems have become available for accurate and rapid mapping. In the airborne method, the transmitter is slung below a plane flying at 150 m, and the receiver is towed behind. The plane generates a magnetic field, and the moving field passsing through the ground creates a secondary electric field whenever it hits something conductive. The receiver trailing behind the plane collects the data (For example, the Australian National Airborne Physics Project).

The Repercussions of Salinity

The Effect of Salinity on Plants

Salts in the soil water may inhibit plant growth for two reasons. First, the presence of salt in the soil solution reduces the ability of the plant to take up water, and this leads to reductions in the growth rate. This is referred to as the osmotic or water-deficit effect of salinity. Second, if excessive amounts of salt enter the plant in the transpiration stream there will be injury to cells in the transpiring leaves and this may cause further reductions in growth. This is called the salt-specific or ion-excess effect of salinity (Greenway and Munns, 1980). The definition of salt tolerance is usually the percent biomass production in saline soil relative to plants in non-saline soil, after growth for an extended period of time. For slow-growing, long-lived, or uncultivated species it is often difficult to assess the reduction in biomass production, so percent survival is often used.

As salinity is often caused by rising water tables, it can be accompanied by waterlogging. Waterlogging itself inhibits plant growth and also reduces the ability of the roots to exclude salt, thus increasing the uptake rate of salt and its accumulation in shoots.

Variation in Salt Tolerance between Species

The three most important crops in the world are wheat, rice and maize. Differences in the growth response of various species. Wheat is one of the more salt-tolerant crop species, and many cultivars that have been selected for yield in water-limited conditions do not suffer a 50% reduction in biomass until salinities reach 15 dS/m (approximately 150 mM NaCl). Rice is more salt-sensitive, and many cultivars suffer a 50% reduction in growth at half this concentration of salts. Maize falls in between these two species in terms of salt sensitivity.

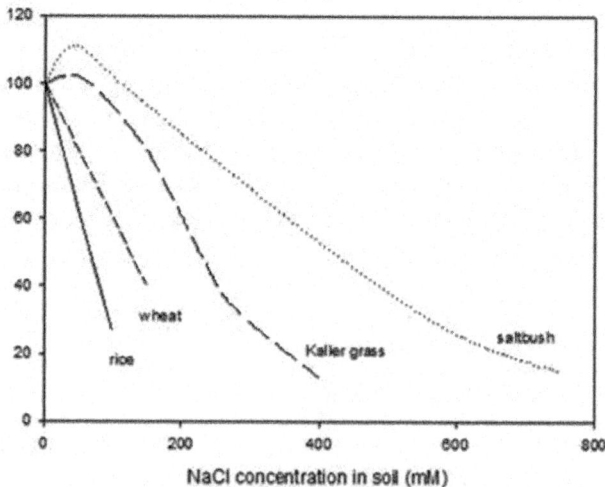

NaCl concentration in soil (mM)

Figure: Biomass production of four diverse and important plant species in a range of salinities. Wheat is one of the more salt-tolerant crops, and rice is one of the more salt-sensitive crops. Two halophytes: a saltbush species *Atriplex amnicola* and a grass *Diplachne* (syn. *Leptochloa*) *fusca* or Kallar grass. Both halophytes show outstanding salt tolerance with high growth rates and are being used in Australia and Asia for grazing on saline land.

Another criterion of salt tolerance of crops is their yield in saline versus non-saline conditions. A survey of salt tolerance of crops, vegetables and fruit trees was made by the USDA Salinity Laboratory. This shows for each species a threshold salinity below which there is no reduction in yield, and then a regression for the reduction in yield with increasing salinity.

Full details are available on line. The data in some cases are for a single cultivar of the species, or a limited number of cultivars at a single site, so they are not necessarily representative of the species. Further, the data are related to an EC_e value, which is not an appropriate reference point for a sandy soil, or for many current soil salinity estimates that based merely on a 1:5 extract. However, the data are useful in that they show the wide range of tolerance across species, and also show that yield has a different pattern of response than does vegetative biomass Yield always shows a threshold in response to a range of salinities, but with young plants a threshold is rarely seen.

With plants exposed to salinity at an early stage of seedling development there are linear reductions in both leaf area expansion and total plant biomass with increasing salinity.

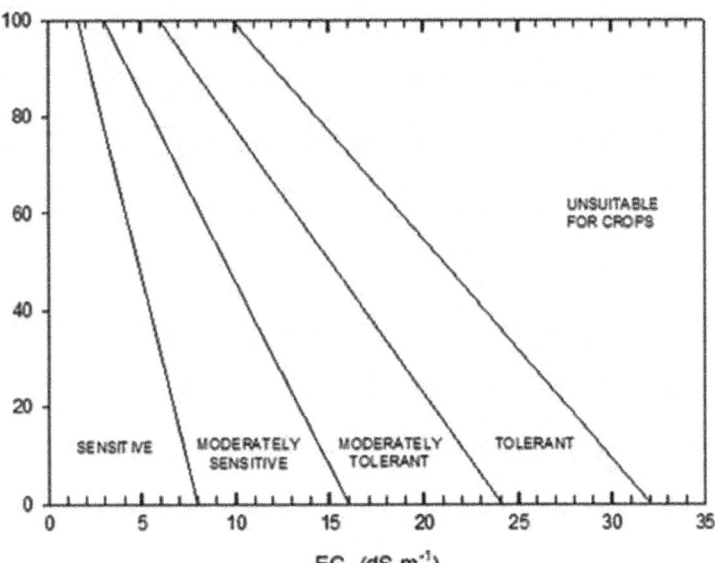

Figure: Categories for classifying crop tolerance to salinity according to the USDA Salinity Lab. Note that the ECe is more applicable to an irrigated than a rainfed field, in the latter the soil moisture content might be 2-4 time less than in a saturated paste.

Causes of the Growth Reduction under Saline Conditions

The effects of a saline soil are two-fold: there are effects of the salt outside the roots, and there are effects of the salt taken up by plants.

The salt in the soil solution (the "osmotic stress") reduces leaf growth and to a lesser extent root growth, and decreases stomatal conductance and thereby photosynthesis (Munns, 1993). The cellular and metabolic processes involved are in common to drought-affected plants. The rate at which new leaves are produced depends largely on the water potential of the soil solution, in the same way as for a drought-stressed plant. Salts themselves do not build up in the growing tissues at concentrations that inhibit growth: meristematic tissues are fed largely by the phloem from which salt is effectively excluded, and rapidly elongating cells can accommodate the salt that arrives in the xylem within their expanding vacuoles. So, the salt taken up by the plant does not directly inhibit the growth of new leaves.

The salt within the plant enhances the senescence of old leaves. Continued transport of salt into transpiring leaves over a long period of time eventually results in very high Na^+ and Cl^- concentrations, and they die. The rate of leaf death is crucial for the survival of the plant. If new leaves are continually produced at a rate greater than that at which old leaves die, then there might be enough photosynthesising leaves for the plant to produce some flowers and seeds. However, if the rate of leaf death exceeds the rate at which new leaves are produced, then the plant may not survive to produce seed. For an annual plant there is a race against time to initiate flowers and form seeds, while the leaf area is still adequate to supply the necessary photosynthate. For perennial species, there is an opportunity to enter a state of dormancy, and thus survive the stress.

The two responses occur sequentially, giving rise to a two-phase growth response to salinity. The first phase of growth reduction is quickly apparent, and is due to the salt outside the roots. It is essentially a water stress or osmotic phase, for which there is surprisingly little genotypic difference. Then there is a second phase of growth reduction, which takes time to develop, and results from internal injury. The two-phase growth response. The experiment was conducted with two genotypes with contrasting rates of Na^+ uptake, and known differences in salt tolerance; previous experiments had shown that the genotype with the low Na^+ uptake rate had a higher survival of high salinity. That during the first 3-4 weeks after the soil was salinised, there was a large growth reduction in both genotypes. This is called the 'Phase 1' response, and is due to the osmotic effect of the salt. Then after 4 weeks, the genotypes separated; the one with the low Na⁺ uptake rate continued to grow, although still at a reduced rate compared to the controls in non-saline solution, but the one with the high Na⁺ uptake rate produced little biomass and many individuals died. This is the 'Phase 2' response, and is due to genotypic differences in coping with the Na⁺ or Cl⁻ ions in the soil, as distinct from the osmotic stress.

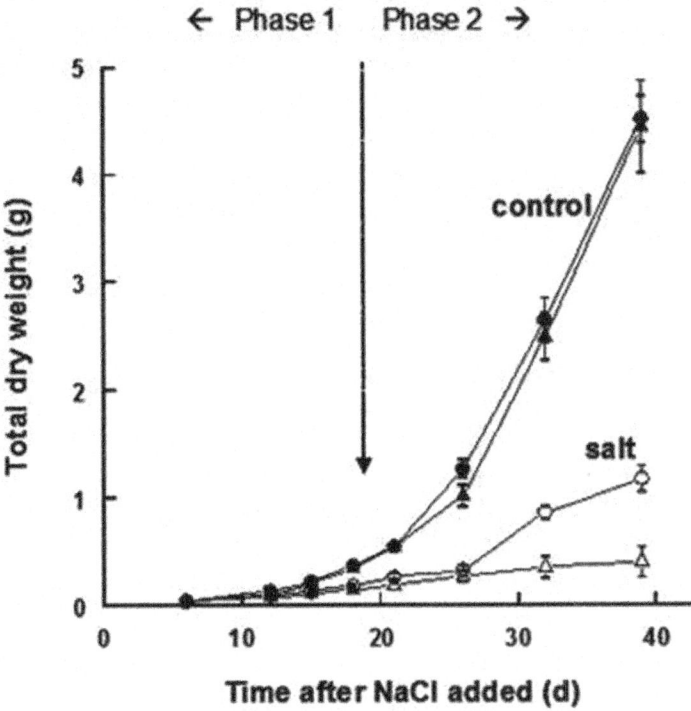

Figure: Two accessions of the diploid wheat progenitor *Ae. tauschii* grown in supported hydroponics in control solution (closed symbols) and in 150 mM NaCl (open symbols). Circles denote the tolerant accession, triangles the sensitive one. The arrow marks the time at which symptoms of salt injury could be seen on the sensitive accession; at that time the proportion of dead leaves was 10% for the sensitive and 1% for the tolerant accession (Munns et al., 1995).

These results illustrate the principle that the initial growth reduction is due to the osmotic effect of the salt outside the roots, and that what distinguishes a salt-sensitive plant from a more tolerant one is the inability to prevent salt from reaching toxic levels in the transpiring leaves, which takes time.

To grow in saline conditions, plants must maintain a high water status in the face of soil water deficits and potential ion toxicity. A plant can only grow or survive in a saline soil if it can both continue to take up water *and* exclude a large proportion of the salt in the soil solution.

The Required Extent of Salt Exclusion

Roots must exclude most of the Na^+ and Cl^- dissolved in the soil solution or the salt will gradually build up with time in the shoot and become so high that it kills it. To prevent salt building up with time in the shoot, roots should exclude 98% of the salt in the soil solution, allowing only 2% to be transported in the xylem to the shoots. This value of 2% can be calculated from the following equation:

The concentration at which NaCl accumulates in the shoot depends on the salt concentration in the soil solution, the percentage of salt taken up by roots, and the percentage of water retained in the leaves:

$$[NaCl]_{shoot} = [NaCl]_{soil} \; x \frac{\text{\% salt taken up}}{\text{\% water retained}} \tag{1}$$

Plants retain only about 2% of the water they transpire, ie they take up about 50 time more water from the soil than they retain in their shoot tissues. The percentage of transpired water that is retained in the shoot can be calculated from the product of the water use efficiency (*wue*; mass of shoot produced per mass of H_2O transpired) and the shoot water content (*wc*; shoot H_2O per shoot mass):

Water use efficiency (WUE) of plants growing at moderate evaporation demand are usually in the range of 3-6 mg g^{-1}, the variation due to extremes of evaporative demand, rather than a peculiarity of the species. For a water use efficiency of 4 mg g^{-1} and a shoot H_2O:DW ratio of 5:1, about 20 mg of water is retained in the shoot for every g of water transpired (Eqn. 2). That is, the shoot retains only 2% of the water transpired. In order to prevent the salt concentration in the shoot increasing above that in the soil, then only 2% of the salt should be allowed into the shoot, *i.e.* 98% should be excluded.

$$\text{\% water retained} = wue \; x \; wc \; x \; 100 \tag{2}$$

A soil salinity of 100 mM NaCl or 10 dS m^{-1} is about as high as most crops will tolerate without a significant reduction in growth or yield, and a concentration of 100 mM NaCl on a whole shoot basis is about as high as is desirable because it will include some old leaves with much higher salt concentrations, as well as younger leaves or other tissues with lower concentrations. So for plants to grow for extended periods of time in soils with salinity of this order of magnitude, roots should ensure that no more than 2% gets to the shoots. Roots themselves do not accumulate excessively high concentrations of salt. The Na+ and Cl- concentration in roots is rarely higher than in the external solution, and often is lower.

The Relationship between Transpiration and Salt Uptake

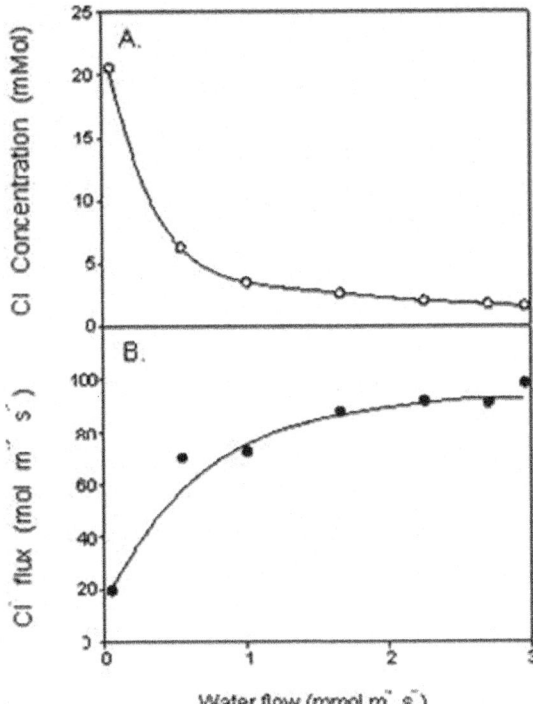

Water flow (mmol m⁻ s⁻)

Figure: The relation between ion concentration in the xylem (A), ion flux to the shoot (B), and transpiration (water flow) (Munns, 1985).

The fundamental processes governing the relationship between water and ion flow through roots are complex and not well understood. NaCl does not move passively with the transpiration stream, neither is its movement entirely independent of it, at least in some species, or over certain ranges of transpiration. The relationship between water and salt flow in the xylem of barley plants (Munns, 1985). As water flow increased from a very low to a moderate rate, there was an increase in Cl- flux, showing that the movement of the ion through the root was enhanced as water flow started to increase. However, when the water flow increased from moderate to high rates, there was little or no further increase in Cl⁻ flux, showing that the movement of the ion was independent of further increases in water flow. This relationship also holds for Na^+ and K^+.

Measurements of ion concentrations in leaves of most plants grown at different humidities are consistent with this pattern, that salt transport to leaves is substantially affected only if transpiration is greatly affected.

An effect might be seen more with species that are very poor excluders, such as lupin as mentioned above, and rice, which carry much more salt in an apoplastic or transpirational "bypass" pathway than other species. In rice, the percentage of water moving through a bypass pathway from roots to shoots was estimated as 5.5% of the total water transpired, and could account for all the Na^+ transported to the shoots, whereas in wheat only 0.4% of the water moved along a bypass pathway and could not account for most of the Na^+ transported (Garcia et al., 1997).

Mechanisms of Control of Salt Transport

Here we look at differences between species in the ability to tolerate the salt-specific component of salinity. Differences in tolerating the osmotic stress itself are in common with drought tolerance.

The rate at which old leaves die depends on the rate at which salts accumulate to toxic levels. Thus, control of the rate at which salt arrives in leaves is essential, as are mechanisms that reduce the toxicity of the salt. For species lacking the ability to compartmentalise salts in the vacuoles to high concentrations, continued transport of salt to the leaves will eventually result in either excessive build-up of salts in the cell walls or in the cytoplasm. The former will cause death through dehydration, and the latter will cause death through poisoning of metabolic systems such as photosynthesis or respiration.

Control at the Whole Plant Level

Control of salt transport into and through the plant takes place at five sites in the plant. Control occurs in the root cortex, at the loading of the xylem, at the retrieval from the xylem in upper parts of the roots. These three processes serve to reduce the transport to the leaves. Control in the shoot occurs by the exclusion of salt from the phloem sap flowing to meristematic regions of the shoot. An additional mechanism occurs in most halophytes: specialised cells to excrete salt from leaves. However, halophytes also rely on the first four mechanisms to reduce the flux of salt to the leaves – excretion is an additional backup for plants growing in very saline site, and for perennial species.

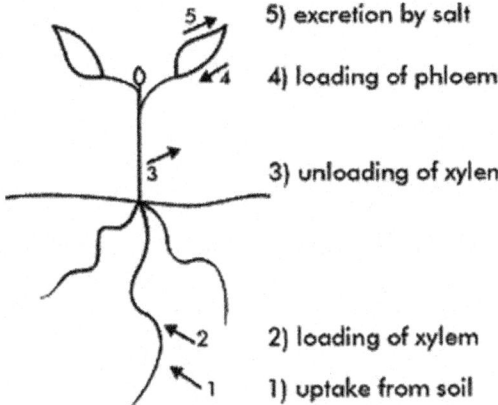

5) excretion by salt

4) loading of phloem

3) unloading of xylem

2) loading of xylem

1) uptake from soil

Figure: Control points at which salt transport is regulated. These are: 1. selectivity of uptake from the soil solution, 2. loading of the xylem, 3. removal of salt from the xylem in the upper part of the plant, 4. loading of the phloem and 5. excretion through salt glands or bladders. For a salt tolerant plant growing for some time in a soil solution of 100 mM NaCl, the root concentrations of Na^+ and Cl^- are typically about 50 mM, the xylem concentration about 5 mM, and the concentration in the oldest leaf as high as 500 mM (Munns et al., 2002).

Exclusion is particularly important for perennial species whose leaves may live for a year or more. For these species there is greater need to regulate the incoming salt load than for annual species whose leaves may live for only one month.

There are contributory features that function to maintain low rates of salt accumulation in leaves. High shoot/root ratios and high intrinsic growth rates (Pitman, 1984), and absence of an apoplastic pathway in roots (Garcia et al., 1997) all will serve to reduce the rate at which salt enters the transpiration stream and accumulates in the shoot.

$$\text{Ion concentration in shoot } \left(\text{mol g}^{-1}\right) = \frac{\text{Ion uptake rate (mol g}^-}{\text{Relative growth rate } \left(\text{g}\right.} \tag{3}$$

Two principles emerge from this equation. First, a fast-growing plant will have a lower concentration in the shoot than a slow growing plant, for the same uptake rate. Note that it is the relative growth rate (RGR) not the size itself that influences the ion concentration in the shoot. The notion of 'vigour' as affecting ion concentration therefore needs to be defined carefully, as RGR is independent of plant size. Second, any increase in shoot ion concentration with increasing salinity may not be due to increase ion uptake rate – it could be due to decreased RGR due to the osmotic effect of the salt. This also means that an additional stress such as waterlogging that depress RGR will also cause an increase in concentration in the shoot for the same uptake rate

Control at the Cellular Level: ion Compartmentation

There is no evidence of adaptations in enzymes to the presence of salt, so mechanisms for salt tolerance at the cellular level involve keeping the salt out of the cytoplasm, and sequestering it in the vacuole. That this occurs in most species is indicated by the high concentrations found in leaves that are still functioning normally, concentrations well over 200 mM, which are known to completely repress enzyme activity *in vitro* (Munns, 2002) Generally, Na^+ starts to inhibit most enzymes at a concentration above 100 mM. The concentration at which Cl^- becomes toxic is even less well defined, but is probably in the same range as that for Na^+. If Na^+ and Cl^- are sequestered in the vacuole of the cell, K^+ and organic solutes should accumulate in the cytoplasm and organelles to balance the osmotic pressure of the ions in the vacuole. The organic solutes that accumulate most commonly under salinity are proline and glycinebetaine, although other molecules can accumulate to lesser degrees. Salt tolerant species have transport systems on the tonoplast that can sequester Na^+ and Cl^- at high concentrations within the vacuoles, while maintaining much lower concentrations in the cytoplasmic compartments. The ion channels and transporters that regulate the net movement of salt across cell membranes were described in several recent reviews.

In summary, roots do most of the work in protecting the plant from excessive uptake of salts, and filter out most of the salt in the soil while taking up water. Even so, there are mechanisms for coping with the continuous delivery of relatively small amounts of salt that arrive in the leaves, the most important being the cellular compartmentalisation of salts in the vacuoles of the mesophyll cells. This strategy allows plants to minimise or delay the toxic effects of high concentrations of ions on important and sensitive cytoplasmic processes. The rate at which leaves die is the rate at which salts accumulate to toxic levels, so genotypes that have poor control of the rate at which salt arrives in leaves, or a poor ability to sequester that salt in cell vacuoles, have a greater rate of leaf death.

Chapter 7

Impact of Mineral Deficiency Stress

A mineral element is considered as essential, when plants cannot complete reproductive stage of life cycle due to its deficiency. Deficiency must be corrected only by supplying the element in question and when the element is directly involved in the metabolism of the plant (Arnon, 1954). Based on these criteria, sixteen elements so far were identified as essential. These are: carbon, hydrogen, oxygen, nitrogen, phosphorus, potassium, calcium, magnesium, sulfur, iron, manganese, zinc, copper, boron, molybdenum and chlorine. Most of the carbon as carbon dioxide enters the plant from the air; hydrogen and oxygen are taken up as water. The rest of the elements are taken up from the soil solution as mineral nutrients. Among these nutrients N, P, K, Ca, Mg, and S are considered major or macronutrients, because they are required in large quantities that range between 1 to 150 g per kg of plant dry matter. Fe, Zn, Mn, Cu, B, Mo and Cl are minor or micronutrient that are required at rates of 0.1 to 100 mg per kg of plant dry matter (Marschner, 1997a). Chloride is essential in micro quantities but can accumulate in the plant in large quantities when present in high concentrations in the soil solution, (Xu *et al.*, 2000).

All the essential nutrients are required by plants in balanced proportions. Deviation from this may result in nutritional disorders. Early detecting of nutritional deficiency stress is important. Stress might extend to the entire plant with loss of yield if relief of stress is not employed. Continuous shortage of a nutrient or nutrients might cause plant death. When two or more elements are deficient simultaneously, the composite picture of symptoms may resemble no single known deficiency. Mineral deficiency symptoms are sometimes confused with other complex field events such as damage caused by insect-pest, disease, salt stress, water stress, pollution, light and temperature injury (Bennett, 1993) and herbicide damage. Toxicity of Mo or Se is similar to P deficiency (Bennett, 1993), Fe deficiency in

Mango is similar to Chloride toxicity (Xu *et al.*, 2000). Therefore, it is necessary to critically observe and define these deficiency symptoms. The deficiency symptoms might be distinguished based on the plant part that shows deficiency symptoms, presence or absence of dead spots and entire leaf or interveinal chlorosis.

Generally, nutrient deficiency in the plant occurs when a nutrient is insufficient in the growth medium and/or cannot be absorbed and assimilated by the plants due to unfavourable environmental conditions. Nutrient disorders limit crop production in all types of soil around the world. Soil conditions associated with nutrient deficiencies of various nutrient elements.

Visible Symptoms of Stress

Nitrogen (N)

The characteristic deficiency symptom of nitrogen is the appearance of uniform yellowing of leaves including the veins, this being more pronounced on older leaves as expressed in rabbit-eye and blueberries (Tamada, 1989); Fescue (Razmjoo, 1997); *Ailanthus triphysa* (Anoop *et al.*, 1998); chili (Balakrishnan 1999) and sugarcane (Nautiyal *et al.*, 2000). The leaves become stiff and erect. In dicotyledonous crops the leaves detach easily under extreme deficiency condition. Cereal crops show characteristics 'V' shaped yellowing at the tip of lower leaves. O'Sullivan *et al.*,(1993) observed relatively small and pale green leaves with dull appearance in sweet potato. If such condition of nitrogen stress do persist, the result is a decreased foliage growth and shoot growth.

Phosphorus (P)

In phosphorus deficiency, leaves remain small, erect, unusually dark green with greenish red in sweet potato (O'Sullivan *et al.*, 1993), bluish green in chili (Balakrishnan 1999), brown in birdsfoot trefoil (Russelle and McGraw, 1986) or purplish tinge in sugar maple (Bernier and Brazeau, 1988); blueberry (Tamada,1989) and sugarcane (Nautiyal *et al.*, 2000). The under side develops bronzy appearance. The root growth is also restricted under phosphorus stress in black pepper (Nybe and Nair, 1986). Anthocyanin pigment increases in leaves of barley (Hamy,1983) and *Arabidopsis thaliana* (Trull *et al.*, 1997) under phosphorus stress.

Potassium (K)

Under potassium stress condition, yellowing of leaves starts from the tips or margins of leaves extending towards the centre of leaf base. The yellowing is interveinal and irregular in the leaves of tomato (Besford, 1978) and blueberry (Tamada, 1989). These yellow parts become necrotic (dead spots) with leaf curling in tobacco (Arnold *et al.*, 1986); sugar maple (Bernier and Brazeau, 1988); sapota (Nachegowda *et al.*,1992) and sugarcane (Nautiyal *et al.*, 2000). There is a sharp difference between green, yellow and necrotic parts.

Calcium (Ca)

Calcium stress in plants results in chlorosis of young leaves along the veins of birdsfoot trefoil (Russelle and McGraw, 1986) and blueberry (Tamada, 1989), if deficiency persist longer, bleaching of upper half leaf followed by leaf tip curling do occur in black pepper (Nybe and Nair, 1987) and sugarcane (Nautiyal *et al.*, 2000). The growing bud leaf becomes chlorotic white with base remaining green, the distortion of the tips of shoots i.e. dieback was observed by Edwards and Hortan, (1997) in peach seedlings. Similarly, Spehar and Galway, (1997) found brown spots on leaves, reduced expansion and premature leaf senescence under Ca stress in soybean crop. Stress during fruiting in tomato increases susceptibility to blossom end rot (Adams and El-Gizawy, 1988; Sonneveld and Voogt, 1991 and Ho *et al.*, 1999). Calcium stress is also responsible for other disorders such as bitter pit in apple (Ford, 1979; Monge *et al.*, 1995 and Silva and Rodriguez, 1996); leaf tip burn in cabbage (Miao *et al.*, 1997) and lettuce; black heart of celery; cavity spot of carrots (Scaife and Clarkson 1978); vitrescence in melons (Jean-Baptist *et al.*, 1999).

Magnesium (Mg)

Magnesium deficiency causes yellowing, but differs from that of nitrogen. The yellowing takes place in between veins of older leaves (Makkanen, 1995) of *Picea abies* and veins remain green, this is followed by necrosis of tissues in birdsfoot trefoil (Russelle and McGraw, 1986), melons (Simon *et al.*, 1986). Black pepper (Nybe and Nair, 1987) and blueberry (Tamada, 1989). Mg deficiency my be induced in tomatoes by high levels of ammonium in the nutrient solution (Kafkafi *et al.*, 1971).

Sulfur (S)

Sulfur deficiency cause leaves to become yellowish in black pepper (Nybe and Nair, 1987); potato (Gupta and Sanderson, 1993) and *Brassica oleracea* (Stuiver *et al.*, 1997) and it appears similar to nitrogen deficiency, but the symptoms are first visible on younger leaves (Russelle and McGraw, 1986). The affected leaves are narrow and the veins are paler and chlorotic than interveinal portion, especially towards the base with marginal necrosis in sugarcane (Nautiyal *et al.*, 2000).

Iron (Fe)

The principal veins remain conspicuously green and surrounding portion of the younger leaves turn yellow tending towards whiteness in chickpea (Mehrotra and Gupta, 1990 and Saxena *et al.*, 1990); groundnut (Reddy *et al.*, 1993); radish, cauliflower, cabbage and sorghum (Preeti *et al.*, 1994); lentil (Zaiter and Ghalayini, 1994) and soybean (Fonts and Cox, 1998). Under sever deficiency, most part of the leaf becomes white (Russelle and McGraw, 1986).

Zinc (Zn)

The leaves become narrow and small in chili (Balakrishnan, 1999), the lamina becomes chlorotic in sweet potato (O'Sullivan *et al.*, 1993), sour orange seedlings (Swietlik, 1995) and chickpea (Khan *et al.*, 1998), while veins remain green. Subsequently, dead spots develop all over the leaf including veins, tips and margins under sever deficiency, shoot growth is reduced (O'Sullivan *et al.*, 1993; Swietlik, 1995 and Yu and Rengel, 1999). Khaira disease in rice results due to zinc deficiency (Gautam and Sharma, 1982; Sharma *et al.*, 1988 and Sahi *et al.*, 1992). Shoot elongation is reduced and a tuft or rosette of distinctly narrow leaves is produced at the shoot terminal in apple and pear. The symptoms are termed 'little leaf' or 'rosette' (Hanson, 1993).

Boron (B)

Boron deficiency causes yellowing or chlorosis of youngest leaves and stems (Yu *et al.*, 1998) which starts from the base to the tip. Rosetting of terminal shoots of potato (Roberts and Rhee, 1990). Leaf tip burn, elongate and become whitish brown in rice (Yu *et al.*, 1998). Death of terminal bud occurs in extreme cases. Boron deficiency causes brown heart in radish (Shelp *et al.*, 1987) and crown choking in coconut (Baranwal *et al.*, 1989).

Manganese (Mn)

The principal veins as well as smaller veins are green, the interveinal portion become chlorotic in *Ailanthus triphysa* (Anoop *et al.*, 1998) followed by necrosis and browning of interveinal tissue in melons (Simon *et al.*, 1986). The affected young leaves remain small and abscise before older leaves in birdsfoot trefoil (Russelle and McGraw, 1986).

Molybdenum (Mo)

The common symptoms of Mo deficiency in plants include a general yellowing, marginal and interveinal chlorosis, marginal necrosis, rolling, scorching and downward curling of margins in poinsettia cultivars (Cox and Bartley, 1987; Cox, 1992) and in various field, horticulture and forage crops (Gupta and Gupta, 1997). The deficiency of molybdenum in cauliflower causes the disorder described as 'Whiptail' (Duval et al., 1991).

Copper (Cu)

In copper deficiency, visible foliar symptoms appear on young leaves as chlorosis changing to necrosis (Conover et al., 1991; Del, 1994); rolling, wilting and twisting of leaves in wheat (Owuoche, 1995). The later affected leaves appear papery and twisted in rice (Nautiyal et al., 1999).

Chlorine (Cl)

The symptoms of chlorine deficiency develop first on the older leaves. Discrete patches of pale green chlorotic tissue appear between the main vein near the tip of the leaf, downward cupping of some of the older leaves of Kiwifruit was observed by Smith et al., (1987). The leaflets of youngest leaves shrivel completely, older leaflets develop a brown necrosis which start near the tip and extend backwards particularly at the margins of red clover (Whitehead, 1985).

Nickel (Ni)

Plant growth is reduced and older leaves turn chlorotic giving plants a nitrogen deficient phenotype, when grown on urea-based nutrient solutions not supplemented with Ni in tomato and soybean (Shimada and Ando, 1980; Krogmeier et al., 1991). Similar results were obtained in oilseed-rape, zucchini and soybean by Gerendás and Sattelmacher (1997).

Heat Stress and its Impact

The Environmental and Physiological Nature of Heat Stress

Heat stress often is defined as where temperatures are hot enough for sufficient time that they cause *irreversible* damage to plant function or development. In addition, high temperatures can increase the rate of reproductive development, which shortens the time for photosynthesis to contribute to fruit or seed production. I also will consider this as a heat-stress effect even though it may not cause permanent (*irreversible*) damage to development because the acceleration does substantially reduce total fruit or grain yield.

The extent to which heat stress occurs in specific climatic zones is a complex issue. Plants can be damaged in different ways by either high day or high night temperatures and by either high air or high soil temperatures. Also, crop species and cultivars differ in their sensitivity to high temperatures. Cool-season annual species are more sensitive to hot weather than warm-season annuals. In Table below there are several examples of cool-season and warm-season annual crop species. I did not include safflower in the table because it is unusual in that during the vegetative stage it grows well in cool conditions and during the reproductive stage it grows well in hot conditions.

Table : Annual crop species adapted to cool and warm seasons (Hall 2001).

Cool-Season Annuals	Warm-Season Annuals
Barley, brassicas, canola, fava bean, flax, garbanzo bean, Irish potato, lentil, lettuce, lupine, mustard, oat, pea, radish, rye, spinach, triticale, turnip, vetch, wheat	common bean, cotton, cowpea, cucurbits, finger millet, grain amaranth, lima bean, maize, mung bean, pearl millet, pepper, pigeon pea, rice sesame, sorghum, soybean, sunflower, sweet potato, tobacco, tomato.

High day temperatures can have direct damaging effects associated with hot tissue temperatures or indirect effects associated with the plant-water-deficits that can arise due to high evaporative demands. Evaporative demand exhibits near exponential increases with increases in day-time temperatures and can result in high transpiration rates and low plant water potentials (Hall 2001). The effects of drought and plant-water-deficits on crop adaptation are discussed in the 'Drought Stress' page.

Air temperatures vary during the day and season. Temperature data from weather stations for many locations in California may be found at the California Weather Databases linked to the Integrated Pest Management Project of the University of California. For many cities in the world, monthly means of daily maximum and minimum air temperatures may be found at the worldclimate web site. Additional world weather data may be found at the Interactive Weather Information Network. Definitions of temperate, subtropical and tropical climatic zones and the average monthly means of daily maximum and minimum air temperatures that can occur in representative locations in these zones may be found in the book by Hall (2001).

The extent of heat stress that can occur in a specific climatic zone depends on the probability of high temperatures occurring and their duration during the day or night. Where global climate change is occurring these probabilities may not be predicted well based only on historical records for specific locations. Heat stress is a complex function of intensity (temperature degrees), duration and rate of increase in temperature. The magnitude of heat stress rapidly increases as temperature increases above a threshold level and complex acclimation effects can occur that depend on temperature and other environmental factors.

High soil temperatures can reduce plant emergence. The maximum threshold temperatures for germination and emergence are higher for warm-season than for cool-season annuals. For example, the threshold maximum seed zone temperature for emergence of cowpea is about 37^0C compared with 25 to 33^0C for lettuce.

During the vegetative stage, high day temperatures can cause damage to components of leaf photosynthesis, reducing carbon dioxide assimilation rates compared with environments having more optimal temperatures. Sensitivity of photosynthesis to heat mainly may be due to damage to components of photosystem II located in the thylakoid membranes of the chloroplast and membrane properties (Al-Khatib and Paulsen 1999). Membrane thermostability has been evaluated by measuring electrolyte leakage from leaf disks subjected to extreme temperatures (Blum 1988). More stable membranes exhibit slower electrolyte leakage. Studies comparing responses to heat of contrasting species indicated that photosystem II of the cool season species, wheat, is more sensitive to heat than photosystem II of rice and pearl millet, which are warm season species adapted to much higher temperatures (Al-Khatib and Paulsen 1999).

Extreme temperatures can cause premature death of plants. Among the cool-season annuals, pea is very sensitive to high day temperatures with death of the plant occurring when air temperatures exceed about 35^0C for sufficient duration, whereas barley is very heat tolerant, especially during grain filling. For warm season annuals, cowpea can produce substantial biomass when growing in one of the hottest crop production environments on earth (maximum day-time air temperatures in a weather station shelter of about 50^0C), although its vegetative development may exhibit abnormalities such as leaf fascinations. For monocotyledons, including both cool-season and warm-season annuals, high daytime temperatures can cause leaf firing which involves necrosis of the leaf tips and this symptom also can be caused by drought.

Reproductive development of many crop species is damaged by heat such that they produce no flowers or if they produce flowers they may set no fruit or seeds. The reviews of Hall (1992, 1993) discuss the detrimental effects of heat stress on reproductive development that has been reported for cowpea, common bean, tomato, cotton, rice, wheat, maize and sorghum. I will examine the detrimental effects of heat stress on cowpea because of the comprehensive information available for this species and the likelihood that many of the same phenomena occur with other warm-season crop species.

Controlled-environment studies in which cowpea plants were subjected to separately controlled root and shoot and day and night temperatures demonstrated that pod set (the proportion of flowers producing pods) was damaged by moderately high night temperature of the shoot (Warrag and Hall 1984a,b). It was surprising that night temperature would have this effect since much hotter day temperatures did not damage pod set of cowpea. Reciprocal artificial pollinations between plants grown under high and optimal night temperatures indicated the low pod set was caused by male sterility and that the pistils did not appear to be damaged by high night temperature. The detrimental effects of high night temperature on pod set also were shown to occur in field conditions (Nielsen and Hall 1985b). In these experiments a unique experimental approach was used in which plots of cowpea plants were subjected to different increments of higher night temperatures during early stages of flowering using enclosure systems placed over the plots only during the night-time (Nielsen and Hall 1985a).

Possible mechanisms for the sensitivity of pod set to high night temperatures have been proposed. Mutters and Hall (1992) demonstrated that there is a distinct period during the 24-hour cycle when pollen development in cowpea is sensitive to high night temperatures. Plants subjected to high temperature during the last six hours of the night exhibited substantially decreased pollen viability and pod set, whereas plants subjected to high temperature during the first six hours of a twelve-hour night exhibited no damage. Mutters and Hall (1992) hypothesized that these results could be explained if a heat-sensitive process in pollen development is under circadian control and only occurs in the late night period. Note that if a heat-sensitive process is under circadian control and if genetic variation exists for the time in the 24-hour cycle when this process occurs, evolution in hot environments would favour plants in which the heat-sensitive process occurs at the coolest time which is just prior to dawn. The damaging effect of high night temperature on pod set was greater in long days than in short days, and red and far red light treatments indicated it is a phytochrome-mediated response (Mutters et al. 1989b).

Studies were conducted in which cowpeas were transferred between growth chambers having high or optimal night temperatures (Warrag and Hall 1984b; Ahmed et al. 1992). They demonstrated that the stage of floral development most sensitive to high night temperature occurs 9 to 7 days prior to anthesis, which is after meiosis and coincides with release of pollen microspores from the tetrads. Damage due to high night temperature was associated with premature degeneration of the tapetal layer that provides nutrients to developing pollen, infertile pollen and in some genotypes anthers did not dehiscence. The transfer of proline from the tapetal layer to pollen was inhibited (Mutters et al. 1989a).

Comparisons of heat-sensitive and heat-tolerant cowpeas showed a genotypic association between sensitivity to heat during pod set and rapid leakage of electrolytes from leaf discs subjected to heat stress (Ismail and Hall 1999). Possibly, the damage to pollen development by high night temperatures may be in some way associated with a heat-induced malfunction in membrane properties.

Floral bud development also can be damaged by heat such that plants do not produce flowers. For cowpea, two weeks or more of consecutive or interrupted hot nights during the first month after germination caused complete suppression of floral bud development (Ahmed and Hall 1993). In extreme cases the floral buds become necrotic and die. In field conditions, the damage occurs under long days but not under short days. However, responses to red and far red light indicated the effect was only partially consistent with the system being mediated by phytochrome (Mutter et al. 1989b). The damaging effect of high night temperature and long days on floral bud development also depended on light quality whereas the damaging effect on pod set did not depend on light quality (Ahmed et al. 1993b). When growth chambers were used with relatively large amounts of fluorescent light and little incandescent light, such that the red/far red ratio was 1.9, floral buds were not suppressed in long-day high night temperature conditions, but pod set was very low. This artificial light system provides a useful experimental method for studying the effects of heat stress on pod set without complications due to heat stress effects on floral bud development. When growth chambers were used with lighting systems providing a red/far red ratio of 1.3 to 1.6, floral bud suppression was observed that was similar to what is obtained under long-day high night temperature conditions in the field where sunlight has a red/far red ratio of about 1.2.

There are two important conclusions from these studies. First, the use of growth chambers with lighting systems that mainly depend on fluorescent lights can result in either serious artifacts or methodological advantages when studying plant reproductive responses to heat stress. Second that intense shading of floral buds could reduce the red/far red ratio below 1.2 in field conditions and intensify the floral bud suppression effect. In densely sown fields of cowpea, individual plants that are suffering from competition and are tall and spindly can exhibit floral bud suppression even though night temperatures are not too hot.

Pods of different cowpea genotypes produce 9 to 20 ovules with many cultivars having 15 ovules, but pods rarely produce these many seeds per pod. Under optimal conditions two-thirds of the ovules may produce seed, whereas with high day or high night temperature (Warrag and Hall 1983) and other stresses, such as drought, fewer seeds are produced per pod. For most cultivars and stresses it is the ovules at the blossom end of the pod that suffer embryo abortion and do not produce seed, resulting in the production of "pinched" pods.

Cowpea seeds produced under high day temperatures can have asymmetrical twisted cotyledons (Warrag and Hall 1984a). Germination of the seed is not influenced and this effect of heat stress may not be a major problem for commercial production. In contrast, heat-induced brown discolouration of cowpea seed coats can occur with some cultivars and be a major problem causing consumers to reject grain. Higher night temperatures resulted in progressively larger numbers of seed with larger areas of brown discolouration on seed coats (Nielsen and Hall 1985b).

The extent to which high-temperature damage to photosynthesis or reproductive development affect fruit or grain yield probably depends on the extent to which the photosynthetic source and the reproductive sink are limiting fruit or grain yield, and this may vary among species and cultivars. Surface and internal tissues of tomato and citrus fruit can be damaged by the combination of high temperatures and intense solar radiation. High tissue temperatures also can damage cambium layers in exposed trunks and branches.

Repercussions of Heat Stress

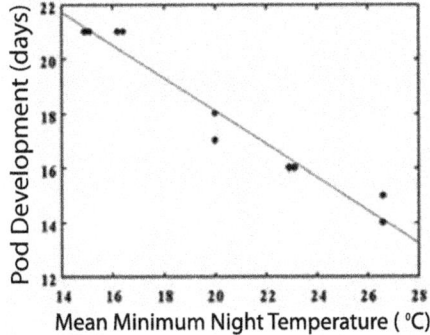

Figure: Days from anthesis to mature dry pod for cowpeas grown under night-time temperatures in the same field (Data from Nielsen and Hall, 1985b)

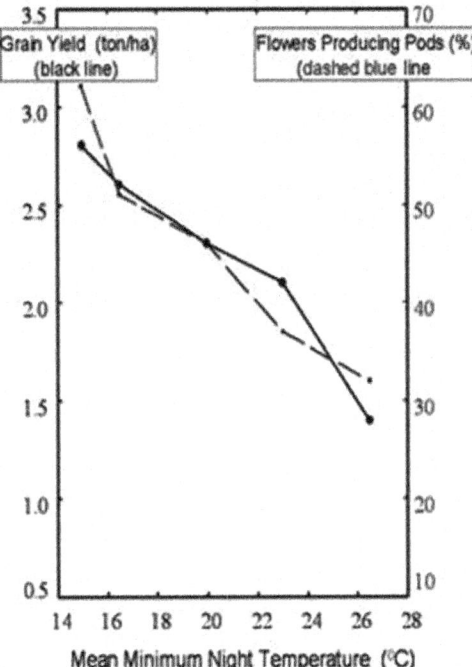

Figure: Grain yield and % flowers producing pods for cowpeas grown under different night-time temperatures in the same field (data from Nielsen and Hall 1985b)

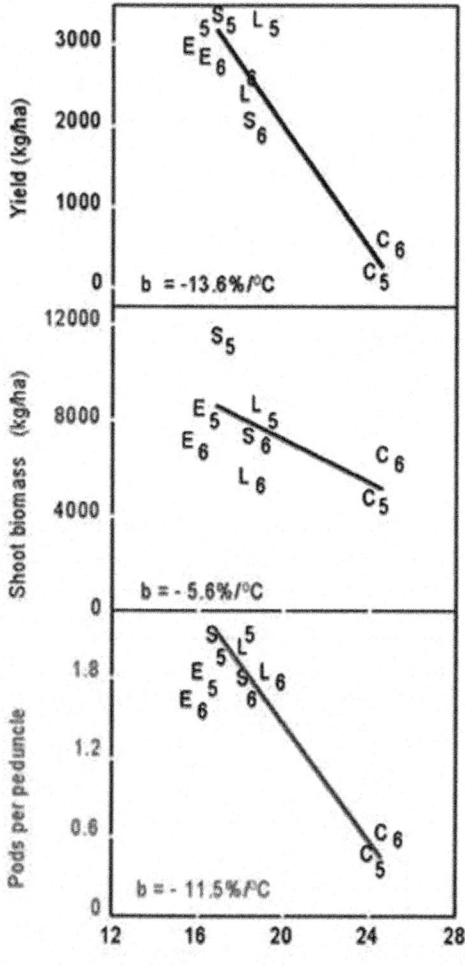

Figure: Plant production of heat-susceptible cowpea lines grown in different fields with contrasting thermal regimes (data from Ismail and Hall 1998) (data from Ismail and Hall 1998)."

The reduction in maximal emergence of annual crops due to hot soils can be so pronounced that yield of the economic product is reduced substantially. This can be a major problem for crops, such as lettuce, whose canopies are not very plastic and cannot compensate for the low emergence. Heat stress at emergence is a major problem for cool-season crops that are sown in the late summer in hot subtropical zones with the objective of producing a crop during the cooler weather in the fall. Warm-season crops also can experience this problem in those tropical zones and seasons where soil temperature can be extremely hot at sowing. The acceleration of reproductive development by high temperatures may partially explain why the potential grain yields of warm season crops, such as rice and cowpea, usually are higher in the subtropics than the tropics. The extent of the acceleration of development of cowpea has been determined by subjecting plants to different night temperatures in field conditions using temporary enclosures imposed only at night with regulated heating systems (Nielsen and Hall 1985a). Under the cool night temperatures of subtropical California (minimum night temperatures of 16°C), individual cowpea pods took 21 days to proceed from anthesis to mature dry pod. In contrast, the pod development period of the same cowpea cultivar was only 14 days with the higher night temperatures that occur under tropical conditions (minimum night temperatures of 26°C).

More rapid pod development may increase the extent of embryo abortion, and individual seed usually are smaller in tropical compared with subtropical conditions for the same cowpea cultivar. Cowpeas subjected to elevated night-temperature treatments produced smaller seed (Nielsen and Hall 1985b). The more rapid development of individual fruits also results in the overall reproductive period of the plant being shorter. Grain yields of cowpea cultivars grown with optimal management are much less in tropical zones than in subtropical zones mainly due to the shorter overall reproductive period caused by the hot nights of tropical zones (Hall et al. 1997). Acceleration of reproductive development also is a problem for cool season crops, such as wheat, growing in environments that are hot during reproductive development (reviewed by Hall 1992, 1993).

Direct evidence for heat-stress effects on grain yield was provided by the studies of Nielsen and Hall (1985b) in which cowpea was subjected to different increments of elevated night temperature under field conditions in a subtropical location in California. For minimum night temperatures greater than 15°C there were linear reductions in both grain yield and the proportion of flowers that set pods with 50% reductions occurring at minimum night temperatures of about 26°C.

Evaluating the same set of cultivars over a range of environments that mainly differ in temperature provides a more indirect approach for evaluating heat stress effects on the performance of crops that is of interest to farmers. Six heat-susceptible cowpea genotypes, including a California cultivar, were grown in eight field environments with contrasting temperatures but similar high levels of solar radiation and optimal management (Ismail and Hall 1998). Grain yield was negatively correlated with mean minimum night temperatures during the three-week period beginning one week before first flowering. For minimum night temperatures greater than 16.5°C grain yield decreased 13.6 % per °C associated with a similar decrease (11.5 % per °C) in number of pods per peduncle but only a small decrease (5.6 % per °C) in shoot biomass production.

The Environmental and Physiological Nature of Stress

Drought (and extreme temperature) is an environmental occurrence that can be defined and measured by indices derived from large historical databases on precipitation and other weather variables. Several drought indices are use, as published by the Drought Mitigation Centre at the University of Nebraska. Among these, the 'Palmer Index' seems to be widely adopted among climatologists. Information on identification and assessment of drought prone environments can be obtained from various world/country climatic maps available on the net.

Geographical Information Systems (GIS) is developing very fast and exciting applications useful for agronomists and environmental scientists are being developed. Even specific applications to assist breeders working in dry environments are being developed. A lightweight GIS data viewer developed by ESRI ("ArcExplorer") is now available online. This freely downloadable software offers an easy way to perform basic GIS functions. ArcExplorer is used for a variety of display, query, and data retrieval applications and supports a wide variety of standard data sources. It can be used on its own with local data sets or as a client to Internet data and map servers. Explore the application of GIS to issues of agriculture development in dry regions as being developed at CIMMYT (search their site for 'GIS'). GIS interfaced with GPS offer great potential in precision agriculture and similar map-based applications. Finally, a free online climate estimator for any point on the globe is available and can be a very useful tool for the agronomist and breeder.

Crop simulation models can reasonably estimate and quantify the impact of specific drought stress conditions on crop productivity. Such models require the input of various water regime and climate variables in order to simulate crop yield with better or lesser level of accuracy.

Crop models and user's guides are available today for:

- Cereals - maize, wheat, rice, sorghum, barley, and millet.
- Grain legumes - soybean, peanut, dry bean, and chickpea.
- Root crops - cassava and potato;
- Sugarcane, tomato, sunflower and pasture,
- And more.

Additional information on crop simulation models and ecological models can be found at Ecobas and at the International Consortium for Agricultural Systems Applications.

The Development of Plant Water Deficit

Plant water deficit develops as the demand exceeds the supply of water. The supply is determined by the amount of water held in the soil to the depth of the crop root system. The demand for water is set by plant transpiration rate or crop evapotranspiration, which includes both plant transpiration and soil evaporation. Evapotranspiration is driven by the crop environment as well as major crop attributes such as plant architecture, leaf area and plant development. The plant functions within a physical system consisting of the soil-plant-atmosphere continuum. During the day the plant is under heavy energy load consisting mainly of the received solar radiation, ambient air temperature and humidity. While some of this radiation energy is important for photosynthesis, most of it is not utilised and it must be dissipated. It is partly dissipated by radiation emitted from the plant in the form of heat, but most of it must be dissipated by transpiration ('latent heat'). Henceforth the term "transpirational cooling" was coined. Transpiration cause leaves to cool relative to ambient temperature when the environmental energy load on the plant is high.

Water is driven through the plant from the soil to the atmosphere by the difference in water potential between the atmosphere (very low potential) and the soil (relatively high potential when wet), analogous to the flow of electrical current under the differences in electric potentials ('Ohm's law'). Flow is also influenced by hydraulic resistances in the plant, such as the resistance regulated by stomata in the leaves or by the conductive system (root and stem xylem elements) of the plant, or by the resistance of cells and cell walls between soil and the root xylem vessel.

As water transpires from the leaf, leaf water potential (LWP) is reduced (becoming more negative). If water is available in the soil (at high soil water potential) then water will flow into the leaf to replenish the loss with only a small reduction in LWP. As soil water potential (SWP) reduces LWP must be further reduced in order to create the necessary gradient differential, which would drive the water up from the drying soil to the leaf.

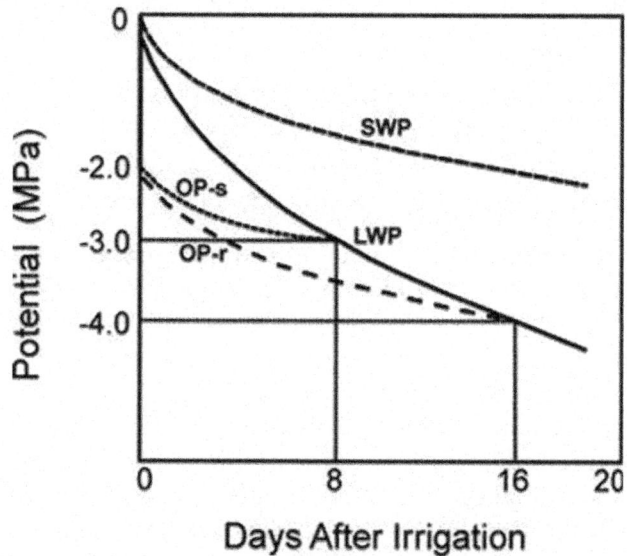

Days After Irrigation

Figure: A schematic representation of the components of leaf water status during a soil drying cycle. SWP– soil water potential; LWP- leaf water potential $(\varnothing_w)$; OP-s and OP-r represent two different cases of change in osmotic potential $(\varnothing_s)$ with the reduction in LWP.

The leaf cells contain various organic and inorganic solutes, which determine the leaf osmotic potential (OP). OP is lower (more negative) than LWP and the difference between the two is turgor potential. Turgor is lost (null value) when LWP=OP (in Fig.1, at about –3.0 and -4.0 MPa on days 8 and 16, respectively).

Leaf turgor is associated with cellular growth and function. When turgor becomes null cells collapse and the leaf wilts, though it is not dead. Stomata are responsive (among other factors) to turgor, closing to reduce transpiration. The reduction in stomatal conductance causes also a reduction CO_2 fixation and photosynthetic assimilation and an increase in leaf temperature. The increased leaf temperature may reach a level causing heat damage to the leaf especially under hot conditions.

Turgor maintenance and transpiration are therefore crucial to plants under drought stress. Turgor can be sustained by keeping a high LWP through water uptake from the drying soil or by reducing

OP through solute accumulation (osmotic adjustment). OP-s represents a variety with little osmotic adjustment (OA), where turgor is lost at a LWP of −3.0 MP on day 8, as compare with variety OP-r where due to greater OA turgor is lost only on day 16 at LWP of -4.0 Mpa. Hence, OA delays turgor loss and wilting. Further details on OA are given in the Drought Mitigation/Drought Resistance page on this section.

Turgor can also be maintained by cell wall hardening during the development of water deficit. While cell wall hardening helps to sustain turgor it impedes cell growth. This is only one of the many examples where the maintenance of plant water status under drought stress might partly be achieved at the cost of reduced growth.

Besides the factors controlling transpiration at the single leaf level, a most dominant factor in controlling whole plant and crop transpiration is total leaf area. When grown in a pot a large plant will require irrigation more frequently than a smaller one for the same pot volume. A major avenue by which plant evolution served to adapt plants to dry environments is by reduced plant size and growth rate, typical of many xerophytic plants. It is also a common observation that when sever water deficit develops lower (older) leaves are desiccated and die first so as to reduce leaf area and water requirement, while upper (younger) leaves retain open stomata and carbon assimilation. This behaviour is typical of sorghum, a relatively drought resistant crop plant.

At the crop level plant size and the demand for water is mainly expressed by leaf area index (LAI), which is the total area of live leaves per unit ground surface ($m^{-2}m^{-2}$). Crop evapotranspiration (ET) increases with LAI until LAI reaches a maximum threshold beyond which ET does not increase. As the crop matures and leaves senesce, LAI is reduced and so does ET.

Cellular turgor is not the only important transducer of whole plant water stress. The growth regulating hormone abscisic acid (ABA) is produced in the shoot in response to desiccation, causing many of the known expressions and consequences of plant water deficit such as arrested growth, stomatal closure, and reproductive failure. ABA is also produced in the root in direct response to the drying soil and its hardness as it dries (e.g. #8058). ABA flows with the transpiration stream to the shoot. Since the soil may dry around only some of the roots (typically in the top-soil) while most roots are in wet soil, root ABA exported to the shoot may cause stomatal closure or arrested growth before any water deficit develops in the shoot. This "hormonal or chemical root signal" may therefore serve as an "early warning system" to the plant. The function and value of this "root signal" in a crop plant subjected to drought stress has not yet been fully resolved. Without entering a long review and discussion it can be postulated that low ABA production or low shoot sensitivity to ABA might be beneficial in most cases of a crop in an agricultural ecosystem.

The overall agronomic role of ABA is still under debate and unresolved. In the evolutionary sense it may well serve as a major signal to place the plant in a dormant and a life conserving state before it enters sever desiccation.

Measuring Plant Water Stress and Water Status

Sensing and estimating plant stress at the whole crop level is difficult mainly because of the need to integrate an estimate based on the whole canopy. Remote sensing technology enabled to develop the Crop Water Stress Index (CWSI). This index is developed mainly from measuring the canopy temperature with an infrared thermometre. Infrared remote measurement of leaf temperature is based on a relationship between leaf temperature and transpiration. Generally, as transpiration rate is reduced due to plant water deficit so does leaf temperature rises relative to air temperature. CWSI is mainly applied to irrigation scheduling, but can be applied to non-irrigated conditions. Infrared thermometry is also being used in selection work towards breeding for drought resistance.

Several direct measurements can be performed on single plants in order to assess their water status, stress status, the repercussions of water deficit and its physiological consequences. Here only the most essential and widely used measures of plant water status are mentioned. For additional information check the section on Methods of Breeding for Resistance. For technical data check the instruments and materials section in the Web Resources Page.

Leaf water potential (LWP) can be measured in detached leaves or tissues by quickly sampling the leaves and putting the sample into the measuring instrument. Measurement can be done by the pressure chamber, which is suitable also for fieldwork. When a leaf (or a stem) is cut off a plant, the sap is sucked back into the xylem, since it is under tension. That tension is broadly equal to LWP. The detached leaf is therefore sealed in a steel chamber with only the cut end (petiole) protruding out. Pressure is applied to the chamber (from a pressure source such as a compressed nitrogen gas). When the sap meniscus appears at the xylem surface the pressure is recorded and taken as the xylem (leaf water) potential. Typical LWP of live transpiring leaves can range from about -0.3 MPa to –2.5 MPa.

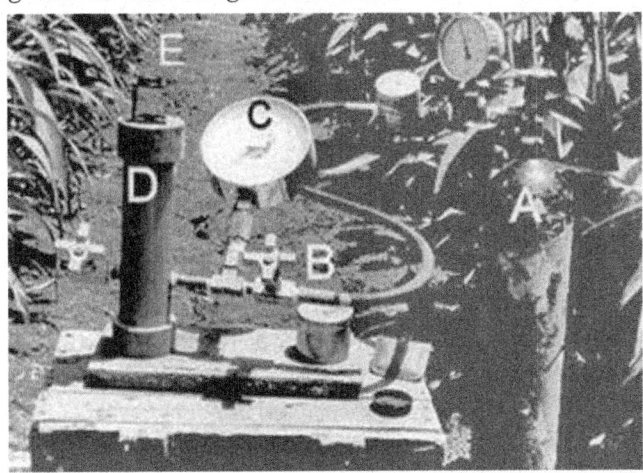

Figure: The pressure chamber; (A) pressurized gas cylinder; (B) the chamber pressurizing valve; (C) pressure gage; (D) the chamber; (E) magnifying glass.

The thermocouple psychrometre had to be initially built by the scientists themselves until commercial variation were put to the market. It is not suitable for fieldwork. With the basic design the tissue sample is sealed in a small chamber containing a thermocouple. After an equilibration period a cooling current is applied to the thermocouple in order to condense water on the junction. The amount of condensed water is proportional to the water potential of the tissue. That water is allowed to evaporate causing a change in the thermocouple output. That output is calibrated for water potential, using salt solutions.

The pressure probe for measuring turgor pressure is limited to work with singular cells. With this method a small capillary tube filled with oil is used to puncture a cell. The oil is pushed back into the tube in proportion to the cellular turgor pressure. Applying and measuring a balancing pressure to the probe estimate pressure.

Relative water content (RWC) is a veteran method that has recently gained favour over LWP as a very relevant physiological measure of plant water deficit. Its advantage is that it accounts for the effect of OA in affecting plant water status. Two plants with the same LWP can have different RWC if they differ for OA. The detailed protocol for RWC is available on this site. Osmotic potential (OP) is determined in freeze-thawed killed tissue, which serves to release all cellular solutes. The potential recorded in such a sample by the thermocouple psychrometre is OP. Alternatively if a drop of solute can be obtained by squeezing the killed tissue or by centrifugation then it can be measured for OP with a standard (micro) osmometre. Turgor is estimated as the difference between LWP and OP. When LWP=OP turgor is null. When this is attempted both LWP and OP should best be measured by the same technique, for example with the thermocouple psychrometre.

Osmotic adjustment (OA) is defined as the net accumulation of solutes after the plant has been exposed to a predetermined rate of water deficit. The reduction in OP during water deficit is not an estimate of OA because it is caused by both cellular water loss (a mere concentration effect) and real cellular solute accumulation. The increase in OP in the latter case over OP in fully hydrated none stressed plant is OA. OA is proportional (non-linearly) to the rate of water deficit (LWP) over time. The method therefore entails careful application of drought stress for an extended period of time to a predetermined level of water deficit, best measured by RWC. OP is then measured in the tissue and OA is calculated as the difference between measured OP and OP estimated for a non-stressed fully turgid state. Alternatively, water deficit is applied to a predetermined level of stress (say 70% RWC) after which the plant is fully rehydrated (typically overnight). OP is then measured and compared with that of a fully hydrated non-stressed plant. The difference between the two measures of OP is OA, measured by the "rehydration method". A more accurate but a resources demanding method involves the derivation of OA from the relationship between OP and RWC during a drying cycle. Typical OA values for crop plants can range from null to around 1.5 MPa.

Measuring Transpiration and Stomatal Conductance

There are numerous methods for measuring transpiration from single detached leaves, whole plants and whole canopies in the field. Methods vary in applicability, accuracy, cost and speed. The review of all methods is beyond the scope of this site. Many methods are simply gravimetric, based on the amount of water lost from a plant grown in a container or from a detached leaf, with known leaf area. Transpiration is expressed as mass of water lost per unit leaf area per unit time. Field methods are often based on measuring the amount of water lost from the soil profile. Various *porometres* were developed to measure transpiration from intact leaves in the field or the laboratory. The most common principle of function is the exposure of a humidity sensor to the transpiring leaves under standard cuvette conditions (the diffusive resistance porometre). Porometres vary in function and specifications, where some constitute only a part of a more elaborate system of monitoring total leaf gas exchange. A pressure-drop (or viscous flow) porometre was designed and used in the 1960' to 1980's. It is based on the relationship between stomatal conductance and the flow of pressurized air across the leaf. It seems that this system was dropped in favour of the diffusive resistance porometre, which is also readily available commercially.

The *sap flow* method is becoming popular with large trees, in which transpiration is difficult to measure by other methods. This method estimates transpiration by the velocity of a heat pulse applied to the trunk and measured above the point of applications.

Measuring Soil Moisture

An overview of soil-water relations is available via PowerPoint presentation (allow time for download).

Soil moisture content (by volume) and status (by potential or tension) is a major variable affecting plant water status and crop water-use. Extractable soil moisture is the amount of water that a given crop can extract from the soil to a given soil water potential and soil depth. Generally, different crops can use 50% to 80% of the extractable soil moisture before crop transpiration is reduced and plants present symptoms of water deficit. These values change with crop, soil and atmospheric conditions. The determination of soil moisture content or status is of major consideration regarding plant water relations.

Methods of testing soil moisture vary from feeling the soil by hand to remote sensing it from aerial or even space platforms. Methods vary in accuracy, cost, convenience and purpose.

The major methods used in farming or in research are detailed in the sites noted below. The basic method is the 'gravimetric' where soil moisture content is determined by weighing the soil before and after drying in an oven. More advanced methods are represented by the 'Time-Domain Reflectometry' (TDR) method, which has developed in recent years into a relatively accurate and convenient method for measuring soil moisture at different depths through soil access tubes. One supplier of the system is UMS, for example. Details on the various measurement methods are available at SOWACS (Soil Water Content Sensors).

The Repercussions of Plant Water Deficit

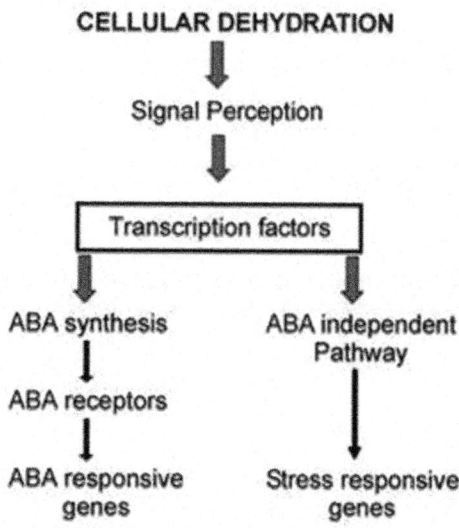

Figure: A consensus scheme of stress perception and gene response.

It is not quite clear which are the primary sensors of cellular dehydration and their order of importance or function, be it cellular water status, pressure, bound water, hormones (mainly ABA), cellular membrane functions or other agents. It is not perfectly clear how cells perceive cellular water deficit and how cellular water deficit is transduced and transcribed into the various consequences of this stress, be it adaptation or mortality. Furthermore, it has not been clearly established which of these responses are stress adaptive and which are expressions of system degradation. The working hypothesis underlying current research in this area recognises multiple signal transduction pathways between stress perception and gene expression. Two major possible pathways transcribe the perception of this signal; one involving ABA production and the other is ABA independent. The ABA independent pathway is not fully resolved. In the ABA dependent pathway ABA induces novel protein synthesis, which regulates numerous "stress responsive", or "ABA responsive" genes. ABA may regulate stress responsive genes without novel protein synthesis. These gene products are either functional (*e.g.* water-channel proteins or key enzymes) or regulatory (*e.g.* protein kinases) and they are involved in mediating various cellular responses some of which are recognised as adaptive. The interface between the molecular domain involving stress responsive gene expression and whole plant response to drought stress is yet to be fully understood. This interface is critical for translating molecular genetics science into advances in crop production under stress conditions. At present there are hundreds of recognised 'drought stress responsive genes' which are upregulated or downregulated under the effect of dehydration and their functions remains to be determined.

At the *whole plant and the crop level*, the important repercussions of water deficit are mediated by effects on plant phenology, phasic development, growth, carbon assimilation, assimilate partitioning and plant reproduction processes. These major effects account for most of the variation in crop yield caused by drought stress. Growth depends on cell expansion and cell division. Cell expansion is probably the most sensitive to water deficit. Cell division might be less sensitive. Cell expansion is dependent on the maintenance of turgor, cell wall extensibility and other factors possibly pertaining to ABA signalling. Reduced cell expansion as a primary response to water deficit serves to reduce plant water use but also lead to reduceed plant productivity. If the reduction in total plant water use is not sufficient to sustain turgor, then transpiration is further reduced by stomatal closure. Initially, stomatal closure reduces transpiration more than it reduces CO_2 assimilation, but at an advance stress both are reduced drastically. Wilting is an expression of turgor loss, which takes up different forms according to plant species, such as leaf rolling in the cereals.

Figure: Drought stressed sorghum plant with rolled leaves

Reduced cell expansion also carries a primary effect on meristematic development of yield components, such as the inflorescence or the tiller initials in the cereals - leading to potentially small reproductive organs and reduced yield. This is an irreversible structural effect that is difficult to amend by re-watering. It can however be amended to some extent by inter-organ compensation following watering, such as renewed tillering in the cereals. The meristematic tissues are generally positioned within the plant in a relatively protected environment as compared with that of a fully expanded leaf and therefore it may take a sever stress for meristem to loose its turgor. However, ABA transported from stress-affected organs can also arrest meristem development even at relatively high meristem water-status.

Water deficit causes advanced or delayed flowering, depending on the species (for example, wheat and rice, respectively). ABA may have a role in this respect as it has been shown to delay flowering in tomato and maize. A delay of up to 50 days has been seen in rice subjected to pre-flowering drought stress. The effect of drought stress on phasic development has been shown to be crucial in affecting maize yield under stress. In maize stress cause a delay in female organ development while the male inflorescence is less affected. Hence, stress causes an increase in the time interval between silking and anthesis. A short anthesis-to-silking interval (ASI) (#3347) has been shown to be a main feature of drought resistance in maize.

It is widely agreed that the reproductive growth stage is the most drought-sensitive stage. Water deficit can cause reproductive failure. Pollen or pollen mother cells are generally more sensitive to desiccation than the ovary so that male sterility is a common result of drought stress during flowering. Reduced grain set of wheat under drought stress has been ascribed to ABA accumulation in the shoot. An ABA responsive gene has been found in the floral parts of tomato. There are some interesting reports (e.g. #3074) showing that grain set of maize subjected to drought stress could be partially improved by the experimental infusion of sugar solution into the stem, leading to conclude that sugar starvation could be an additional factor in affecting grain set and ovary abortion under stress. Short supply of sugar could very well be a generally important factor in the abortion of fruit under drought stress.

Shortage of assimilates and sometimes nitrogen availability is a major cause of arrested grain and fruit growth during drought stress.

Drought stress during cereal grain development reduces the duration of grain filling. If the rate of grain filling is not adjusted upward, final grain weight is reduced. Increased grain growth rate under drought stress depends on the supply of assimilates. This supply is becoming short due to the inhibition of current photosynthesis during stress. An alternative source of assimilates are pre-anthesis stem reserves in the form of sugars, starch or fructans, depending on the species. These reserves are readily utilised for grain filling and their availability may become a critical factor in sustaining grain filling and grain yield under drought stress.

Root/shoot dry weight ratio increase as plant water stress develops. The increase is mostly due to a relative reduction in shoot dry weight. However there were rare cases where an absolute amount of root dry weight increase was observed under drought stress. ABA may have a role in promoting root growth under drought stress. Osmotic adjustment has also been found to improve deeper root growth under stress. Most certainly root distribution within the soil profile changes as stress develops, in a way that helps the plant to explore deep soil moisture. In the cereals, dry topsoil inhibits the formation and establishment of new roots in the topsoil while assimilates partitioned to the root are used in furthering the growth of existing roots into deeper soil. In the small grains and rice, tillering is associated with the development of new roots from tillers. Therefore, extensive tillering is generally associated with dense and shallow roots while limited tillering tends to associate with sparser and deeper roots. This is one of the reasons why most cereal crop cultivars developed in dry regions tend to have a limited tillering habit.

Chapter 8

Impact of Cold Stress

Successful adaptation of a crop species is dependent upon the programming of critical growth stages so that the plant can capitalize on favourable weather periods during the growing season. Plants have evolved a variety of adaptive mechanisms that allow them to optimise growth and development while coping with environmental stresses. These mechanisms include seed and bud dormancy, photoperiod sensitivity, and low-temperature response. Seed dormancy delays germination until after the embryo has gone through an after-ripening period. The over-winter survival of buds of many temperate zone trees and shrubs is dependent on a dormancy stage that starts in the late summer or early fall and ends after exposure to an extended period of cold or increasing day length in the spring. In addition to trees, many other dicots and grasses have a photoperiod response that can advance or delay flowering. Vernalization is a requirement for growth at low temperatures before a plant will flower. Most winter annual and biennial plants have a vernalization requirement. Low-temperature acclimation is an ability of plants to cold acclimate when exposed to gradually decreasing temperatures below a specific threshold. This is the most common mechanism that plants have evolved for adapting to low-temperature stress and examples of plants with the capacity to cold harden can be found in most species.

Types of Low Temperature Injury

There are two types of injuries a plant can sustain through exposure to low temperatures (Stushnoff et al. 1984). The first is chilling injury that occurs from approximately 20 to 0°C. The resultant injuries may include a variety of physiological disruptions in germination, flower and fruit development, yield, and storage life. Minor chilling stress at non-lethal temperatures is normally reversible. Exposure to gradually decreasing temperatures above the critical range can also result in hardening of plants that may reduce or eliminate injury during subsequent exposure to low temperatures. The second type of injury is called freezing injury. This type of injury occurs when the external temperature drops below the freezing point of water. Some varieties of plants that are susceptible to chilling injury can be killed by the first touch of frost. At the other extreme, many plants that are native to cold climates can survive extremely low temperatures without injury (Levitt 1980). Plants may experience intracellular freezing and/or extracellular freezing. Intracellular freezing damages the protoplasmic structure and the ice crystals kill the cell once they grow large enough to be detected microscopically. In extracellular freezing, the protoplasm of the plant becomes dehydrated because a water-vapour deficit is created as cellular water is transferred to ice crystals forming in the intercellular spaces. In some cases, water can remain liquid as low as -47°C without nucleating and forming ice. When nucleation of this supercooled water does occur, intracellular ice forms suddenly resulting in death of the plant.

Types of Plants

Plants can be grouped into three different classes according to their low-temperature tolerance (Stushnoff et al. 1984). The first group includes frost tender plants that are sensitive to chilling injury and can be killed by short periods of exposure to temperatures just below freezing. They cannot tolerate ice in their tissues and readily exhibit frost injury symptoms that include a water soaked flaccid appearance with loss of turger followed by rapid drying upon exposure to warm temperatures. Beans, corn, rice, and tomatoes are examples of plants in this category.

Low-temperature acclimation of plants in the second group allows them to tolerate the presence of extracellular ice in their tissues. Their frost resistance ranges from the broad-leafed summer annuals, which are killed at temperatures slightly below freezing, to perennial grasses that can survive exposure to -40°C. As temperatures decrease the outward migration of intracellular water to the growing extracellular ice crystal causes dehydration stress that will eventually result in irreversible damage to the plasma membrane, which is the primary site of low-temperature injury. If ice nucleation does not occur at -3 to -5°C, supercooling may result in intracellular freezing and death of individual cells.

The final group is made up of very cold hardy plants that are predominantly temperate woody species. Like the plants in the previous group, their lower limits of cold tolerance are dependent on the stage of acclimation, the rate and degree of temperature decline, and the genetic capability of tissues to accommodate extracellular freezing and the accompanying dehydration stress. Deep supercooling allows certain tissues in plants from this group to survive low temperatures without the formation of extracellular ice. However, the most cold hardy species do not rely on supercooling and can withstand temperatures of -196 °C.

Plant Chilling Stress and Its Repercussions

Most crops of tropical origin as well as many of subtropical origin are sensitive to chilling temperatures. This limits production areas and causes potential damage during storage if they are exposed to low temperatures. The temperature below which chill injury can occur varies with species and regions of origin, ranging from 0 to 4°C for temperate fruits, 8°C for subtropical fruits, and about 12°C for tropical fruits such as banana (Lyons 1973). Amongst the highest volume world food crops, maize (*Zea mays*) and rice (*Oryza sativa*) are sensitive to chilling temperatures. Their growth and development can be adversely effected by temperatures below 10°C resulting in yield loss or crop failure. Christiansen and St. John (1981) estimated annual losses of $60 million to the cotton industry due to chilling temperature immediately following field planting. Chilling during the seedling stage in cotton can reduce plant height, delay flowering and adversely affect yield and lint quality. Seedlings can also suffer water stress and leaf desiccation at chilling temperatures, floral initiation is inhibited at 7°C and seed set is inhibited at 15°C. Other crops suffering stand loss, delayed maturity, and reduced yield as a result of chilling after planting include soybean (*Glycine max* L.), lima bean (*Phaseolus lunatus* L.), cucurbits (*Cucurbita* sp.), tomato (*Lycopersicon esculentum* Mill.) pepper *Capsicum annuum* L.), eggplant (*Solanum melongena* L.), okra (*Abelmoschus esculentus* L.), and various cereal crops.

Physiological age, seedling development, and pre-harvest climate can also influence chilling sensitivity. Freshly imbibed seeds of chill-sensitive species tend to be very sensitive, as does the pollen development stage. Fruits maturing at high temperature are more susceptible than those maturing at lower temperatures. Post-harvest storage at lower temperatures is commonly used to extend the storage life of fruits and vegetables. Tropical and subtropical plants however are often subject to physiological damage and loss of quality due to chill injury under these storage conditions. The severity of injury to chill-sensitive tissues tends to increase with decreasing temperatures and with length of low-temperature exposure. Chilling has been found to change the entire metabolic system of the cell with some processes recovering quickly and others only slowly. Chilling affects the entire internal environment of each cell and each molecule within the cells. Enzymatic reactions, substrate diffusion rates, and membrane transport properties are all affected. Chilling injury is therefore likely a direct consequence of these effects (Kratsch and Wise 2000).

Amelioration of Chilling Injury

Avoidance: To avoid chilling injury, planting dates can be altered though this is often difficult because of its effect on later development of the plant. To overcome this problem, cultivars have been bred for early vigour and maturity. In the case of stored fruits and vegetables, maintenance of appropriate storage temperatures is essential to avoid chilling injury. Investigations have also been undertaken to examine synthetic plant growth regulators for the protection of chilling sensitive crops (Li 1989).

Temperature conditioning: Low-temperature 'hardening' allowing tolerance to chilling temperatures appears to have little effect although some sensitivity to 'slight chilling' can be reduced by exposure to temperatures slightly above the chilling range. It also appears that chilling injury to stored fruits and vegetables can be ameliorated by warm temperatures if they are imposed before tissue degeneration becomes advanced. Other treatments such as waxing, fungicides, hormones, and antioxidants have produced variable results that have been dependent upon the species and treatment conditions (Lyons 1973).

Duration: Ultrastructural-chilling injury increases with time and with prolonged exposure the injury becomes irreversible. It is therefore important to minimise the time of chilling temperature exposure.

Relative humidity: High (100%) relative humidity has been found to protect chloroplasts from chill injury, an effect that is enhanced by darkness.

Theories of Chilling Injury

Early research focused on chilling causing an imbalance in plant physiological processes. Chilling was found to affect O_2 evolution, organic acids, sugars, polyphenols, phospholipids, protein, and ATP. Research indicates that chilling stress in sensitive plants changes most chemical entities. There is evidence of accumulation of toxins such as ethanol and acetaldehyde. Although many altered processes involve key metabolites; it is difficult to separate the critical chilling-sensitive metabolic processes from those that are byproducts of metabolic disruptions or of ultrastructural breakdown. Ion leakage due to membrane permeability changes has often been reported in chill sensitive plants. Phase transition of the lipid portion of the cellular membranes has also received considerable attention as the *primary* response to chilling temperatures (Lyons 1973).

Ultrastructural changes: On an ultrastructural level, several changes have been associated with chilling injury. Although there are a number of variables affecting chill injury, the ultrastructural symptoms are very similar across species. Ultrastructural symptoms of chilling injury become evident before obvious physical symptoms are visible. These include changes to chloroplasts, mitochondria and membranes associated with these organelles and the vacuoles (Christiansen and St. John 1981). The symptoms include swelling and disorganisation of the chloroplasts and mitochondria, reduced size and number of starch granules, dilation of thylakoids and unstacking of grana, formation of small vesicles of chloroplast peripheral reticulum, lipid droplet accumulation in chloroplasts, and condensation of chromatin in the nucleus (Kratsch and Wise 2000).

Chloroplasts are the first and most severely affected organelle. Irradiance during chilling greatly exacerbates the resulting injury. Chilled plants in darkness have been found to remain green and, except for starch depletion, chloroplasts appear normal. In the presence of light, however, chlorophyll becomes bleached, lipid droplets accumulate, and thylakoids degenerate. Mitochondria appear more resistant to chilling temperature but an immediate effect of low temperature on chilling-sensitive species is a suppression of mitochondrial activity. Electron micrographs of chilled sweet potato roots revealed that the mitochondria had a swollen appearance due to the release of phospholipids from the inner and outer membranes during storage at chilling temperatures. The capacity to bind phospholipids was also greatly decreased.

Membrane permeability and phase transition: Measures of solute leakage or ion permeability have provided evidence of increased membrane permeability in response to chilling. The plasma membrane is often considered the primary site of freezing injury and electrolyte leakage. Early work indicated that plants originating in warm climates tend to have more saturated fatty acids in their membrane lipids. More recent work on mitochondrial membranes has shown that membranes do undergo a physical phase transition from a flexible liquid-crystalline to a solid-gel structure at 10 to 12°C, which coincides with the temperature sensitivity range of species of tropical origin. Fruits of several apple cultivars have been observed to undergo phase transition in the 3 to 10°C range suggesting the same mechanism of chilling injury as found in tropical species. The correlation between fatty acid composition and temperature induced phase transition is, however, not precise. It may be that other membrane components such as sterols also play a role.

It is possible that the phase transition of cellular membranes could account for the entire range of physiological and metabolic changes associated with chilling injury. Increased membrane permeability could lead to an altered ion balance and also to the ion leakage observed from chilling of sensitive tissues. Phase transition could result in conformational changes in membrane bound enzymes and account for the observed discontinuities in the function of many enzyme systems. This may cause an imbalance between membrane bound and non-membrane bound systems. Over time the cells inability to cope with increased concentrations of metabolites could result in injury. Different tolerances to these metabolites could explain why some cultivars are more resistant to damage while still undergoing phase transition. Imbalances in metabolism, accumulation of toxic compounds, and increased permeability could all be the result of temperature-induced phase transition (Lyons 1973).

The contribution of unsaturated fatty acids in cell membrane lipids has been discussed for many years in relation to chilling sensitivity. Nishida and Murata (1996) have shown that chilling injury can be manipulated by modulating levels of unsaturation of fatty acids by the action of acyl-lipid desaturases and glycerol-3-phosphate acyltransferase. Lyons (1973) proposed that temperature induced phase transition of membrane lipids may play a primary role in chilling sensitivity of plants. Continued exposure to chilling temperatures would result in phase separated membranes becoming incapable of maintaining ionic gradients resulting in metabolic disruption and eventual cell death. A positive correlation has been found between chilling sensitivity of herbaceous plants and the level of saturated and trans-monounsaturated molecular species of phosphatidylglycerol (also termed high-melting-point molecular species) in thylakoid membranes. However, there is still a question of how directly these high-melting-point molecular species relate to chilling sensitivity in plants. Growth at low temperature generally increases the degree of unsaturation of membrane lipids, which compensates for the decrease in fluidity caused by the lower temperature. This increased unsaturation is also correlated with the sustained activity of membrane-bound enzymes at low temperature. The unsaturation of membrane lipids is therefore considered critical for the functioning of biological membranes and the survival of plant cells at low temperature. However, since low temperature is also known to induce or alter the expression level of a large number of genes it is not clear if the association between membrane lipid unsaturation and chilling tolerance is a cause or effect relationship.

Recently the role of unsaturation of membrane lipids in chilling tolerance and in response to low temperature has been reexamined using mutant and transgenic lines (Nishida and Murata 1996). In this way unsaturated fatty acids can be manipulated independent of temperature so that their individual effects can be evaluated. Tobacco was transformed with squash and *Arabidopsis* phosphatidylglycerol (PG) species found in thylakoid membranes. Squash has low levels of *cis*-unsaturated PG while *Arabidopsis* has relatively high levels of *cis*-unsaturated PG. It was found that tobacco transformed with squash PG was more chilling sensitive and tobacco transformed with *Arabidopsis* PG was the most chilling resistant, as measured by photosynthesis at 1°C under strong illumination. These results indicate that chilling sensitivity can be manipulated by altering the level of unsaturated PG in the chloroplasts. These and other experiments have shown that unsaturation of membrane lipids protect the photosystem II complex from low-temperature photoinhibition by accelerating recovery from the photoinhibited state. However, it is likely that other factors such as accumulation of polyols and amino acids, or their derivatives, contribute to chilling sensitivity in plants. Some specific proteins may also be responsible for chilling tolerance.

Alteration of intracellular pH: Yoshida et al. (1999) noted that intracellular pH was, in part, actively controlled by H^+-transport from the cytoplasm to the vacuole catalysed by H^+-ATPase located on the vacuolar membrane in mung bean (Vigna radiata L.), which is a very chilling-sensitive species. The vacuolar H^+-ATPase is extremely sensitive to low temperature and is preferentially inactivated upon exposure to chilling temperatures. This inactivation occurs much earlier than the symptoms of cell injury and the decrease in enzyme activity associated with plasma membranes, endoplasmic reticulum, and mitochondria. Cold-induced inactivation of H^+-ATPase also occurs in chilling sensitive rice. Cold-induced suppression of proton transport disrupts cytoplasmic homeostasis and causes a change in the pH. The chilling sensitivity of cultured mung bean cells changed markedly during the growth cycle and a close relationship was found between sensitivity of the cells and of H^+-ATPase to the cold. Cold-induced inactivation of the vacuolar H^+-ATPase was closely linked to acidification of the cytoplasm and the corresponding alkalization of the vacuoles suggesting a passive release of H^+ ions across the vacuolar membrane. The susceptibility of vacuolar H^+-ATPase to low temperature in vivo was found to be markedly different between chilling-sensitive and chilling-resistant species. In contrast to the H^+-ATPases of chilling-sensitive species like mung bean and kidney bean (Phaseolus vulgaris), the H^+-ATPases of the chilling-tolerant species such as pea (Pisum sativum) and broad bean (Vicea faba) were very stable over long periods of low-temperature exposure. The molecular structures of the 16 kDa proteolipids from the two types of H^+-ATPase appeared to be very different. Low-temperature-induced pH reduction of the cytoplasm caused by inactivation of vacuolar H^+-ATPase may therefore be the cause of extreme chilling-sensitivity.

Plants are uniquely adapted to their native environment through developmental programming and the particular composition and conformation of their molecular components is optimised within each species for maximum competitive ability. These differences in adaptation result in the wide range of cellular disturbances that have been observed when these plants are moved to cooler environments. Changes in enzyme reactions, substrate diffusion rates, membrane properties, and cytoplasmic pH affect the entire metabolic system of cells subjected to chilling stress. The resulting injury depends on the duration of exposure and on the individual species, or variant, being observed.

The Physiological and Agronomic Repercussions of Freezing Stress

Low-Temperature Response Mechanisms

Plants have adapted two mechanisms to protect themselves from damage due to below freezing temperatures. Supercooling is a low-temperature tolerance mechanism that is usually associated with acclimated xylem parenchyma cells of moderately hardy woody plants. When sources of ice nucleation are absent, pure water can supercool or remain unfrozen to its homogeneous nucleation point of approximately -40°C. The initiation of freezing at the limit of supercooling occurs suddenly and is accompanied by an exotherm that can be detected by thermal analyses of plant tissues. Plant tissues suffer irreversible damage once ice nucleation of supercooled water occurs and the distribution in nature of tree species with the ability to deep supercool is normally restricted to regions where winter temperatures are warmer than -40°C (George et al. 1982).

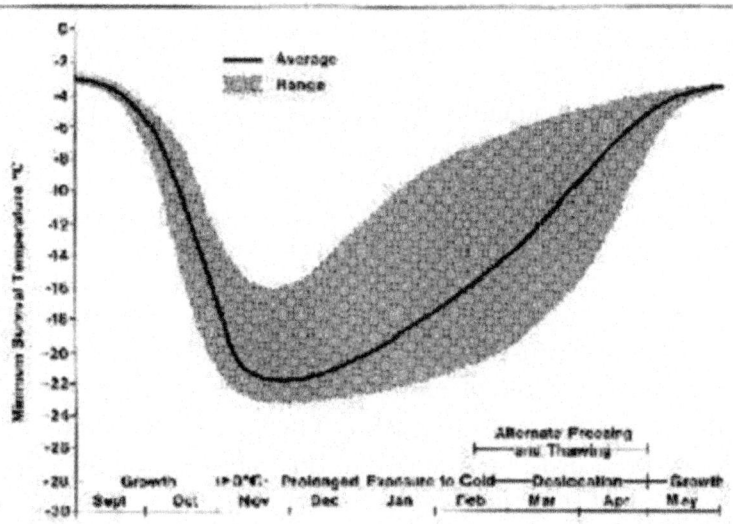

Figure: Changes in cold hardiness of Norstar winter wheat for the period September to May. The primary factors responsible for these changes are shown at the bottom of the graph.

The second and most common low-temperature response mechanism is acclimation. Low-temperature acclimation is a gradual process during which there are changes in just about every measurable morphological, physiological, and biochemical characteristic of the plant. These changes are determined by genotype x environment interactions that are quite complex and not clearly understood. They have been studied most extensively in cereals where a wide range in genetic potential and the availability of unique cytogenetic stocks has allowed for novel approaches to investigations at the molecular and whole plant level. Potential gene donors have been evaluated for use in interspecific transfers and the control of alien (donor species) low-temperature gene expression has been studied in a variety of backgrounds. A survey of the published research in these areas has allowed us to construct a field validated winter survival model that successfully simulates the over winter changes in low-temperature tolerance of a wide range of genotypes (Fowler et al. 1999).

Consequently, this review will focus mainly on the genetic systems that winter cereals have evolved for low-temperature adaptation, the regulation of these systems, and their complex interaction with the environment.

Low-Temperature Acclimation in Winter Cereals

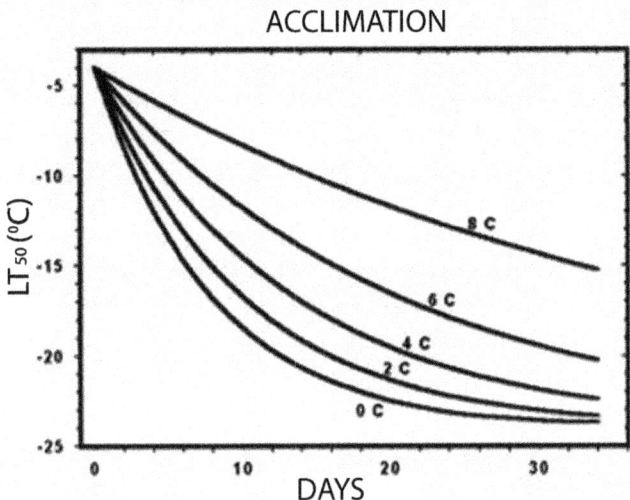

Figure: Relationship between average daily crown temperature (0 to 8oC) and low-temperature tolerance of Norstar winter wheat during acclimation at a constant temperature before vernalization saturation (LT50 for Norstar = -24oC when fully acclimated).

Under field conditions in western Canada, eight to 12 weeks of fall growth is usually required for the full development of cold hardiness in winter cereals. The first four to five weeks is a period of active growth that takes place when average daily soil temperatures at crown depth are above 9°C. Both the cold acclimation process and winter survival require energy and this period of warm temperature allows for the establishment of healthy vigorous plants. Plants with well-developed crowns before freeze-up are in the best position to withstand the rigours of winter and regenerate roots and leaves in the spring. However, plants that enter the winter with two to three leaves are usually not seriously disadvantaged.

Cold acclimation of winter wheat plants begins once fall temperatures drop below approximately 9°C. A translocatable substance that promotes cold acclimation is not produced when winter wheat plants are exposed to acclimating temperatures (Limin and Fowler 1985). Consequently, the cold-hardiness level of different plant parts, such as leaves, crowns and roots, is dependent upon the temperature to which each part has been exposed. Because the crown contains tissues that are necessary for plant survival, it is the soil temperature at crown depth that determines critical cold-acclimation rates.

Plant growth slows considerably at temperatures that promote cold acclimation. In the field, soil temperatures gradually decrease as winter approaches and four to seven weeks at temperatures below 9°C is usually required to fully cold-harden plants. Cold acclimation during this period is dependent upon crown temperatures and the rate of acclimation increases dramatically as temperature drop from 9 to 0°C. Exposure of winter wheat crowns to soil temperatures above 9°C during this period results in a rapid loss of cold hardiness. The rate of dehardening is dependent upon the temperature to which the crown is exposed. At this stage, plants that have been exposed to crown temperatures above 9°C will resume cold acclimation once they return to temperatures below 9°C.

Winter wheat normally does not realise its maximum cold hardiness potential until after the soil is frozen in the late fall. In Saskatchewan, full acclimation is usually achieved by the middle to the end of November. Once cold acclimation has been completed, winter wheat can maintain a high level of cold hardiness provided crown temperatures remain below freezing. In the fall, winter wheat will cold acclimate when exposed to crown temperatures colder than 9°C. However, prolonged exposure of acclimated plants to winter temperatures above freezing results in the transition of the plant from the vegetative to the reproductive phase and a gradual loss of cold hardiness. The warmer the crown temperature during the winter, the shorter the period that maximum levels of cold hardiness can be maintained and, once started, the more rapid the rate of decline in cold hardiness.

Death of the crown tissue will result if the soil temperature falls below the plants minimum survival temperature. Exposure of winter cereal plants to crown temperatures that are 2 to 3°C warmer than their minimum survival temperature will cause immediate damage and a reduction in cold hardiness (Fowler et al. 1999). Longer periods of exposure to temperatures approaching the minimum survival temperature can quickly reduce the plant's ability to tolerate cold stress. The expected LT_{50} for different exposure times (T) to constant temperature can be calculated from Equation 1 (Fowler et al. 1999).

$$LT50_{(T)} = LT50_{(0)} + 5.72 + 1.53 * \ln(T) \qquad [1]$$

Where T is the number of days that plants are exposed to a constant low-temperature stress. $LT50_{(0)}$ is determined using a series of test temperatures where the low-temperature stress is removed as soon as the crown samples are exposed to a predetermined minimum temperature. For example, fully acclimated Norstar winter wheat will normally survive to -24.0°C in a controlled-freeze test where plant samples are gradually cooled at a rate of 2 to 6°C hr^{-1} and removed as soon as they reach a predetermined temperature. However, two days exposure to -17.2°C in a controlled environment will reduces the minimum survival temperature of fully hardened Norstar winter wheat from -24.0 to -17.2°C, a cold hardiness loss of 6.8°C.

Once vernalization saturation is complete and the plant enters the reproductive stage, it loses its ability to cold acclimate (Fowler et al. 1996a) and it will start to deharden at temperatures warmer than approximately -4°C. This means that winter wheat will eventually completely deharden once plant growth resumes in the spring. Growth rate and rate of dehardening are both temperature dependent and because frozen soils warm slowly in the spring, several weeks of warm air temperatures are required to re-establish and completely deharden winter cereal plants that have survived without serious winter damage.

Chapter 9

The Impact of Mineral Toxicity Stress

Soil pH or soil reaction is an indication of the acidity or alkalinity of soil and is measured in pH units. Soil pH is defined as the negative logarithm of the hydrogen ion concentration. The pH scale goes from 0 to 14 with pH 7 as the neutral point. As the amount of hydrogen ions in the soil increases the soil pH decreases thus becoming more acidic. From pH 7 to 0 the soil is increasingly more acidic and from pH 7 to 14 the soil is increasingly more alkaline or basic.

Descriptive terms commonly associated with certain ranges in soil pH are:

- extremely acid, < than 4.5; lemon=2.5; vinegar=3.0; stomach acid=2.0; soda=2–4
- very strongly acid, 4.5–5.0; beer=4.5–5.0; tomatoes=4.5
- strongly acid 5.1–5.5; carrots=5.0; asparagus=5.5; boric acid=5.2; cabbage=5.3
- moderately acid, 5.6–6.0; potatoes=5.6
- slightly acid, 6.1–6.5; salmon=6.2; cow's milk=6.5
- neutral, 6.6–7.3; saliva=6.6–7.3; blood=7.3; shrimp=7.0
- slightly alkaline, 7.4–7.8; eggs=7.6–7.8
- moderately alkaline, 7.9–8.4; sea water=8.2; sodium bicarbonate=8.4
- strongly alkaline, 8.5–9.0; borax=9.0
- very strongly alkaline, > than 9.1; milk of magnesia=10.5, ammonia=11.1; lime=12

Measuring Soil pH

Soil pH provides various clues about soil properties and is easily determined. The most accurate method of determining soil pH is by a pH metre. A second method which is simple and easy but less accurate then using a pH metre, consists of using certain indicators or dyes.

Phosphorus is never readily soluble in the soil but is most available in soil with a pH range centred around 6.5. Extremely and strongly acid soils (pH 4.0-5.0) can have high concentrations of soluble aluminum, iron and manganese which may be toxic to the growth of some plants. A pH range of approximately 6 to 7 promotes the most ready availability of plant nutrients.

But some plants, such as azaleas, rhododendrons, blueberries, white potatoes and conifer trees, tolerate strong acid soils and grow well. Also, some plants do well only in slightly acid to moderately alkaline soils. However, a slightly alkaline (pH 7.4-7.8) or higher pH soil can cause a problem with the availability of iron to pin oak and a few other trees in Central New York causing chlorosis of the leaves which will put the tree under stress leading to tree decline and eventual mortality.

The soil pH can also influence plant growth by its effect on activity of beneficial microorganisms Bacteria that decompose soil organic matter are hindered in strong acid soils. This prevents organic matter from breaking down, resulting in an accumulation of organic matter and the tie up of nutrients, particularly nitrogen, that are held in the organic matter.

Changes in Soil pH

Soils tend to become acidic as a result of: (1) rainwater leaching away basic ions (calcium, magnesium, potassium and sodium); (2) carbon dioxide from decomposing organic matter and root respiration dissolving in soil water to form a weak organic acid; (3) formation of strong organic and inorganic acids, such as nitric and sulfuric acid, from decaying organic matter and oxidation of ammonium and sulfur fertilizers. Strongly acid soils are usually the result of the action of these strong organic and inorganic acids.

Lime is usually added to acid soils to increase soil pH. The addition of lime not only replaces hydrogen ions and raises soil pH, thereby eliminating most major problems associated with acid soils but it also provides two nutrients, calcium and magnesium to the soil. Lime also makes phosphorus that is added to the soil more available for plant growth and increases the availability of nitrogen by hastening the decomposition of organic matter. Liming materials are relatively inexpensive, comparatively mild to handle and leave no objectionable residues in the soil. Some common liming materials are: (1) Calcic limestone which is ground limestone; (2) Dolomitic limestone from ground limestone high in magnesium; and (3) Miscellaneous sources such as wood ashes. The amount of lime to apply to correct a soil acidity problem is affected by a number of factors, including soil pH, texture (amount of sand, silt and clay), structure, and amount of organic matter. In addition to soil variables the crops or plants to be grown influence the amount of lime needed.

Causes and Effects of Soil Acidity

Soil acidity is a crop production problem of increasing concern in central and western Oklahoma. Although acid soil conditions are more widespread in eastern Oklahoma, the more natural occurrence there has resulted in farm operators being better able to manage soil acidity in that part of the state. However, in central and western Oklahoma the problem appears to grow with time. This fact sheet explains why soils become acid and the problems acid soils create for crop production. OSU Extension Facts No. 2229 explains how soil acidity and the lime requirement are determined by soil testing. A subsequent fact sheet discusses managing wheatland soils in Oklahoma.

Why Soils are Acid

The four major causes for soils to become acid arelisted below:

1. Rainfall and leaching
2. Acidic parent material
3. Organic matter decay
4. Harvest of high yielding crops.

The above causes of soil acidity are more easily understood when we consider that a soil is acid when there is an abundance of acidic cations (pronounced cat-eyeon), like hydrogen (H+) and aluminium (Al+++) present compared to the alkaline cations like calcium (Ca++), magnesium (Mg++), potassium (K+), and sodium (Na+).

Rainfall and Leaching

Excessive rainfall is an effective agent for removing basic cations over a long time period (thousands of years). In Oklahoma, for example, we can generally conclude that soils are naturally acidic if the rainfall is above 30 inches per year. Therefore, soils east of I-35 tend to be acidic and those west of I-35, alkaline. There are many exceptions to this rule though, mostly as a result of item 4, intensive crop production. Rainfall is most effective in causing soils to become acidic if a lot of water moves through the soil rapidly. Sandy soils are often the first to become acidic because water perco-lates rapidly, and sandy soils contain only a small reservoir of bases (buffer capacity) due to low clay and organic matter contents. Since the effect of rainfall on acid soil development is very slow, it may take hundreds of years for new parent material to become acidic under high rainfall.

Parent Material

Due to differences in chemical composition of parent materials, soils will become acidic after different lengths of time. Thus, soils that developed from granite material are likely to be more acidic than soils developed from calcareous shale or limestone.

Organic Matter Decay

Decaying organic matter produces H+ which is responsible for acidity. The carbon dioxide (CO_2) produced by decaying organic matter reacts with water in the soil to form a weak acid called carbonic acid. This is the same acid that develops when CO_2 in the atmosphere reacts with rain to form acid rain naturally.

Several organic acids are also produced by decaying organic matter, but they are also weak acids. Like rainfall, the contribution to acid soil development by decaying organic matter is generally very small, and it would only be the accumulated effects of many years that might ever by measured in a field.

Crop Production

Harvesting of crops has its affect on soil acidity development because crops absorb the lime-like elements, as cations, for their nutrition. When these crops are harvested and the yield is removed from the field, then some of the basic material responsible for counteracting the acidity developed by other processes is lost, and the net affect is increased soil acidity. Increasing crop yields will cause greater amounts of basic material to be removed. Grain contains less basic materials than leaves or stems. For this reason, soil acidity will develop faster under continuous wheat pasture than when grain only is harvested. High yielding forages, such as bermudagrass or alfalfa, can cause soil acidity to develop faster than with other crops.

The approximate amount of lime-like elements removed from the soil by a 30 bushel wheat crop. Note that there is almost four times as much lime material removed in the forage as the grain. This explains why wheat pasture that is grazed out will become acidic much faster than when grain alone is produced. Using 50 percent ECCE lime, it would take about one ton every 10 years to maintain soil pH when straw (or forage) and grain are produced annually at the 30 bushel per acre level.

The use of fertilizers, especially those supplying nitrogen, has often been blamed as a cause of soil acidity. Although acidity is produced when ammonium containing materials are transformed to nitrate in the soil, this is countered by other reactions and the final crop removal of nitrogen in a form similar to that in the fertilizer.

What Happens in Acid Soils

Knowing the soil pH helps identify the kinds of chemical reactions that are likely to be taking place in the soil. Generally, the most important reactions from the standpoint of crop production are those dealing with solubilities of compounds or materials in soils. In this regard, we are most concerned with the affects of pH on the availability of toxic elements and nutrient elements.

Toxic elements like aluminium and manganese are the major causes for crop failure in acid soils. These elements are a problem in acid soils because they are more soluble at low pH. In other words, more of the solid form of these elements will dissolve in water when the pH is acid. There is always a lot of aluminium present in soils because it is a part of most clay particles.

Element Toxicities

When the soil pH is above about 5.5, the aluminium in soils remains in a solid combination with other elements and is not harmful to plants. As the pH drops below 5.5, aluminium containing materials began to dissolve. Because of its nature as a cation (Al^{+++}), the amount of dissolved aluminium is 1000 times greater at pH 4.5 than at 5.5, and 1000 times greater at 3.5 than at 4.5. For this reason, some crops may seem to do very well, but then fail completely with just a small change in soil pH. Wheat, for example, may do well even at pH 5.0, but usually will fail completely at a pH of 4.0.

The relationship between pH and dissolved manganese in the soil is similar to that just described for aluminium, except that manganese (Mn^{++}) only increases 100 fold when the pH drops from 5.0 to 4.0.

Toxic levels of aluminium harm the crop by "root pruning." That is, a small amount of aluminium in the soil solution in excess of what is normal causes the roots of most plants to either deteriorate or stop growing. As a result, the plants are unable to absorb water and nutrients normally and will appear stunted and exhibit nutrient deficiency symptoms, especially those for phosphorus.

The final affect is either complete crop failure or significant yield loss. Often the field will appear to be under greater stress from pests, such as weeds, because of the poor condition of the crop and its inability to compete.

Toxic levels of manganese interfere with the normal growth processes of the above ground plant parts. This usually results in stunted, discoloured growth and poor yields.

Desirable pH

The adverse effect of these toxic elements is most easily (and economically) eliminated by liming the soil.

Liming raises the soil pH and causes the aluminium and manganese to go from the soil solution back into solid (non-toxic) chemical forms. For grasses, raising the pH to 5.5 will generally restore normal yields. Legumes, on the other hand, do best in a calcium rich environment and often need the pH in a range of 6.5 to 7.0 for maximum yields.

A soil pH in the range of 6.0 to 7.0 is also desirable from the standpoint of optimum nutrient availability. However, the most common nutrient deficiencies in Oklahoma are for nitrogen, phosphorus and potassium, and availability of these elements will not be greatly changed by liming. Nutrients most affected by soil pH are iron and molybdenum. Iron deficiency is more likely to occur in non-acid (high pH) soils. Molybdenum deficiency is not common in Oklahoma, but would be most apt to occur in acid soils and could be corrected by liming.

Metals occur naturally in soils which may be beneficial or toxic to the environment. Although excess of metals may produce some common effects on plants in general, there are many cases of specific effects of individual metals on different plants (i.e. both macro- and micro-flora). The biota requires some of these elements in trace quantities but may be sensitive to higher concentration of metal. Metal toxicity in plants has been reported by many workers. Aluminium (Al) is not regarded as an essential nutrient, but low concentrations can sometimes increase plant growth or induce other desirable effects. Aluminium toxicity is an important growth-limiting factor for plants in acid soils below pH 5.0 but can occur at pH levels as high as 5.5 in minespoils. Generally, Al interferes with cell division in root tips and lateral roots, increases cell wall rigidity by cross linking pectins, reduces DNA replication by increasing the rigidity of the DNA double helix, fixes phosphorous in less available forms in soils and on root surfaces, decreases root respiration, interferes with enzyme activity governing sugar phosphorylation and the deposition of cell wall polysaccharides, and the uptake, transport, and also use of several essential nutrients (Ca, Mg, K, P and Fe). Excess Al even induces iron (Fe) deficiency symptoms in rice (*Oryza sativa* L.), sorghum and wheat. Al is present in all soils, but Al toxicity is manifested only in acid conditions, in which the phytotoxic form Al3+ predominates. Recent progress in the study of toxic metals and their interactions with essential elements has greatly increased our understanding of the mechanism of toxicity at the biochemical level.

The Impact of Flooding Stress on Plants and Crops

The complex and highly developed land-based biology with which we are most familiar, is a relatively new phenomenon in evolutionary terms. Its existence is predicated on a successful invasion from the sea by photosynthetic macrophytes ~ 400 million years ago (Corner, 1964). This gave rise to organisms with the then novel ability to operate photosynthetically in air while securing water and minerals using a non-photosynthetic, foraging root system. Present day representatives of the more than 300,000 plant species presently known to science now occupy almost every terrestrial niche. However, although, their progenitors were aquatic, the land plants derived from them are relatively intolerant of free water in their surroundings, especially if it is slow moving or immobile. The resulting effect is so severe that biochemical and morphological adaptations have emerged many times during evolution (Cook, 1999) to allow a sizeable minority of species to succeed in sporadically or permanently flooded areas on land. These include areas prone to ice encasement (Andrews, 1996). The ability of excess water to damage plants may seem counter-intuitive since water is chemically benign. However, certain physical properties of water, most notably its ability to interfere with free gas exchange, can injure and kill plants when they are totally submerged (Jackson and Ram, 2003) or even when only the soil is waterlogged (Vartapetian and Jackson, 1997).

The non-frozen permanently wet places are known variously as bogs, mires, marshes, fens, peatlands, bottomlands, wetlands etc. These can be natural or man-made, static or flowing, fresh, brackish or salty, seasonal or permanent. They are widespread and contain a highly adapted and characteristic flora that is under threat from drainage, peat extraction and redevelopment. These destructive activities conflict with an increasing recognition of the considerable economic, ecological, social and amenity value of many wetlands. They have great significance as wildlife sanctuaries, as buffer zones that reduce flooding intensity of surrounding areas and detoxify the drainage water. This recognition gave rise to the Convention on Wetlands, an intergovernmental agreement adopted at Ramsar, Iran on 2 February 1971. Its remit is to work for the protection and wise management of wetlands world-wide. To-date (2003), the Ramsar Convention had 136 signatory countries, with 1287 wetland sites included in the Ramsar List of Wetlands of International Importance.

Figure: Wetlands in Okavango, Botswana. (*Photograph by Jim Thorsell*)

The extent of more or less permanently wet places is more readily quantified than areas subjected to sporadic waterlogging or deeper flooding. In reality, almost all the land surface becomes flooded at some time. This is true even in deserts, such as those of central Australia. Worldwide, it has been estimated that approximately 10 % of all irrigated farmland suffers from frequent waterlogging, which may decreases crop productivity 20 %. But, in addition, many rainfed regions are also susceptible to temporary flooding. Satellite imaging has the potential to locate these areas of temporarily flooded land but the technique is under-utilised compared with similar work assessing the extent of water deficient soils.

Clearly, stress from flooding has, and remains, a major influence on species distribution worldwide. It can be the dominant determinant of species success in wet areas (Lenssen et al. 2003). Furthermore, along with drought, salinity and mineral deficiency, flooding also has serious economic consequences for productivity of much arable farmland. This review assesses the impact of excess water on plant growth and development and on the underlying biochemistry and molecular biology. Particular attention is given to responses that appear to enhance tolerance or survival. The major part deals with waterlogging of the soil, and its impact on root systems, the aerial shoot and on farm crops. This is preceded by an assessment of the factors that influence the scale and occurrence of waterlogging and submergence.

Factors Favouring Occurrence of Waterlogging and Submergence

Increasing Water Input

It is self-evident that episodic waterlogging of the soil or deeper submergence (referred to collectively as flooding when a distinction is not necessary) occur when water enters soil faster than it can drain away under gravity. There is mounting evidence that, in several parts of the World, inputs of water are growing. One cause may be climate change. For example, records show an increased incidence of flooding and rainfall in much of northern and Western Europe during the last century. Modelling studies indicate that this shift to more intense rainstorms in this region may raise both the frequency and magnitude of flooding from river overflow over the next 50 years (Prudhomme C., Jakob, and Svensson, 2002).

A factor contributing to increased rainfall will be faster seawater evaporation at the warmer temperatures that, in turn, may produce more rain. Associated with these trends is an increase in sea level predicted to be up to 20 cm over the next half century. This will principally be the result of thermal expansion and melting of polar ice. Greater fluxes of river water resulting from mountain deforestation are also being experienced in many parts of the World, with loss of mountain forest and wetlands playing a major part in heightening peaks of river outflow. The overall outcome is an increasing frequency of flooding of lowland regions such as the lower reaches of the River Rhine in Europe and the Euphrates delta of Bangladesh and West Bengal. These are highly populated areas but also contain much productive farmland where satellite imaging has recorded many major flooding events. While it is damage to human life and property that attracts most media coverage, flash flooding can devastate vegetation of poorly adapted species especially farm crops. In developing countries especially, this threatens the well-being of many people who depend on locally produced food.

Intensive and large-scale irrigation of farmland can also increase the incidence of waterlogging of the soil. Water tables can rise as a result. This is especially likely in heavily irrigated dry regions such as Sindh Province in the Indus valley of Pakistan. Here, 50 to 60 years ago, the water table was 4 m below ground. By 1984, it was less than 1.6 m over most of the irrigated region, the rising water being laden heavily with salts. The resulting environmental catastrophe has led to a multi-million dollar drainage project (the Left Bank Outfall Drainage Project) of immense scale but bringing with it much controversy and environmental concern. The problem is exacerbated by the flatness of the topography that inevitably slows the rate of lateral drainage.

A third contributory factor can be change of land use. For example, conversion of meadow land to arable farming in Germany since the 1950s, has contributed to increased surface run-off and exacerbated flooding problems elsewhere in the landscape (Van Der Ploeg *et al.*, 2002). Expanding urbanisation of the landscape also creates large expanses of non-absorbing hard surface that concentrates rainwater to its periphery via surface run-off or underground drainage systems.

Slow Drainage Through the Soil Profile

The duration and severity of flooding or deeper submergence can be influenced not only by the rate of water input but also by the rate of water flow out from the rooting zone and by the water absorbing capacity of the soil. Topography plays an important part in determining the speed of lateral flow within and above the soil. Obviously, it will be slower on the plains than on sloping land. However, the impact of rate of vertical drainage through the soil profile is critical and strongly affected by soil structure, which is highly variable. Soil structure is a complex subject, involving both macro and microstructural components. It is well-described in Soils An Introduction (5th edition) by Michael J. Singer and Donald N. Munns. In brief, it can be said that approximately 40 – 60 % of the soil volume is made of solid material (mostly minerals and organic matter) that is permeated by spaces filled with water, with gas or with roots and other living organisms. The total volume of these spaces (pores), the size range of the pores, their interconnectivity, stability and the relative proportions of each size class all have a major impact on how much water is held by the soil and how readily it drains through the profile. Small pores hold water more strongly by capillary forces than do larger ones. Adopting the classification of (Greenland, 1977), interconnected pores with a diametre range larger than 50 μm (transmission pores) drain under gravity.

This allows air to enter (critically oxygen) to support aerobic respiration and also gives space for root exploration. Pores with diametres in the range of approximately 50 – 0.5 μm (storage pores) can hold water against the force of gravity but weakly enough for roots to extract it using driving forces of up to -1500 kPa. However, they are not large enough to allow roots to penetrate. Pores smaller that 0.2 μm hold water so strongly that neither gravity nor roots can extract their contents. These are, therefore, permanently filled with water and are termed residual pores. These classifications which contrasts a sandy loam with a more clayey soil. It is readily apparent that it will take relatively little extra water for a clay soil to become waterlogged from field capacity compared to a sandy loam soil. Pores in clay soils are also unstable when emptied of water. Thus, drainage is not always followed by the entry of air because some pores collapse. A further consideration is the interconnectedness of transmission pores. The ease of movement of water under gravity from pore to pore is quantified as the soil's hydraulic conductance (mm d^{-1}). The pores of clay soils are less well connected than those of sandier soils and thus drain more slowly because hydraulic conductance is low. Drainage rates are also affected by a soil's macro-structure. In clayey soils, the small intrinsic hydraulic conductance can be offset to some extent by a tendency to crack thereby opening up fissures through which water may move readily. Channels formed by earthworms, decayed root axes, also improve the rate of drainage. The tendency for soil particles to form aggregates (crumbs and clods) with relatively wide channels between them also influences drainage rate. On a larger scale, the rate of soil drainage can be affected indirectly by the hydraulic conductivity of the sub-soil. An impermeable layer at sub-soil depth such as that created by the soles of ploughs, or imposed naturally as in so-called duplex soils, can cause saturation and flooding of the topsoil that, in itself, has good drainage properties. In duplex soils, rainwater moving readily through the sandy topsoil layers then encounters an impermeable layer, typically rich in clay that may have been further compacted by the use of heavy agricultural machinery. In wet weather, the outcome is a perched water table that limits rooting depth and saturates much of the overlying soil. In Australia, large areas used for growing wheat and other crops in south west of Western Australia, and in Victoria and Queensland are especially susceptible to flooding by such perched water tables.

Soil Waterlogging

Oxygen Shortage and Other Damaging Features of Waterlogged Soil

In waterlogged soil, diffusion of gases through soil pores is so strongly inhibited by their water content that it fails to match the needs of growing roots. A slowing of oxygen influx is the principal cause of injury to roots, and the shoots they support (Vartapetian and Jackson, 1997). The maximum amount of oxygen dissolved in the floodwater in equilibrium with the air is a little over 3 % of that in a similar volume of air itself. This small amount is quickly consumed during the early stages of flooding by aerobic micro-organisms and roots. In addition to imposing oxygen shortage, flooding also impedes the diffusive escape and/or oxidative breakdown of gases such as ethylene (Arshad and Frankenberger, 1990) or carbon dioxide that are produced by roots and soil micro-organisms. This leads to accumulations that can influence root growth and function. For example, accumulated ethylene may slow root extension, while carbon dioxide in the soil can severely damage roots of certain species e.g.,

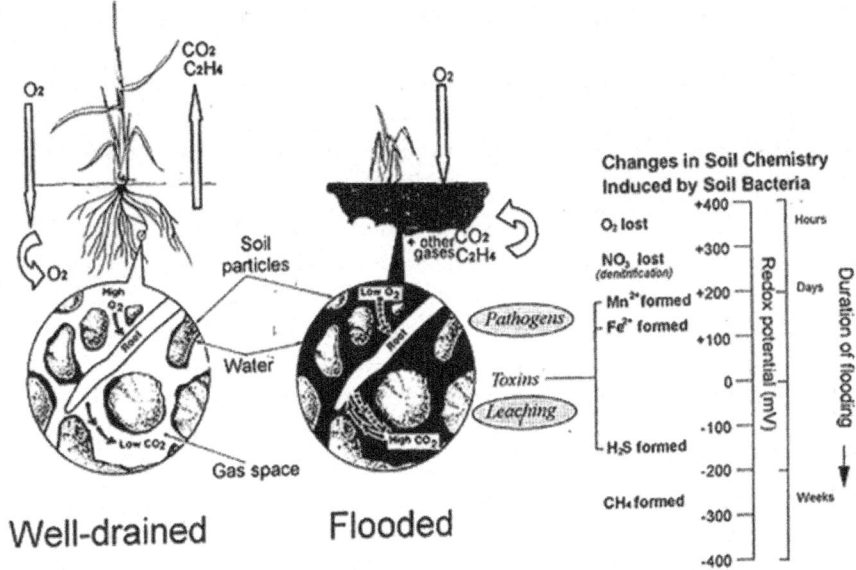

Figure: Effect of flooding on (i) the displacement and exclusion of aerial oxygen from the soil, entrapment of metabolically generated gases in the soil and (ii) the consequences, over time, of bacterial respiration for soil redox potential, loss of free nitrate and subsequent generation of chemically reduced end-products. (Developed from (Setter and Belford, 1990)

Trapped carbon dioxide may form bicarbonate ions that can accentuate the effect of high lime content, leading to iron unavailability and chlorosis. Warm temperatures and ample supplies of organic matter will inevitably accelerate the development of these potentially damaging soil conditions. If root tips survive oxygen shortage *per se*, they may be injured or killed by subsequent changes in soil biochemistry (Ponnamperuma, 1972). These changes come about because of microbial respiration that utilises inorganic ions as alternative electron acceptors to oxygen in order to sustain energy generation. The changes are associated with measurable decreases in redox potential. Facultative anaerobes first chemically reduce nitrate, converting it to nitrite, nitrous oxide and nitrogen gas (denitrification) rendering nitrate unavailable to roots. As the reducing intensity of the soil increases further obligate anaerobes chemically reduce oxides of Mn^{4+}, and Fe^{3+} to form highly soluble Mn^{2+} and Fe^{2+} (Laanbroek, 1990) that may enter roots and interfere with enzyme activities and damage membranes. Ferrous ion toxicity can be a particular problem for rice farming on acidic soils. If flooding is prolonged, further, anaerobic bacteria may then convert SO_4^{2-} to H_2S, a poison of respiratory enzymes and non-respiratory oxidases. Acidic soils that are low in iron are especially likely to contain free and undissociated H_2S (Ponnamperuma, 1972). In the most severely reducing soils, methogenic bacteria reduce carbon dioxide to methane. Although the gas is harmless to plants it is second in importance to carbon dioxide as a greenhouse gas contributing to global warming. Rice paddies are globally significant methane sources.

Flooding may also increases the incidence of soil-borne fungal diseases (Yanar, Lipps, and Deep, 1997). Germinating seeds are particularly vulnerable to fungal colonisation (e.g., *Gliocladium roseum*). Infection of alfalfa, vegetables and trees by phytophthora (wilting), pythium (damping-off) and anaerobic bacteria (e.g. *Pseudomonas putida*) are common problems in practical farming (Walker, 1991). However it is not always clear whether injury is principally the result of the microbial infection or of the direct affects of flooding (Davison, 1997) and if infection follows rather than precedes injury.

ATP Supply and Demand

Anaerobic roots generate ATP mainly by glycolysis. This pathway also feeds pyruvic acid into ethanolic fermentation (and also into lactic acid fermentation, especially in the first hours of anoxia before the cytosol acidifies). Glycolysis is a cytoplasmic pathway that forms pyruvic acid from glucose, yielding 2 ATPs from each glucose molecule. This is only about 6 % of the ATP generated by mitochondria-based aerobic respiration. Thus, glycolysis is highly inefficient while still requiring a plentiful supply of glucose. It also requires pyridine nucleotide coenzyme in its oxidized form (NAD$^+$). The required NAD$^+$ is generated anaerobically from chemically reduced NADH during ethanolic or lactic acid fermentation. Unless metabolic processes that consume ATP are simultaneously suppressed, the small yield of ATP in anaerobic cells is insufficient for survival beyond a few hours. Suppression can be brought about if the roots are 'trained' beforehand by a few hours of partial oxygen shortage (e.g., 0.04 mol m^{-3} in solution or 3 % v/v in the air phase) before the supply of oxygen is finally extinguished (Xia, Saglio, and Roberts, 1995). Thus, an inability to restrict ATP utilisation to essential life-support processes may be at least as important a reason for death of flooded root tips as slow ATP production. Greenway and Gibbs (2003) conclude that early cell death can only be avoided if the small amounts of available energy are successfully re-diverted to permit synthesis of certain critical 'anaerobic' proteins (e.g., alcoholic fermentation enzymes), that support glycolysis and fermentation and help to prevent excessive acidification of the cytoplasm and vacuole and maintain membrane integrity. A marked decrease of membrane integrity may well be one of the most critical consequences of ATP imbalance for the viability of root cells. It is a consequence of lipid hydrolysis that is probably mediated by lipolytic acyl hydrolase (Rawyler et al., 1999). When membrane integrity is lost, the cell is irrecoverably damaged (Zhang et al. 1992). Unlike in some animal tissues, there is little evidence that ion channels and carrier systems can become sealed in response to anoxia thereby helping to retain soluble cell contents.

The modest ATP generation capability of glycolysis/fermentation depends on a ready supply of glucose and its precursors. Sugar shortage caused by anaerobic arrest of starch breakdown and sugar unloading in roots can thus shorten the duration of survival. This is illustrated by the ability of rice seeds to germinate without oxygen. Such ability is due, in part, to its possession of an anaerobically inducible gene coding for á-amylase, the enzyme principally responsible for degrading starch to a range of sugars (Loreti et al. 2003). In *in vitro* studies of anaerobic roots, hexose feeding is a prerequisite for long periods of anaerobic survival of the cultures. Over shorter time scales too (hours or several days), seedling roots have also been found to survive longer and ferment more vigorously when given external glucose (Webb and

Armstrong, 1983), Tadege et al., 1998). But, even when anoxic roots are given extra hexose they die eventually, indicating causes of death other than simply substrate-starved arrest of glycolysis. This may be because rates of glycolysis cannot speed-up sufficiently to satisfy demand or because ATP demand is not sufficiently down-regulated. But, other factors also come into play. These include the absence of molecular oxygen to support essential non-respiratory oxidative and oxygenation reactions (e.g., synthesis of polyunsaturated fatty acids used in membrane formation (Vartapetian, Mazliak, and Lance, 1978).

Self-Poisoning

Anaerobic roots may also die from self-poisoning by products of anaerobic metabolism; the most notable toxin being excess protons that acidify the cytoplasm and vacuole (Gerendás and Ratcliffe, 2002). In support of this notion, roots of pea (*Pisum sativum*), black eyed peas and navy beans, which collapse particularly quickly when anoxic, acidify their cytoplasm more rapidly than do longer-lived anoxic maize, soybean or pumpkin root tips. The sources of the extra protons within the cell have proved difficult to identify (Gerendás and Ratcliffe, 2002). Another possible toxin is acetaldehyde. In alcoholic fermentation, activity of the enzyme that converts acetaldehyde to ethanol (alcohol dehydrogenase - ADH) usually exceeds that of the enzyme that promotes acetaldehyde production from pyruvic acid (pyruvate decarboxylase - PDC).

Normally, this state of affairs ensures low sub-toxic concentrations of acetaldehyde in anoxic cells. However, after such tissue is returned to air, this control is sometimes lost and plant tissue typically generates a burst of acetaldehyde that could be damaging (Boamfa et al. 2003). Another potential toxin is nitric oxide (Dordas et al. 2003). This is a free radical gas that can be formed by the action of nitrate reductase (coded for by an anerobically inducible gene) and possesses the ability to kill cells, as is well-known in mammalian tissues. However, the roles of this molecule are still very unclear and it has even been suggested that the beneficial impact of nitrate on survival of anoxia may be an outcome of increases in nitric oxide arising from the reduction of nitrate to nitrite (Stoimenova et al. 2003). The conventional view is that death of the root-tip caused by one or more of the above-mentioned factors threatens the vigour and survival of the entire plant since fully functional root tips are required for soil exploration, uptake of water and mineral nutrients. However, Subiah and Sachs, 2003 take the view that rapid death of anoxic root tips is an adaptive response. They consider that loss of the root tip allows the remainder of the root (with its dormant lateral root primordia) to survive for longer, an interpretation supported by their finding that prior removal of root tips prolongs the life of anaerobic maize seedlings.

Although not strictly self-poisoning, cell death arising from oxidative reactions following the re-introduction of oxygen cannot be excluded as a cause of waterlogging injury and death. Underlying this notion is that an absence of oxygen harms the ability of plant cells to protect themselves against the formation and action of active oxygen species (e.g. superoxide radicals) when the floodwater recedes and free oxygen returns to the cells. Roots of soybean are thought to suffer in this way (Van Toai and Bolles, 1991). Most information on this phenomenon comes from work with rhizomes and cultured cells. However, recent studies with cell cultures do not strongly support to the view that post-anoxic damage is a major cause of death from anoxia (Rawyler et al. 2003).

How Roots Survive Anaerobic Conditions

During natural waterlogging of the soil, anoxia will be preceded by a period of partial oxygen shortage (hypoxia). This will last for as long as it takes for dissolved oxygen in the floodwater to be consumed by roots and other aerobic soil organisms. This hypoxic interlude can act as a training period by improving the ability of the roots to survive subsequent anoxia by inducing biochemical acclimation or anatomical acclimation.

Biochemical Acclimation

As little as 6 h prior exposure to partial oxygen shortage (typically 3 - 5 %, v/v in the gas phase) can lengthen survival time of anoxic maize root tips from 8 h to 72 h (Saglio, Drew, and Pradet, 1988). The mechanism by which cells initially sense the partial oxygen shortage that triggers this acclimation is unknown. One possibility is that sensing works through a binding of oxygen to non-leguminous haemoglobin, which is ubiquitous in plants. However, this mechanism has been ruled out on the grounds that binding is too tight for sensitive detection of partial oxygen shortage (Dordas et al. 2003). Despite this, haemoglobin is undoubtedly important for anoxia tolerance in other as yet undiscovered ways (Hunt et al. 2002) and petunia plants transformed with a Vitreoscilla haemoglobin gene, show remarkably enhanced tolerance to submergence (Imao et al., 2003).

Following sensing of partial oxygen deficiency, genes coding for so-called anaerobic proteins (actually hypoxically-induced proteins or HIPs) are up-regulated at transcriptional and post- translational levels while others coding for many aerobic proteins remain expressed and translated up to the point when the cells finally become anoxic. (Chang et al. 2000; Dolferus et al. 2003; Fennoy and Bailey-Serres, 1995; Subiah and Sachs, 2003; Baxter-Burrell et al. 2003). The anaerobic proteins (or HIPs) are necessary for the acclimation. The up-regulation of these genes is effected by the action of proteinaceous transcription factors (e.g., Myb factors, G-box factors, 14-3-3 proteins) that bind to promoter regions of target genes and influence their expression. The base sequences of these regions determine the susceptibility of a gene to any given transcription factor. One such region is the so-called 'anaerobic response element' characterised by a GT/GC-rich motif. This has been associated with several genes, such as an alcohol dehydrogenase (*ADH1*), that are strongly upregulated by oxygen shortage (Dolferus et al., 1994). HIPs can be divided into (1) enzymes involved in cytosolic energy metabolism especially those involved in starch breakdown and the glycolytic and fermentative pathways upon which anoxic energy generation depends; (2) enzymes implicated in pH regulation; (3) enzymes involved in aerenchyma formation; (4) enzymes with protective functions such as scavenging for potentially damaging active oxygen species generated when anoxic roots are returned to air; (5) proteins involved in signal sensing and transduction (e.g., the ethylene receptor ETR), (6) others of unidentified function. The complexity of the picture is being increased as sophisticated methods of protein analysis such as MALDI-DE-TOF mass spectrometry (Chang et al. 2000) become employed. Thus, HIPs play a major part in prolonging the life of anoxic roots that previously experience at least some hours of partial oxygen deficiency. If their synthesis is interfered with by applying protein synthesis inhibitors or through mutations, the effect of partial oxygen shortage on prolonging survival of subsequent anoxia is suppressed (reviewed by (Jackson and Ricard, 2003). It is notable that translation mRNA of many 'aerobic' genes is suppressed in anoxic cells.

Expression of HIP genes and production of survival protein is complimented by a co-ordinated down-regulation of demand both for oxygen, respirable substrates and for ATP. This suppression of demand is instigated by modest decreases in oxygen supply and well in advance of the onset of fermentation or any increase in NADH relative to oxidized NAD^+ (Geigenberger et al., 2000). Thus, while HIP production can be seen as a preparation for anoxia, the early down regulation of ATP and oxygen demand appears to delay the point at which tissue anoxia actually sets in. This down regulation of ATP- and oxygen-consuming pathways is probably underpinned by changes in gene expression. As previously mentioned, roots that are partially deprived of oxygen (typically 5 %) undergo marked changes in expression (Andrews et al. 1994; Klok et al. 2002). The kinetics of the up- and down-regulation of various groups of genes is complex with transcription factors and signal transduction pathway genes changing soonest. The identification of gene clusters bearing common regulatory sequences is a priority since by this means the overall co-ordination of events will become clearer (Dolferus et al. 2003). It will be important to understand both the down-regulation of genes coding for energy-consuming steps and the up-regulation of genes coding for survival proteins. Post-transcriptional processing of survival proteins and their patterns of association will be a further priority

Anatomical Acclimation through Aerenchyma Formation

The presence of large interconnected intercellular gas-filled spaces that often extend from the shoots to near the root tip (aerenchyma) is feature shared by most (Justin and Armstrong, 1987b) although not all species (e.g. *Caltha palustris* – Seago et al. 2000) that grow well in wet places. The spaces are created by cell separations resulting from differential rates of division or expansion by neighbouring cells or from the death of certain cells. The radially orientated spaces of the rice root is an example of aerenchyma formed by the death of particular cells that takes place mostly as a part of normal constitutive development. Mathematical modelling and direct experimentation have demonstrated that sufficient oxygen can diffuse through such tissue from the aerial shoot to satisfy the respiratory needs of root axes up to 30-cm-long at growing temperatures. In certain aquatic species, pressure-driven gas flow may aerate even longer lengths of rhizomes to which roots are attached (reviewed in (Jackson and Armstrong, 1999). Tests with wheat plants have shown that, without aerenchyma, roots longer than 100 mm are fatally damaged by an O_2-free medium while roots of this length equipped with 12 % porosity continue elongating. The effectiveness of oxygen transport is increased in species like rice where oxygen losses to the soil are inhibited by an inducible barrier to outward radial diffusion (Colmer et al., 1998). Any oxygen leaking radially out of the roots into the anaerobic soil can oxidise the rhizosphere thus decreasing injury from chemically reduced toxins such as ferrous ions.

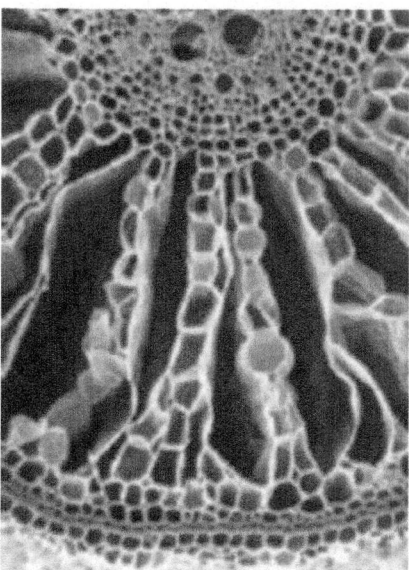

Figure: Scanning electron micrograph showing the cortex of a young rice root where radial lines of intact living cells alternate with gas-filled space created by cell death. Photograph by Stewart Young.

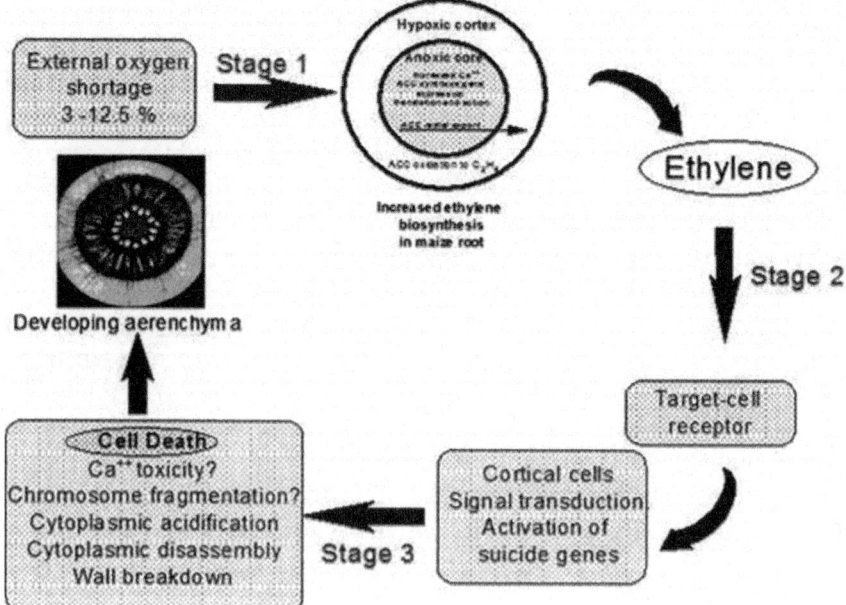

Figure: Summary of possible stages in aerenchyma formation in roots of *Zea mays* induced by partial oxygen shortage external to the root and mediated by increased synthesis of ethylene that in turn induces a form of programmed cell death in target cells of the cortex.

However, most species are intolerant of soil waterlogging and do not possess extensive aerenchyma (e.g., arabidopsis and tomato) but, in some (e.g. *Zea mays*), aerenchyma in roots can be stimulated during the early stages of waterlogging by the associated partial shortage of oxygen. This morphological acclimation is the result of spatially targetted process of programmed cell death that is localized in the root cortex and is stimulated by increases in ethylene concentration (Drew et al. 1979) brought about by entrapment and by faster biosynthesis; the latter being favoured by low but not extinguished oxygen supply e.g. 0.68 mol m^{-3} or 5 % v/v oxygen). The additional ethylene is trapped within the flooded roots by the floodwater and acts on files of target cells in the cortex resulting in their disassembly to create longitudinally interconnected gas spaces. Processes of cellular disassembly that destroy the cell are only partially understood and the state of knowledge is summarised. This so-called lysigenous aerenchyma is the outcome of an ordered set of structural degradative changes (lysis) that commence within 6 h and destroy the cell in 2-3 d. These changes include, acidification of the cytoplasm, possible loss of control of cytoplasmic Ca^{++}, a loss of microtubule orientation, plasmamembrane invagination, the formation of membrane-bound vesicles and, the appearance of membrane bound organelles, internucleosomal DNA fragmentation, chromatin condensation and changed patterns of pectin methylation in cell walls (Gunawardena et al. 2001a) and (Gunawardena et al., 2001). The features of those cortical cells that confer target status for ethylene induction of programmed cell death remain elusive.

How Above-ground Shoots Survive Soil Waterlogging and Submergence

It is inevitable that, because of the close functional interdependence between roots and shoots, stress on roots from soil waterlogging also threatens the shoot system. One example of this is the arrest of nitrate uptake that arises from microbial denitrification and damage to uptake mechanisms from an absence of oxygen. Young leaves then take this nutrient from older leaves leading to premature senescence in the latter (Drew and Sisworo, 1977). A second example of this knock-on effect on shoots is the tendency of waterlogged plants to wilt severely in bright light. This paradoxical response to waterlogging of the soil is the outcome of a lowered conductivity to water uptake that typifies oxygen-deficient roots and is brought about by proton induced conformational changes to water channel proteins.

Root to Shoot Communication and Shoot-Water Conservation

The tendency for leaves to dehydrate irreversibly in response to increases in root hydraulic conductivity is ameliorated by rapid signalling from roots to shoots that results in a slowing of water loss from the foliage. This is achieved by a prompt decrease of stomatal apertures, leaf expansion and, in tomato at least, by marked petiolar epinasty. Epinasty is the outcome of a downward re-orientation of whole leaves and leaflets that involves growth promotion on the upper (i.e., adaxial) surface. Epinasty reduces the amount of energy incident on the foliage and thus will slow the rate of water loss.

Figure: Sunflowers suffering from wilting in the sun during an episode of summer waterlogging. (Photograph supplied by LACJ Voesenek)

The signals inducing these effects in the shoots have been reviewed recently (Jackson, 2002) and, in the main, comprise increases (positive messages) or decreases (negative messages) in the flux of regulating hormones, or related substances, carried to the shoot in xylem sap. The signalling processes controlling epinasty in tomato are well understood, involve ethylene and can be summarised as follows. Severely hypoxic or anaerobic roots were shown to generate a positive message that (Bradford and Yang, 1980) showed later to be the ethylene precursor 1-aminocyclopropane-1-carboxylic acid (ACC). The amounts have been estimated using mass spectrometry and are demonstrably sufficient to enrich the shoot tissues with enough ACC to support the faster shoot ethylene production needed to induce epinasty (Else and Jackson 1998). The release of large amounts of ACC into xylem sap by anaerobic roots has two probable causes. The first is a blocking of ACC oxidation to ethylene by the absence of oxygen. This promotes a build-up of ACC. Some of this enters the transpiration stream and is drawn in to the shoot by transpiration steam as positive message. This enrichment has been shown in tomato and in xylem sap of *R. palustris*. The second putative cause is an up-regulation, and presumed translation of one member (*LE-ACS7*) of the six-gene family of ACC synthases responsible for forming ACC from its precursor *S*-adenosyl-L-methionine. Up-regulation occurs within 1 h and takes place along with a rise in ACC concentration in roots and enhanced ethylene production in the leaves (Shiu et al. 1998). Changes in gene expression and enzyme levels also enhance the oxidation of ACC on its arrival in the shoot since the ability of petioles to oxidase ACC to ethylene increases within 6 h (English et al. 1995) along with a rise in ACC oxidase mRNA. However, it remains unfortunate that detailed time courses of ethylene production using modern sensors such as photoacoustics, are not available to confirm the link between the onset of epinasty and increased shoot ethylene production.

Fast stomatal closure is often seen in many species soon after waterlogging begins but has proved difficult to explain. In tomato, many potential signalling messages such as increases in delivery of the hormone abscisic acid (ABA), sharp decreases in mineral nutrient delivery or increased xylem sap alkalinity have received little support, although in *Ricinus communis*, temporary dehydration of the leaves induced by severely decreased root hydraulic conductivity appears to trigger stomatal closure that is mediated by ABA (Jackson et al., 2003). Other less well-researched possibilities for root to shoot signalling in waterlogged plants involve decreases in pH and in the delivery of cytokinin or gibberellin hormones. In addition to possibly contributing to closing the stomata these changes may also help explain other shoot responses to soil waterlogging such as depressed stem elongation rates and faster leaf senescence.

Root to shoot signalling is best seen as a stop-gap mechanism that rapidly reduces the need for root-sourced supplies such as water and minerals at a time when root activities are suppressed by inadequate oxygen. But, if the roots remain inactive, or are killed and not succeeded by functional replacements, the shoot will not survive and the plant will then fail in its entirety. For longer-term survival of the plant, oxygen must be introduced into the interior of roots in amounts that support respiration and activities such as mineral and water uptake. Alternatively, replacement roots must develop that are better aerated than their predecessors. In species with adaptive capacity, shoots respond to flooding in ways that favour such changes. These are discussed briefly below.

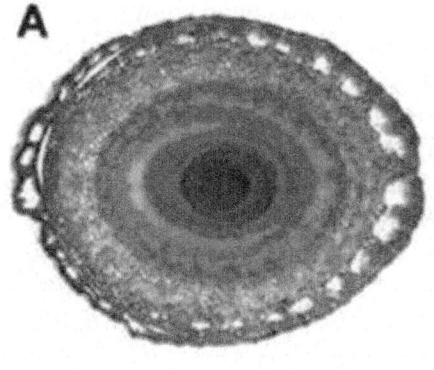
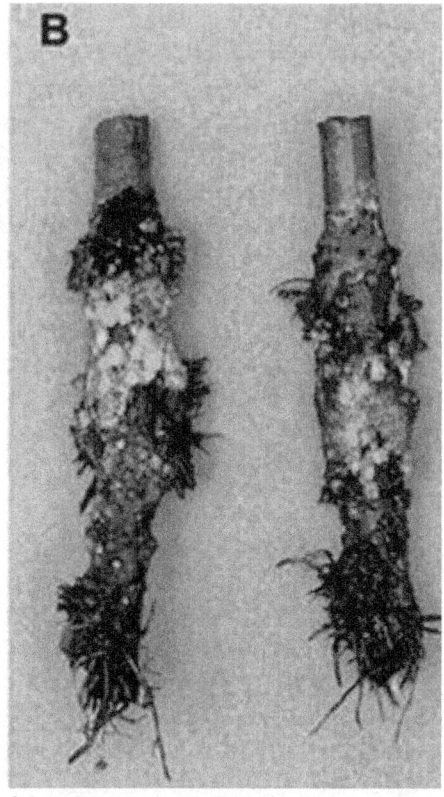

Figure: Two examples of reactivation of development at the shoot base caused by soil waterlogging. A. Aerenchyma formation in the leaf base of *Zea mays*. B. Hypertrophic lenticels at the stem base of young apple plants (*Malus domestica*) waterlogged for 40 days (adventitious roots removed for clarity). Photographs by the author.

Increased Porosity of the Shoot Base

In many species the shoot base become morphogenetically activated when plants are waterlogged, especially if the shoot base itself becomes water- covered. In *Z. mays*, the base of outer leaves form aerenchyma and studies in rice reveal a link with transverse veins and a distinct interconnectedness between the cells that eventually die to create the aerenchyma (Matsukura et al. 2000). In many dicots, swelling of submerged portions of the lower shoot and the development of swollen stems and hypertrophied lenticels are common. These effects probably enhance flooding tolerance by promoting tissue gas exchange, although experimental data demonstrating this are elusive. Swelling and hypertrophic lenticels are seen in many herbaceous dicots and woody species (e.g., apple) (Hook and Brown, 1973). Experimental blocking hypertrophic lenticels with lanolin grease has been shown to decrease oxygen entry into nearby roots via intercellular spaces and aerenchyma. This oxygen is then potentially available to support root functioning and also the formation and emergence of new adventitious roots nearby. Hypertrophic lenticels may also allow dissipation of metabolically generated volatiles such as ethanol, ethylene and acetaldehyde, although the physiological significance of this for plant performance and survival has not been examined. The leaf-base aerenchyma, and swelling of stem-base and lenticels are probably consequences of cell expansion promoted by endogenous ethylene trapped in the submersed tissue by the water covering (Kawase, 1974).

Figure: Effect of approximately two weeks waterlogging on the formation of replacement adventitious roots at the soil surface by plants of [A] maize (*Zea mays*) and [B] sunflower (*Helianthus annuus*). In A, the roots are outgrowths of pre-existing primordia. In B, the original root system has been killed and replaced by newly initiated roots formed at the shoot base and hypocotyl. Both mechanisms generate a root system with improved access to aerial oxygen

Replacement Rooting

Species with inherently surface-inhabiting root systems are notably tolerant of prolonged waterlogging (Justin and Armstrong, 1987a). However, if those species with deeper and thus more vulnerable roots are to revive, they must form replacement roots positioned near or at the better-aerated soil surface. There are three mechanisms for generating these replacement root systems. One is a stimulation of the outgrowth of root primordia already present within the shoot base (e.g., in Zea mays). A second is the induction of a new root system that involves initiation of root primordia and their subsequent outgrowth (e.g. in *Helianthus annuus* and *Rumex palustris*). Ethylene seems to be involved in both these processes. Applications of the gas to *Z. mays* strongly promotes primordial emergence (Jackson et al. 1981) although subsequent elongation is inhibited in association with aerenchyma formation. In nature, this inhibition may be mitigated by internal ventilation of ethylene out through the aerenchyma (Visser et al. 1997). Where replacement rooting involves initiation of new root primordia, auxin and ethylene are both thought to interact to bring this about (Visser et al., 1996). A third mechanism of placing roots at the soil surface involves a re-orientation of the root extension. Lateral roots of certain species grow upwards in waterlogged soils (Pereira and Kozlowski, 1977); Gibberd et al. 2001). When they reach the surface, they offer a replacement pathway for aeration of other attached roots provided there is adequate internally interconnected aerenchyma.

Fast Upward Shoot Elongation

Another developmental effect of flooding, that supplements the aerenchyma system, is the promotion of shoot extension and/or increased uprightness of submerged leaves (Grimoldi et al. 1999; Cox et al. 2003,). These responses are mostly found in aquatic or amphibious species well-adapted to periodically flooded areas or to rising water levels (e.g., rice and water lilies). The resulting increase in effective plant height, that can begin after a delay of less than 2 h of inundation, improves access to aerial or dissolved oxygen or to light for the generation of photosynthetic oxygen. In appropriate species, this oxygen can diffuse readily to the root elongation zone and elsewhere through aerenchyma (Waters et al., 1989). Internodes of the highly compressed stem in the base of the shoot of *Z. mays* are also stimulated to elongate when water levels cover just the shoot base. This raises the height of the ring of replacement roots emerging from a basal node. For the most part, the promotion of elongation by shoot or leaf are primary responses to ethylene entrapped in the growing tissue by the floodwater (Musgrave, Jackson, and Ling, 1972). The hormone is known to act in conjunction with gibberellins and/or with auxin to stimulate cell extension. This effect is in marked contrast to the effect of ethylene on less-well adapted species, where the gas most often strongly inhibits shoot elongation. In stems of deepwater rice (Kende et al., 1998) and petioles of *Rumex palustris* (Voesenek et al., 2003), faster underwater elongation is strongly dependant on a prior degradation of the growth-inhibiting hormone abscisic acid. Faster extension is also associated with the expression of a gene encoding a member of the family of cell-wall loosening enzymes known as expansins and with the up-regulation of an ethylene receptor gene (Voesenek et al., 2003). The range of species that utilise ethylene in this depth-accommodation response is wide and includes dicots, monocots and at least one liverwort and a tropical fern (*Regnellidium diphyllum*). However, the mechanism is not universal. In *Potamogeton pectinatus,* submerging the stem promotes elongation even though the plant is unable to synthesise ethylene itself because of ACC oxidase activity is lacking (Summers et al., 1996). Leaves of rice seedlings may also respond to a signal other than ethylene (Ella et al., 2003). Clearly, some other signal can be generated by submergence that is responsible for stimulating elongation. That signal may comprise accumulated carbon dioxide or partial oxygen shortage since rice coleoptiles elongate faster underwater in response to their collaborative effects as well as to ethylene. In the stems of deepwater rice, these signals act more independently (Kende, 1987). Young rice seedlings are readily submerged in water too deep for them to escape by means of fast upward elongation. In these circumstances, the expenditure of energy imposed by the faster growth rates may prejudice their survival. Survival is also challenged by more extensive leaf senescence that is promoted by accumulated ethylene (Jackson and Ram, 2003; Ella et al. 2003). In lowland rice types, those most tolerant to several days of total submergence as small seedlings (e.g. FR13A) are usually those that elongate very little underwater while retaining a full set of green leaves.

The most severe stress that flooding can impose on the shoot is total immersion in dark, anaerobic surroundings. Green leaves of some species tolerant of total submergence and of icing-over in arctic environment have the ability to survive such conditions, as do perennating organs of many temperate wetland species (Crawford and Brändle,1996; Schlüter and Crawford, 2001). A variety of biochemical adaptations reminiscent of those discussed earlier for roots are involved although studies with the dicot seedlings of *Arabidopsis thaliana* (Ellis et al. 1999) indicate that the biochemical pathways involved in adaptation to very low oxygen supply (approx. 0.1 %) may be different in roots and shoots. In addition, a select group of species including *Potamogeton pectinatus, Potamogeton distinctus, Sagittaria pygmaea* and rice can respond to a complete absence of oxygen by elongating their stem (or coleoptile in rice) much more quickly in an upward, gravity sensitive manner for many days. The biochemical basis of this remarkable achievement includes being able to degrade starch anaerobically and to sustain a fast rate of glycolysis in the growing parts (Summers et al. 2000). The faster growth is principally the outcome of enhanced cell extension linked to the promoting action of low pH and to the action of auxin that is implicated in anaerobic gravity sensing (Summers and Jackson, 1996). This stimulation of upward shoot elongation by a complete absence of oxygen is coupled with a marked constitutive aerenchyma. The two features together constitute a mechanism of survival by escape from flooding stress by means of upward elongation. But in this case the signal promoting fast elongation is, surprisingly, the complete absence of oxygen, even when this most severe stress is prolonged for many days.

Waterlogging and Crop Production

- Remedial measures in crop production
  - o *Improving land drainage*
  - o *Direct treatment to crops*
  - o *Plant breeding.*

Based on soil typing, (Dudal, 1976) estimated that 12 % of the World's soils are likely to suffer from excess water while (Boyer, 1982) indicated that about 16 % of US soils experience excess water. Clearly, considerable transient and more persistent waterlogging of the soil and deeper submergence of crops occurs in much rainfed farmland world-wide. The extent of its occurrence remains speculative. This is because of a lack of useful definitions of what constitutes excess water content. The transient nature of most farmland flooding also frustrates accurate estimation of the extent of farmland waterlogging. One potentially useful rule of thumb is the SEW_{30} value. This value (as centimetre days) combines all the days in a growing season when excess water is present in some or all of the top 30 cm of the soil profile (i.e. the zone in which most roots are usually to be found). Values of about 100 – 200 cm d are often sufficient to depress crop growth in temperate growing conditions.

Despite the uncertainties of quantification, it is possible to give examples to indicate the scale of the problem. Through such an approach we can appreciate that although statistically, floods are amongst the most common and widespread of all natural disasters and the effects are sometimes catastrophic (e.g., Mozambique in 2000), it is the more mundane persistent inadequacies in soil drainage in the face of near-average or modestly excessive rainfall that constrain farm productivity year by year. North Dakota State University

Extension Service reports that throughout the 1990s, excess rain affected every area of the State. The impact of flooding on cropping has been particularly well studied on the duplex soils of Western Australia (McFarlane and Cox, 1992). Each year, wherever rainfall exceeds approximately 400 mm, waterlogging occurs in a minimum of 8 % of the total cropped area i.e., 1.3 million ha of pastures and approx. 500,000 ha of cultivated land. The area of the latter may be increasing as arable farming expands because of poor returns on animal husbandry. For the 250,000 ha of barley grown in Western Australia, waterlogging is thought to reduce yields by 20-25 %. Assuming waterlogging in 2 out of 5 years, the value of the lost yield has been calculated at 6.84 million Australian Dollars each year. In 1988, crop losses due to waterlogging in a 27000 ha catchment area around Yornanning were approximately 1.1 million Australian Dollars fro barley alone. In a 700,000 ha sandplain area, approximately 400,000 ha are subject to transient waterlogging, making financially viable lupin production almost impossible. Furthermore, these waterlogged soils are believed to be sites of groundwater recharge, contributing to serious salinization of the sandplain. In the state of Victoria, 1.8 million ha of duplex soils are affected by waterlogging, with another 2.3 million ha are susceptible. This is estimated to cost the state of Victoria 36 million Autralian Dollars in lost crop production each year. Setter and Waters (2003) give other examples such as the wheat-growing regions of northern India.

Figure: Left - typical late spring waterlogging of poorly drained field of peas (*Pisum sativum*) in Cambridgeshire, UK. Right – close-up of the injury sustained by leaves of a pea plant after several days soil waterlogging.

Waterlogging damage to crops may sometimes occur when crops are irrigated. The dividing line between adequate and excessive irrigation is not well defined and it is probable that some of the benefits can be lost by heavy-handed irrigation. For example, ponding on the soil surface for more than a day is known to be harmful to wheat watered by flood-irrigation. Prolonged irrigation over many years in dry regions can cause waterlogging problems created by raising the water table. A well-known example, already mentioned, is to be found in the Sindh Province in the Indus Valley of Pakistan. Here, 50 to 60 years ago, the watertable was 4 m below ground. By 1980s it had risen to between 1 – 2 m over most of the irrigated region and the water being highly saline. A multimillion-dollar drainage project (the Left Bank Outfall Drainage project) intended to reverse the situation and funded by the World Bank and the Asian Development Bank is in place but not without causing controversy and additional environmental concerns.

The extent of damage to yield depends heavily on the stage of development as well as on more obvious factors such duration of waterlogging and temperature. For most crops, seed germination is probably the most vulnerable, reflecting both the fast metabolic rate of germinating seeds being coupled with complete inundation. Studied examples include peas (Jackson, 1983) and temperate cereals (Lynch et al., 1981), (Setter and Waters, 2003). Problems are exacerbated if germination takes place in association with decomposing organic matter such as straw residues from the previous crop. Microbial colonisation of such residues depletes the soil of oxygen and nitrate while discharging potentially harmful substances such as unsaturated fatty acids and the growth-inhibiting hormone abscisic acid. Leaking of soluble carbohydrates from seeds is stimulated by anaerobiosis and this also encourages fungal pathogens such as *Gliocladium roseum*. Seeds of rice are exceptional in being able to germinate in flooded soil without oxygen. But the germination is abnormal since only the coleoptile but not the root or the mesocotyl or leaves emerge from the embryo without oxygen. Elongation by the emerging coleoptile is stimulated under these conditions by a combination of lack of oxygen, carbon dioxide and ethylene. Underpinning this is the ability of rice seeds to hydrolyse endosperm starch to sugars by means of anaerobically transcribing genes, such as the *Ramy3D* á-amylase gene, that are up regulated by the associated low sugar levels (Loretti et al. 2003).

Once germinated, stages in subsequent development of crop plants influence susceptibility to flooding injury. Small cereal seedlings with their shoots below ground are highly susceptible (e.g., winter wheat – (Cannell et al. 1980) but thereafter tolerance rises until early reproductive stages when susceptibility again increases (e.g. in soybeans - Linkemer et al. 1998). Heightened vulnerability at or just before flowering has also been noted for several other crops including peas, wheat, sorghum, maize and cow peas (Cannell and Jackson, 1981) when inundated for one or more days. By contrast, several weeks of waterlogging in winter of young plants of crops such as autumn sown wheat or oil seed rape causes little yield loss because of compensatory growth in the following spring and the slower metabolic demands created by cool temperatures during the period of waterlogging.

Unlike other major crop species, rice can yield heavily when grown in waterlogged soil (paddy rice production) and specialised ecotypes are able to yield usefully even when partially submerged in several metres of water (Catling, 1992). This ability is based partly on ethylene-mediated stem elongation that is induced when the shoot base becomes submerged. However, it is less widely recognised that these do not readily escape total submergence. Small vegetative rice plants (even of the so-called deepwater ecotypes) are totally submerged for only a few days as a result of uncontrolled flooding in lowland areas, they are severely damaged or killed. Part of the problem is thought to be that stimulated leaf elongation and associated ethylene-promoted leaf senescence quickly deprive the young plants of starch and sugars, thus prejudicing their survival and re-growth potential.

Remedial Measures in Crop Production

A comprehensive treatment of remedial measures is outside the scope of this review. However, a brief account of various approaches is given. It pre-supposes that, where appropriate, measures such as installing protective barriers against water inflow from rivers or the sea are installed in association with high-volume pumping systems. The creation and maintenance, by these means, of highly productive farmland in reclaimed coastal areas of The Netherlands represents the pinnacle of achievement in this regard. A common feature in the management of this and other riverine farmland (e.g. The Mississippi floodplain) is the building of dykes or levees (long high banks) along rivers to prevent them from overflowing or at least displacing the flooding to lower reaches of the river. The approach has its dangers.

Improving Land Drainage

Estimates of the areas of drained farmland that have been published for several major arable-cropped regions. While highly approximate, they serve to illustrate the enormous scale of the problem of farmland waterlogging and the key role of drainage systems in supporting economically viable agriculture. Nosenko and Zonn (1976

- quoted by Cannell and Jackson 1981) estimated that 155 million ha carried subsurface land drains. Much draining has been installed in the last 50 years (e.g. 6 % of UK farmland received new drains between 1940 and 1970). Most of these systems have been installed to arbitrary drainage targets, reflecting the scarcity of drainage experiments on representative soils.

According to Trafford (1975), installation of drainage systems may be motivated by four rather different sets of circumstances. (1) Where reclamation of flooded land is needed before conventional agriculture is possible; (2) where farming of a given crop may have to be abandoned unless drainage is installed (e.g., white lupin production on the south coast of Western Australia); (3) where improved drainage is needed to support more intensive and higher-value cropping and (4) drainage to improve the yields and profitability of an existing cropping system. In situations (3) and (4) the availability of subsidies to fund part of the cost of installing land drains may affect the financial viability of installing land drains. On a clay soil in the UK, even a 28 % improvement in yields of winter wheat reported over a six year period after installing pipe and mole drains would not have been an economic investment unless installation was subsidized by government grants (Hunter and Trafford, 1979). Such subsidies are no longer available in the UK and much of western Europe. However, less financially accountable benefits such as a marked increase in the number of days the drained land can bear heavy farm machinery without damage must not be ignored.

Mention has already been made of Pakistan's Left Bank Outfall Drain (LBOD) project, one of world's largest drainage schemes. It highlights the installation of land drainage schemes as one of the most widely adopted means of decreasing the incidence of farmland waterlogging. At the individual farm level, measures for improving drainage range from simple and historically interesting ridge and furrow constructions (basically creating raised beds for the crop - system sometimes used in Australia) to the installation of complex subsurface land drains designed to meet target drainage flows and installed using laser-guided drain layers that insert perforated plastic piping to precise depths and gradients. The use of such systems is illustrated by guidance given by the USDA-Agricultural Research Service to sugarcane growers in Lower Mississippi River Valley (LMRV). The machinery used in such work can be formidable. In contrast the simplest subsurface land drains are those created in clayey soils using a 'mole' that is dragged through the soil at a depth of about 50 cm. Despite its seeming simplicity, powerful tractors are still required for its execution. A combination of the two forms of drainage installation is especially effective. Other means of lowering the water table include the judicious planting of waterlogging tolerant trees such as certain eucalyptus species, guava and mango that also produce a useful crop for cash-poor farmers in poorly developed tropical regions. *Eucalyptus aggregata, E. camphora, E. crenulata, E. gunnii & E. gunnii divaricata* are especially tolerant of waterlogging. In New Zealand and Tasmania they grow naturally in undrained peat moors where surface water is present for at least six months of the year. They also tolerate stagnant water just below the surface (unlike poplar and willow). They are, however, stunted on permanently flooded soils. Another approach aimed at reducing the rate of water input into the lower reaches of floodplains such that of the River Rhine in northern Europe, is to retain as much water as possible upstream. This can be improved by preserving and improving what remains of upstream flood plains (so-called soft engineering).

Direct Treatment to Crops

Remedial effects can be observed after applying nitrogen fertilizers as a foliar spray. There is evidence of beneficial effects of this on the yield of cotton. Cereals may also benefit from soil applications of N provided the plants are only moderately damaged. In Western Australia, 100kg N ha[-1] or more raised yields of small grain cereals waterlogged for 3-7days but little benefit is seen in more severely damaged crops. The use of additional nitrogen to offset waterlogging damage has support from basic physiological investigations. These show that applied nitrate may enter anaerobically damaged roots by passive means and be translocated to the shoot. The nitrate may also act simply to replace that leached in the drainage water or destroyed by anaerobic denitrifying bacteria (Trought and Drew, 1981).

Germinating seeds have been identified as being especially vulnerable to damage from soil waterlogging and the associated anaerobiosis and injury from chemically reduced metabolites such as organic acids and respiratory intermediates such as acetaldehyde. It is therefore not surprising that pre-treating seeds with an oxidizing coat has been tried as an insurance against such effects. Coatings include calcium hydroxide and calcium peroxide (Sladdin and Lynch, 1983), although the effects on rice have not always been positive. There are companies selling products that promise to improve plant growth by using calcium peroxide to 'inject' oxygen in to the soil. A more mundane approach in situations where germinating seeds risk contact with undegraded crop residues in wet conditions, is to rotovate and dice these residues or remove them before sowing, possibly by burning if legislation allows this.

Plant Breeding

As with other major abiotic stresses, breeding and selecting successful tolerant cultivars have not yet met with notable commercial success although Setter and Waters (2003) report that a waterlogging tolerant wheat cultivar suitable for Australian conditions is close to being released. This has been derived from doubled haploid (DH) lines generated by crosses with tolerant wheat originally selected by the International Maize and Wheat Improvement Centre (CIMMYT) in Mexico. Despite this limited progress, there are numerous indications that the production of cultivars with improved tolerance that also retain their desirable agricultural traits is a realistic prospect for several major food crops. The utility value of much published work depends on the criteria used to assess tolerance. Not all reports compare final yields, or include non-waterlogged controls in their trials. Nevertheless, a basis for future successful breeding can be found in reports of greater than normal tolerance in a number of major crop species, or allied species that could perhaps be employed to introduce tolerance traits through 'wide hybridization'. For example, waterlogging-tolerant accessions have identified in seven of the eight taxonomic sections in *Trifolium* when compared in terms of relative growth rates. Several lines of soybean and of winter wheat have also been shown to posses unusually high tolerance on the basis of injury levels and final yield. Comprehensive tests in the glasshouse and in the field over several seasons and involving yield comparisons of over 20 cultivars (Musgrave and Ding, 1998) revealed notable tolerance in two lines of winter wheat that was positively correlated with iron-rich surface deposits on the roots, implying a link with increased amounts of aerenchyma. Based on levels of chlorosis in waterlogged spring wheat, (Boru et al. 2001) concluded that tolerance was largely controlled by four genes with beneficial effects that could be additive. (Setter and Waters, 2003) have assessed, in detail, many other reports of inheritable variation amongst the major small-grained cereals. Examples of variability in tolerance in a variety of other crops are given by (Cannell and Jackson, 1981).

Bibliography

Adhikari M.K. : *Research on Plant Tissue Culture*, Department of Plant Resources National Herbarium and Plant Laboratories, 2004.

Alka Rani Upadhyay: *Aquatic Plants for the Waste Water Treatment*, Daya, Delhi, 2004.

Ansari, Tariq M : *Molecular Plant Pathology*, Pearl Books, Delhi, 2008.

Bahar A. Siddiqui and Samiullah Khan: *Plant Breeding Advances and in vitro Culture*, CBS, Delhi, 1997.

Bailey, L.H. : *Manual of Cultivated Plants*, The Macmillan Company, New York, 1949.

Bandyopadhyay, P C : *Breeding and Crop Production*, Gene Tech Books, Delhi, 2007.

Bodeker Gerard : *Medicinal Plants for Forest Conservation and Health Care*, Daya, Delhi, 2005.

Boyer , J.S.: *Measuring the Water Status of Plants and Soils,* Academic Press, N.Y., 1995.

Brooker. S. G., Cambie. R. C. : *Economic Native Plants of New Zealand*, Oxford University Press 1991.

Brown R G : *Dictionary of Plant Tissue Culture*, Ivy Pub, Delhi, 2004.

Byrd, Graf: *Advances in Plant Physiology*, Rajat Pub, Delhi, 2008.

Chadha K. L. and Pareek O. P.: *Advances in Horticulture: Fruit Crops,* New Delhi, Malhotra Publishing House, 1993.

Chandola, R.P. : *Dictionary of Plant Pathology*, Daya, Delhi, 2010.

Chevallier, A. : *The Encyclopedia of Medicinal Plants,* Dorling Kindersley, London, 1996.

Christopher B. Johnson : *Breeding Research on Aromatic and Medicinal Plants,* New York, Haworth Herbal Press, 2002.

Coste R.: *Coffee: the Plant and the Product*, London, MacMillan, 1992.

Degras L.: *Yam: a Tropical Root Crop*, Wageningen, CTA/MacMillan, 1993.

Devi, C.R. Sudharmai : *Analytical Procedures in Soil Science and Agricultural Chemistry*, Agrotech, Delhi, 2004.

Evans, M. : *Herbal Plants*, Studio Editions, London, 1991.

Featherly H. I.: *Taxonomic Terminology of the Higher Plants*, USA, Iowa State College Press, 1954.

Gerard, J. : *The Herbal or General History of Plants*, Dover Publications, New York, 1975 .

Ghosh Biswajit : *Plant Tissue Culture : Basic and Applied*, Universities Press, Delhi, 2005.

Gibbs, D.R. : *Chemotaxonomy of Flowering Plants*, McGill-Queen's University Press, Montreal, 1974.

Gilkey & Dennis : *Handbook of Northwestern Plants*, Corvallis: Oregon State University, 1980.

Gour, H.N.: *Integrated Plant Pathology*, Scientific, Delhi, 2009.

Graf, Alfred Byrd: *Advances in Plant Physiology*, Rajat Pub, Delhi, 2008.

Gupta, O.P. : *Water in Relation to Soils and Plants : With Special Reference to Agriculture*, Agrobios, Delhi, 2002.

Heims, Dan : *Heucheras and Heucherellas : Coral Bells and Foamy Bells*, Portland: Timber Press, 2005.

Herminie Broedel Kitchen: *Soils and Crops : Diagnostic Techniques*, Satish Serial Publishing, Allahabad, 2004.

Jackson, E. : *Crop Management and Soil Conservation*, Biotech Books, Delhi, 2011.

Jacquat Christiane: *Plants from the Markets of Thailand*, Bangkok, Duang Kamol, 1990.

Janardhan K Reddy : *Advances in Medicinal Plants*, Universities Press, Delhi, 2007.

Jha Timir Baran and Ghosh Biswajit : *Plant Tissue Culture : Basic and Applied*, Universities Press, Delhi, 2005.

Jones, R. M.: *Plant Resources of South-East Asia*, Wageningen, Pudoc Scientific Publishers, 1992.

Kala C P : *Medicinal Plants of Indian Trans-Himalaya : Focus on Tibetan Use of Medicinal Resources*, BSMPS, Delhi, 2003.

Kapoor, R.L. and M.L. Saini: *Plant Breeding and Crop Improvement*, CBS, Delhi, 1997.

Kataria, T N : *Plant and Crop Physiology*, Pearl Books, Delhi, 2008.

Khilare V.C. : *Molecular Biology of Plant Pathogens*, Daya, Delhi, 2010.

Kumar M Reddi : *Recent Trends in Rapid Diagnosis of Plant Pathogens*, BS Publications, Delhi, 2008.

Kumar Shailesh : *Plant Tissue Culture—Theory and Techniques*, Scientific, Delhi, 2009.

Kumar, N. : *Breeding of Horticultural Crops : Principles and Practices*, New India Pub, Delhi, 2006.

Mabberley D. J.: *The Plant-Book : a Portable Dictionary of the Vascular Plants*, Cambridge, Cambridge University Press, 1997.

Mannetje, L. T. & Jones, R. M.: *Plant Resources of South-East Asia*, Wageningen, Pudoc Scientific Publishers, 1992.

McPherson. A. and S. : *Wild Food Plants of Indiana*, Indiana University Press 1977.

Meenu Bhatnagar: *Groundwater Management : Sustainable Approaches*, Icfai Books, Delhi, 2012.

Nobel, P. S.: *Physicochemical and Environmental Plant Physiology*, Academic Press, San Diego, 1999.

Oldham P.: *Cost of Production of Major Tree Crops in Dominica*, Roseau, Ministry of Agriculture, 1991.

Pareek L.K. : *Trends in Plant Tissue Culture and Biotechnology*, Agrobios, Delhi, 2005.

Patel, K.P. : *Heavy Metals in Soils and Plants*, Agrotech Pub, Delhi, 2011.

Percival John: *The Wheat Plant - A Monograph*, London, Duckworth & Co., 1921.

Premjit Sharma: *Agricultural Drainage and Water Quality*, Gene Tech Books, Delhi, 2007.

Rani, K.P. : *Advances in Soil Borne Plant Diseases*, New India Pub, Delhi, 2008.

Rao K. Manibhushan : *Recent Developments in Biocontrol of Plant Pathogens*, Today & Tomorrow, Delhi, 1996.

Rob Jenkins and C.K. Jain: *Advances in Soil-Borne Plant Diseases*, Oxford Book Company, Delhi, 2010.

Sambamurty, A V S S : *Textbook of Plant Pathology*, I K Pub, Delhi, 2006.

Samuel W. Johnson: *Crops Feed from Air and Soil*, Reprint Pub, Delhi, 2005.

Sathyanarayana B N : *Plant Tissue Culture : Practices and New Experimental Protocols*, I K International, Delhi, 2007.

Sharma, Priya Vrat : *Classical Uses of Medicinal Plants*, Chaukhambha Visvabharati, Delhi, 1996.

Shekhawat N.S. : *Plant Tissue Culture and Molecular Markers: Their Role in Improving Crop Productivity*, I.K. International, Delhi, 2009.

Shukla, R.S. and P.S. Chandel: *A Textbook of Plant Ecology : Including Ethnobotany and Soil Science*, S. Chand Publisher, Delhi, 2009.

Singh M. P. : *Neem Herbal Medicinal Plant*, Enkay Publishing House, Delhi, 2011.

Swarup, Ram : *Elements of the Nature and Prospectus of Soil*, Manglam Pub, Delhi, 2011.

Tripathi, D P : *Plant Pathology at a Glance*, Scientific, Delhi, 2008.

Tyagi B.K. : *Biodiversity and Conservation of Medicinal Plants*, Swastik Pub, Delhi, 2009.

Tyagi, I.D. : *Plant Breeding and Genetics at a Glance*, South Asian, Delhi, 2005.

Vanangamudi, K.: *Principles and Methods of Plant Breeding*, International Book, Delhi, 2005.

Verma J.P. : *Detection of Plant Pathogens and their Management*, Angkor, Delhi, 1995.

Vijai Pal : *Research Methods in Plant Sciences : Allelopathy : Plant Pathogens*, Scientific, Delhi, 2004.

Wackerman A E : *Harvesting Timber Crops*, Biotech, Delhi, 2002.

Whealy K.: *The Garden Seed Inventory*, Decorah, Seed Saver Publications, 1988.

Whistler, W.A. : *Polynesian Herbal Medicine*, National Tropical Garden, Kauai, Hawaii, 1992 .

Wickens, C. and Hollands, J.: *Industrial and Agriculture Performance*, Prentice Hall, NY, 1996.

Wilde, S.A. : *Forest Soils and Forest Growth*, Periodicals, Delhi, 1991.

Index

□□□